THE SCIENCE OF QUALITATIVE RESEARCH

This book is a unique examination of qualitative research in the social sciences, raising and answering the question of why we do this kind of investigation. Rather than offering advice on how to conduct qualitative research, it explores the multiple roots of qualitative research – including phenomenology, hermeneutics, and critical theory – in order to diagnose the current state of play and recommend an alternative. The diagnosis is that much qualitative research today continues to employ the mind–world dualism that is typical of traditional experimental investigation. The recommendation is that we focus on constitution: the relationship of mutual formation between a form of life and its members. Michel Foucault's program for "a historical ontology of ourselves" provides the basis for a fresh approach to investigation. The basic tools of qualitative research – interviews, ethnographic fieldwork, and analysis of discourse – are reforged in order to articulate how our way of living makes us who we are and so empower us to change this form of life.

Martin Packer is Associate Professor of Psychology at Duquesne University in Pittsburgh and at the University of the Andes in Bogotá. He received his BA at Cambridge University and his PhD at the University of California, Berkeley. He has previously taught at the University of California, Berkeley, and the University of Michigan. His research has explored interactions between neonates and their mothers, early childhood peer relations, conflict among adolescents, and the way schools change the kind of person a child becomes. Packer is coeditor of *Entering the Circle: Hermeneutic Investigation in Psychology* (with Ritch Addison) and *Cultural and Critical Perspectives on Human Development* (with Mark Tappan) and author of *The Structure of Moral Action* and *Changing Classes: School Reform and the New Economy*. He is one of the founding coeditors of the journal *Qualitative Research in Psychology*, and he has published articles in *American Psychologist*, *Educational Psychologist*, and *Mind, Culture and Activity*.

The Science of Qualitative Research

Martin Packer

Duquesne University
University of the Andes

CAMBRIDGE
UNIVERSITY PRESS

CAMBRIDGE UNIVERSITY PRESS
Cambridge, New York, Melbourne, Madrid, Cape Town, Singapore,
São Paulo, Delhi, Dubai, Tokyo, Mexico City

Cambridge University Press
32 Avenue of the Americas, New York, NY 10013–2473, USA

www.cambridge.org
Information on this title: www.cambridge.org/9780521148818

First published 2011

Printed in the United States of America

A catalog record for this publication is available from the British Library.

Library of Congress Cataloging in Publication Data
Packer, Martin J.
The science of qualitative research / Martin Packer.
     p.   cm.
Includes bibliographical references and index.
ISBN 978-0-521-76887-0 (hardback) – ISBN 978-0-521-14881-8 (pbk.)
1. Social sciences – Research.   2. Qualitative research.   I. Title.
H62.P223   2011
001.4′2–dc22   2010028296

ISBN 978-0-521-76887-0 Hardback
ISBN 978-0-521-14881-8 Paperback

It is not easy to say something new; it is not enough for us to open our eyes, to pay attention, to be aware, for new objects suddenly to light up and emerge out of the ground.

Michel Foucault, *The Archeology of Knowledge*, 1969/1972, pp. 44–45

CONTENTS

LIST OF FIGURES AND TABLES

FIGURES

TABLES

LIST OF BOXES

ACKNOWLEDGMENTS

Acknowledgment is due to The Johns Hopkins University Press for permission to use as a chapter epigraph an excerpt from *The Act of Reading: A Theory of Aesthetic Response* by Wolfgang Iser © 1978 The Johns Hopkins University Press. Acknowledgment is due to *The Review of Metaphysics* for permission to use as chapter epigraphs excerpts from two articles by Charles Taylor (*Interpretation and the Sciences of Man* and *Understanding and Human Science*) to which they hold the copyright. Thanks are due to Duquesne University for support of my writing in the form of an Endowed Faculty Development Award and for assistance in index preparation; to the students who have taught me much; and to my wife, Leticia, for her support and understanding.

Introduction

This is an exciting time to be writing about the character of inquiry in social science for there is a growing interest in and openness to new forms of inquiry. Researchers throughout the social sciences are increasingly working with qualitative data – interview transcripts, verbal reports, videos of social interactions, drawings, and notes – whether they view these as "soft data" (Ericsson & Simon, 1984), "messy data" (Chi, 1997, p. 271), or "the 'good stuff' of social science" (Ryan & Bernard, 2000, p. 769). Research projects that include such empirical material are becoming increasingly popular. In addition to self-styled "qualitative researchers," investigators in the learning sciences, developmental psychology, cultural psychology, and even in survey research, as well as many other areas, have turned to nonquantitative material and are exploring ways to collect, analyze, and draw conclusions from it.

At the same time, a strong backlash has developed against this kind of inquiry. In the United States, as in England and Australia, the funding priorities of government agencies emphasize "evidence-based" research. We are told repeatedly that there is a "gold standard" for research in the social sciences, the randomized clinical trial. Other kinds of research – typically cast as naturalistic, observational, and descriptive – are viewed as mere dross in comparison, good only for generating hypotheses, not for testing them. They are seen as lacking the rigor necessary for truly scientific research and as failing to offer practical solutions to pressing problems. Clinical trials, in contrast, are seen as relevant because they test treatments and interventions, and as rigorous because they involve direct manipulation, objective measurement, and statistical testing of hypotheses. Any suggestion that there might be inquiry that follows a logic of inquiry different from that of traditional experimental research is dismissed. The possibility that complex human phenomena might require a kind of investigation that traces them in time and space and explores how they are constituted is not considered.

In the 1980s, there was general agreement that the "paradigm wars" had ended (Gage, 1989). For many, the correct way to proceed seemed to be with "mixed methods" that combined qualitative techniques with aspects of

1

traditional experimental design and quantification. Arguments against mixing "qual" and "quant" are often dismissed as an unnecessarily belligerent perpetuation of the conflict. But now the "science wars" are being fought over much the same territory (Howe, 2005; Lather, 2004). It seems we need to revisit the arguments against applying a naive model of the natural sciences to human phenomena. Today we are in a much stronger position than at any time in the past to articulate the logic of a program of research that explores a more fundamental level of phenomena than can be studied using clinical trials. Important theoretical and empirical work across the social sciences but also in the humanities – in history, philosophy, linguistics, and literary theory – now enables us to define a program of investigation that is focused on "constitution," a term I shall define in a moment.

Researchers must bear some responsibility for the evidence-based movement. There is, for example, a bewildering variety of types of qualitative research. For some this is a potpourri to be savored and celebrated, but for others social science research has "become unhelpfully fragmented and incoherent," divided into "specialist domains . . . that are too often treated in isolation" (Atkinson, 2005). This plurality makes it difficult to establish criteria for evaluating research or to design curricula for teaching research methods. It creates the impression that nonexperimental research cannot provide genuine knowledge. The enormous number of "how to" books currently published is one indication of the profusion of approaches to social scientific research and also the huge appetite for guidance. But, at the same time, the sheer number suggests that this appetite hasn't been satisfied. Readers find themselves left with fundamental confusions and buy book after book in a search for clarification.

In the face of all this, the student who wants to learn how to do qualitative research, or the more experienced researcher who wants to try something new or better, could be forgiven for being confused. This book is an attempt to bring some clarity to the subject. It is not a book on how to do qualitative research – it is not a "how to" book at all. Instead it raises the question that must come first: *why* are we doing qualitative research? Once we have figured out why we are doing research, we will have much more clarity about how research should be conducted because in any activity we can't really know what to do if we don't know what we're aiming for. Only when we are clear about *what* we are doing and *why* can we figure out *how* to do it well.

Qualitative research is, in my view, frequently misunderstood. It is often equated with any kind of investigation that doesn't use numbers, but we will discover that quantification has its place, in the *descriptive* phase of qualitative inquiry. It is often defined as the objective study of personal experience, but we will see that such a view – for example, in empirical phenomenology, interpretative phenomenological analysis, and grounded theory – gets helplessly tangled in the opposition of subjectivity and objectivity. Finally, qualitative research is often seen as the ethnographic study of culture and intersubjectivity, but here the problem turns out to be the uneasy combination of participation and observation.

How then should we understand qualitative research? It seems to me that fundamental mistakes are made today in many approaches to qualitative inquiry and that important opportunities are being missed. Researchers are not asking the right questions. We are not asking sufficiently difficult or interesting questions – we are not aiming high enough. At the same time, we are not digging deep enough; we are not questioning our basic assumptions about human beings and the world in which we live, our assumptions about knowledge and reality. I have been practicing and teaching qualitative research for almost 30 years, working to make it accessible and comprehensible, and although it is gratifying to see this kind of research becoming increasingly widespread, at times I find myself frustrated that the potential of qualitative research is not being realized. This potential is, I believe, profound. Attention to human forms of life, to the subtle details of people's talk and actions, to human bodies in material surroundings, can open our eyes to unnoticed aspects of human life and learning, unexplored characteristics of the relationship between humans and the world we inhabit, and unsuspected ways in which we could improve our lives on this planet.

I will try to demonstrate this potential by introducing the reader to debates that often do not cross the boundaries between disciplines and to historical, conceptual, and ethical aspects of qualitative research that have frequently been forgotten or ignored. I will examine the central practices of qualitative research – interviewing, ethnographic fieldwork, and analysis of interaction – in order to tease out the assumptions embedded in these practices and suggest new ways to think about, collect, and analyze qualitative material. I will suggest new kinds of questions we should set out to answer and outline the general form of a program of qualitative inquiry. Qualitative research is sometimes viewed merely as a set of techniques – a toolbox of procedures for the analysis of qualitative materials – but in my view it is something much more important, the basis for a radical reconceptualization of the social sciences as forms of inquiry in which we work to transform our forms of life.

An important part of this reconceptualization is a new sense of who we are. Humans are products of both natural evolution and history. As products of evolution, we are material beings, one kind of biological creature among many others, participants in a complex planetary ecological system. The long-standing belief that we are somehow not only different from but also better than other animals has been complicit in an attitude toward our planet as merely a vast repository of raw materials, resources that we can exploit for profit. We are witnessing the dire consequences of this attitude and are running up against the limits of this lifestyle of "development."

A change in attitude will require a change in our understanding of our place in nature and our responsibilities as stewards of the planet, a role that we have forced on ourselves as a consequence of our efforts to satisfy a craving for power over nature.

As products of history – of cultural evolution – we are cultural beings, and in this regard we *do* differ from other living creatures. We share 99.5% of the genetic material of the Neanderthals who lived 30,000 years ago, but our lives are 100% different. We can shape our environment in ways that Neanderthals never dreamed of and that other animals are unable to compete with, and our environments have changed us in return. Our continuing naive beliefs in "human nature" fly in the face of important cultural differences and the deep penetration of our being by cultural practices, and they serve to justify our dangerous tendency to demonize people whose way of life is different from our own. Each human group tends to presume that it is internally homogeneous and identical and that the only significant differences are those that distinguish it from others. This attitude fosters a simplistic conception of good and evil and a destructive impulse to "civilize" other peoples and impose our values on them. A change in this attitude will require the recognition that humans are not identical, that there is no universal mental apparatus, and that different traditions, customs, and ways of living have created a variety of ways of living: ways of thinking, seeing, and being.

Thirty years ago, proponents of qualitative research (e.g., Dallmayr & McCarthy, 1977; Rabinow & Sullivan, 1979) wrote of a crisis in the social sciences that they linked to an underlying human crisis – the lack of meaning that the failure of Enlightenment rationality had exposed. In the 18th century, thinkers such as the Austrian philosopher Immanuel Kant – still sometimes described as the most influential philosopher ever – proclaimed the existence of a universal capacity for reason, the same for all cultures and all times, that could provide an objective foundation for knowledge, morality, and ethics. Every book needs a villain, and mine will be Kant. The model of human beings that he defined has caused many more problems than it has solved. It is a model in which each individual constructs personal and private representations of the world around them. It separates people from one another and divides mind from world, value from fact, and knowledge from ethics. It is a big mistake!

Today we are facing a crisis more profound than a loss of meaning, the crisis of mounting environmental damage and escalating war between civilizations. It would be naive to suppose that qualitative research alone could provide a solution to worldwide crises. But we can at least ask that qualitative inquiry counter, rather than bolster, the attitude of seeking to dominate not only other peoples but the planet as a whole. I will argue that qualitative research has the potential to change our attitude of domination because it is sensitive to human forms of life in a way that traditional research cannot be. It can draw on powerful new conceptions of human rationality, alternatives to Kant's model. In this book, I will trace a line of theoretical and empirical work that has developed the proposal that the basis for rationality and order of all kinds is the hands-on know-how, the embodied practical and social activity, of people in a form of life. This line of work leads to new ways of conceptualizing social inquiry.

It might seem strange to link a form of research to a moral imperative. Yet traditional social science has just this kind of linkage, although it is disguised. As we shall see, the German philosopher Jurgen Habermas (1971) has argued that scientific knowledge is never disinterested and that the sciences, both natural and social, are generally motivated by a "technical" interest, an interest in fostering our instrumental action in the world and increasing our mastery of our planet. To some degree, qualitative research has succeeded in adopting a different attitude, one that Habermas calls (rather misleadingly) a "practical" interest: an interest in understanding other people. This is certainly an admirable goal, but one of the points I will make in this book is that too often this understanding has been based on the reduction of others to the status of objects for objective observation. Studying humans as objects – albeit complex and sophisticated objects – is not the same as studying humans as beings who live in particular cultural and historical forms of life and who are made and make themselves as specific kinds of subjects. What we need is a human science that is able to grasp this "constitution." Such a science would not abandon objectivity in favor of relativism, either epistemological or cultural. Rather, it would adopt a moral and epistemological pluralism resting on what has been called a "plural realism" (Dreyfus, 1991, p. 262). Such a science, I suggest, is exactly what qualitative inquiry is, properly understood.

What is needed is a kind of inquiry that is motivated by neither a technical interest nor a practical one but rather what Habermas called an "emancipatory" interest. How can we create this? The imperatives to change our paradigm – to assume a new ontology, to adopt a new view of understanding and knowledge – emerge *within* qualitative inquiry as much as they are demanded by the crisis we face. Much qualitative research is stuck in contradiction and anxiety, and it is crucial to understand why. By refusing to abandon a posture of detached neutrality, much qualitative inquiry today continues to bolster the attitude of domination. Neutrality is equated with objectivity and viewed as genuine knowledge. This kind of research promotes a way of knowing other people that leaves them feeling misunderstood and treated as objects, and fails to recognize either the political and ethical dimensions of understanding or its own transformative power. When we understand another person, we don't merely find answers to our questions about them (let alone test our theories about them) but are challenged by our encounter with them. We learn, we are changed, we mature. Contemporary qualitative research, with a few welcome exceptions, fails to recognize these things or even allow space for such recognition in its repertoire of techniques and its methodological logic.

I believe that if we think carefully about what we are doing, if we examine our own conduct carefully, we will see the inconsistencies in our current research practices and will start to notice where new possibilities lie. We will start to ask new kinds of questions, become able to see different kinds of connections and different kinds of causality, and perhaps view ourselves and our planet in a new light. This book, then, is a wide-ranging review and overview of types and varieties of qualitative research throughout the social sciences. It is selective rather than exhaustive; indeed, the qualitative research literature is now so extensive that trying to cover it comprehensively would be impossible. But in this literature certain issues and dilemmas recur. Studying them can help us envision a new program for qualitative research.

WHAT IS QUALITATIVE RESEARCH GOOD FOR?

So what *is* qualitative research good for? I will be making the case that qualitative research is good for *historical ontology*. I am adopting here a phrase that Michel Foucault coined in an article – "What Is Enlightenment?" – written toward the end of his life (Foucault, 1984b). Foucault sketched "a historical ontology of ourselves" that, he proposed, would involve "a critique of what we are saying, thinking, and doing." It would attend to the complex interrelations of knowledge, politics, and ethics. It would foster personal and political transformation without resorting to violence. It would be an investigation that could create new ways of being.

Foucault was, in my view, describing the kind of inquiry that many of us have been looking for. He viewed it as a form of investigation, even a particular attitude or ethos, that would be scientific without being disinterested, because we need knowledge that is relevant, not knowledge that is disengaged. In Foucault's terms, it would include both "genealogical" and "archeological" components and have an "ethical" aim. That is to say, it would include a historical dimension, attentive to genesis and transformation without reducing them to the linear unfolding of a unidimensional "progress." It would include an ethnographic dimension that would be sensitive to power and resistance. It would carefully examine practical activities – "discourse" – to discover how we human beings are made and how we make ourselves. And it would foster social change not through violent revolt but by promoting "a patient labor giving form to our impatience for liberty" (Foucault, 1975/ 1977, p. 319), working to change who we are.

Such a program of investigation defines what qualitative research can do and organizes its tools (interviews, ethnographic fieldwork, and analysis of interactions) and tasks (to offer knowledge, provide critiques, and foster transformation) in powerful ways. But before we can grasp what such a program involves, we need to reexamine how these tools have been used. The first part of this book explores how the qualitative research interview has become a tool with which researchers try to study subjective experience objectively and suggests that it is better understood as an interaction between two (or more) people, a tool better employed to discover how a person has been constituted in a particular form of life. The second part examines the theory and practice of ethnographic fieldwork, uncovering its tacit ontological assumptions. I explore the popular notion that reality is a "social construction" and distinguish two forms of this claim, one radical, the other not. I suggest that ethnographic fieldwork is an important tool for investigating how a form of life has been constituted and that interaction analysis is a tool for exploring how this constitution continues. The third part turns to the ethical dimension of research, understood as a critical and emancipatory or enlightening practice. I define the tasks of a research program of historical ontology employing these three research tools to answer questions about constitution.

The natural sciences have investigated how the natural world works in order to enable us to manipulate and control it. In doing so, they have created the means for great destruction as well as, hopefully, instruments with which we may undo the damage we have caused. The traditional social sciences have investigated how humans operate as information-processing organisms and have helped design better manipulation in the form of advertising and spin. We desperately need a program of inquiry that can ask questions whose answers would empower us to transform our forms of life, our moral paradigms, and our discursive practices for the better. Qualitative inquiry could overcome its current confusion and fragmentation by adopting a program such as this.

Changing the attitude of seeking to dominate the planet, exploiting its raw materials and exporting one way of life to those who do not share it, will be no simple matter. It is a matter not merely of changing what we believe but changing who we are. Finding the freedom to do this will require that we engage in a critique of how we became who we are, to identify the limits placed on us by history and culture and step beyond them.

OVERVIEW OF THE BOOK

In Chapter 1, I frame what follows by showing how our thinking about science is still influenced by the logical positivism of the early 1900s. The positivists tried to outlaw talk about "ontology" – the kinds of entities that exist – because they considered such talk untestable and unscientific. Science, in their view, should be a solely logical process. The prohibition of ontology is still prevalent today, and proponents of randomized clinical trials have the same vision of science. Yet, as Thomas Kuhn showed, the natural sciences operate within qualitatively distinct paradigms, and a central component of any paradigm is the ontological commitments embedded in its practices. Science is not a purely logical process but a social practice in which some aspect of the world is explored systematically. The lesson is that what we need to do is not *avoid* ontology but adopt an ontology that is appropriate.

Part I. The Objective Study of Subjectivity

The first part of the book examines two of the most common practices of contemporary qualitative research, the semistructured interview and the analysis of interview material by coding. Chapter 2 compares the qualitative research interview with both the traditional survey interview and everyday conversation. The semistructured interview is more flexible than the survey and makes use of the resources of everyday interaction. But compared with a typical conversation, the interview is asymmetric in its use of these resources, shining all the light on the interviewer and encouraging a particular kind of self-disclosure. This would make sense if an interview provided an expression of the interviewee's subjective experience. But this way of thinking about interviewing rests on a common but misleading metaphor about language – that it is a "conduit" through which "meaning" is transferred from one individual to another. This metaphor clashes with the belief that an interview is always a joint production.

Chapter 3 finds the conduit metaphor at work again in the analysis of interviews by coding. Coding involves practices of abstraction and generalization that divide an interview transcript into separate units, remove these units from their context, identify abstract and general categories, extract the content of these categories, and then redescribe this content in formal terms. Language is treated as a collection of words that are labels for concepts, and coding as a process that "opens up" these words and "takes out" the meanings they contain.

Grounded theory (Glaser & Strauss, 1967; Strauss & Corbin, 1990) is a key example of this approach to analysis. But a paradox underlies coding, for although it celebrates individual subjectivity, it tries to eliminate the researcher's subjective experience. It ignores context, the diversity of participants, and the influence of the researcher. People are assumed to be separate from one another and separate from the world we live in. Experience is assumed to be internal and subjective, distinct from an eternal, objective reality.

The anxiety behind the insistence on coding and the confusion over how to do it stem from the conundrum that it seems impossible to obtain objective knowledge from subjective experience. Scientific knowledge is assumed to be abstract, general, and formal, so coding must eliminate what is concrete, specific, informal, and personal. Particular things are treated merely as exemplars of general concepts. Specific experiences are viewed merely as cases of general knowledge that can be formally expressed. But both philosopher Ludwig Wittgenstein and sociologist Harold Garfinkel have questioned the central assumption in coding: that the meaning of a general term is what is *common* to all its exemplars. They recommend instead that the meaning of a word is to be found in its *use*. And, certainly, in practice coders inevitably rely on their tacit understanding of the material they are coding, especially their everyday understanding of how words are used.

There is a gap between the theory of coding and how it is practiced. Coding doesn't do what is claimed. But what is the alternative? Chapter 4 turns to hermeneutics – the theory of interpretation – and the 200-year debate over what it means to understand a text. Wilhelm Dilthey and Friedrich Schleiermacher assumed – like many modern researchers – that to understand a text one needs to reconstruct the author's subjectivity. But Hans-Georg Gadamer argued convincingly that understanding a text always involves its active "application" to a current situation. Meaning is an *effect* of reading a text, and this will be different for each reader. This means that there can be no single correct interpretation of any text but multiple readings, each of which has relevance to a specific time and place.

In Chapter 5, I explore the implications of Gadamer's argument for the analysis of interviews. To cut the Gordian knot of subjectivity-objectivity, we need to attend closely to the language of an interview transcript, its rhetorical structures, techniques, and strategies. Any text – written or spoken discourse – engages its readers and invites them to see the world in a new way. The work of literary critic Wolfgang Iser and historian Hayden White helps us understand how to study the *effects* of reading an interview transcript. An interview has *ontological* power, the power to change how the world is understood. Analysis should focus on how an interviewee crafts a way of *saying* to invite a way of *seeing*.

Our understanding of what someone tells us in an interview builds unavoidably on factors that are not personal or individual but *intersubjective*. Language itself is an intersubjective phenomenon, and the researcher's knowledge of language plays a crucial role in both the conduct and the analysis of an interview. The interview, which seemed a personal, individual source of data, turns out to be based on shared, public linguistic conventions and practices. At the same time, seemingly simple notions such as "subjectivity," "experience," and "meaning" turn out to be surprisingly slippery. This suggests that we should ask whether qualitative research should be the study of *intersubjective* phenomena, such as language, culture, and society.

Part II.   Ethnographic Fieldwork – The Focus on Constitution

The second part of the book turns to how intersubjective phenomena have been studied. Chapter 6 begins with three calls for a new kind of interpretive social science that were made in the 1970s. Charles Taylor argued that political science cannot avoid interpretation. Anthony Giddens pointed out that the logic of sociology involves a double hermeneutic. Clifford Geertz proposed that a culture should be viewed as a collection of texts that requires an interpretive anthropology. In each case, immersion in the social practices of a community – ethnographic fieldwork – was considered crucial, rather than surveys, questionnaires, or even interviews. In each case, interpretation – hermeneutics – was regarded as a central aspect of inquiry. In each case, the new approach was expected to resolve core dualisms that plagued the discipline. And in each case it was said that we would study the key relationship of *constitution* between humans and the world.

The term constitution is rarely defined, but it can be traced back to Aristotle's recognition, more than 2,000 years ago, that there is a relationship of mutual "constitution" between human beings and our forms of life. He argued in his *Politics* and *Ethics* that a human is naturally a societal animal, *zoon politikon*, whose nature it is to live in communities, and that "the natural outcast ... may be compared to an isolated piece at draughts" (Aristotle, 1995, p. 5). Outside of society, a human being has no game to play. The state is prior to the individual, as the whole is prior to the parts, but society doesn't just regulate and direct its members' conduct; it is concerned with their flourishing as humans. The ultimate end of the state, for Aristotle, is the well-being of its citizens, enabling them to develop, to live the good life. At the same time, citizens also play an active role, for it is in participation that they find out the human good.

So the citizens of a community "constitute" it: they decide, formally and informally, how they will live together. Sometimes there is an explicit "constitution," but often the decision emerges tacitly. At the same time, a community doesn't just regulate its citizens' activity but fosters their flourishing. Only by living together with others can humans actualize their capacities, both intellectual and moral. Communities "constitute" the people who live in them. Constitution, then, is this relationship of mutual formation between people and their forms of life.

What is the best way to grasp this interconnectedness and study it adequately? The notion of constitution is developed in Chapters 7 and 8 by tracing the history of two distinct treatments. One has been to make the *epistemological* claim that a human being's *knowledge* of the world they live in is constituted by social practices. I trace this first approach from Kant to philosopher Edmund Husserl and sociologists Alfred Schutz, Peter Berger, and Thomas Luckmann, and propose that ultimately it fails to escape from Kant's individualistic model of human being. With such a model, we can only explore how the world can *appear* objective to an individual subjectivity. This kind of "social construction of reality" can never establish a distinction between what is mere opinion and what is valid knowledge.

The second approach has been to make the *ontological* claim that social practices constitute *real* objects and subjects. This approach is much more powerful and has far-reaching implications. In Chapter 8, I begin with Georg Hegel's response to Kant and then trace the work of philosophers Martin Heidegger and Maurice Merleau-Ponty and sociologist Harold Garfinkel. Their work has articulated a *nondualist ontology* and shifted the focus from *conceptual* knowledge, studied with a detached, theoretical attitude, to practical, embodied *know-how*, studied in an involved way. They have shown how we can see reason and thinking as cultural and historical, as grounded in practical know-how, and how we can see research as thinking that doesn't take itself for granted.

Chapter 9 returns to the debates in cultural anthropology over the manner and purposes of ethnographic fieldwork. Traditional ethnography was wedded to the image of the researcher "alone on a tropical beach close to a native village," as Bronisław Malinowski (1922/1961) put it, and to the ontological presuppositions that culture is bounded, systematic, and integrated (Faubion, 2001). These imply that a fieldworker must enter a culture and participate as a member, describing a member's point of view of *their* world. A more adequate ontology presumes that a culture is a dispersed, dynamic, and contested form of life. Ethnographers need to find and trace this form as newcomers who are representatives of what is "elsewhere." Rather then try to describe structures *behind* everyday life, they need to focus on the order that has been constituted in a form of life: the *regional ontology*, how people and things "show up." And this is not a matter of mere description: ethnographers write accounts to have an *effect* on their readers, inviting new ways to see the world. Malinowski was surely right to see fieldwork as a way of understanding other people in order to better know ourselves and grow a little in our wisdom.

One of the implications of this second treatment of constitution is that the processes of what I call "ontological work" can be studied by researchers, and Chapter 10 compares two approaches to the study of practical activity, in particular discourse practices: critical discourse analysis and conversation analysis. The former turns out still to assume a dualism between person and form of life and tries to bridge the gap with representation. In contrast, conversation analysis pays attention not to *what* people say so much as to what they *do* by saying. It also attends to the way participants in a form of life *display* their understanding of what they and others are doing.

Part III. Inquiry with an Emancipatory Interest

But participation in the practices of a form of life can provide *mis*understanding, and this means that inquiry needs to have a *critical* dimension. Part III explores different approaches to critical inquiry. Chapter 11 traces the origins of critique back to Kant, whose exploration of the conditions for the possibility of knowledge in his *Critique of Pure Reason* defined one aspect of the term critique. When Karl Marx's analysis of capitalism showed that the exploitation of workers, their labor squeezed to extract value, is the condition for the possibility of capital accumulation, the term came to mean both the exploration of the conditions that make a phenomenon possible and the exposure of exploitation.

Marx anticipated the new ontology of ethnography when he proposed that capitalism is open, dynamic, and contested. He argued that the notion that knowledge and research can be disinterested is a myth, an ideology, and he practiced instead a critical and emancipatory kind of inquiry. He did this by seeking a *historical* perspective that people lack in everyday life.

Marx drew his conception of history from Hegel. The following three chapters explore three attempts to base critical and emancipatory investigation on a different kind of history. Chapter 12 focuses on the German philosopher Jurgen Habermas, who has considered what is needed for emancipatory research. He suggests that a researcher needs the know-how of a member but also a historical perspective in the form of a rational, theoretical reconstruction of ontogenesis and societal history. Such a history provides a lens through which a form of life can be studied and critiqued. Habermas has accepted Kant's conception of enlightenment, but he looks for the source of rationality not in transcendental reason but in communicative practices. Research, for Habermas, involves articulating what participants in a form of life presuppose unquestioningly and questioning what they recognize unthinkingly. In doing so, the researcher "deepens and radicalizes" the context of communication that is being investigated.

Chapter 13 turns to French sociologist Pierre Bourdieu, for whom research is a reflexive enterprise that objectifies its own techniques of objectification. Bourdieu's "reflexive sociology" centered around the relational concepts of "habitus" and "social field." Whereas Habermas focused primarily on people's intellectual judgments, Bourdieu emphasized their embodied and situated practical know-how and how this often serves to reproduce an inequitable social order. Bourdieu was more radical than Habermas in his insistence that reason is historical and embodied, and that each of us has acquired bodily dispositions to produce strategic action in a social field that is the site of a game, a struggle, that only the researcher can grasp as a whole. It is the reflexive aspect that gives social science its special status among such games and its ability to produce knowledge that transcends a specific time and place.

In Chapter 14, the central figure is French historian Michel Foucault. Foucault criticized the human sciences for adopting the view that humans are at one and the same time objects and subjects, assuming paradoxically that people are both determinate and uniquely free. When we examine the historical record, he insisted, we find a *variety* of kinds of human beings, in multiple forms of life. Foucault developed a way to study how humans are formed that had three aspects. First is an *archaeology*: a form of investigation that excavates not bones, pottery, and metalwork but official theories or concepts. The second is *genealogy*: tracing the family tree of these official pronouncements to write "histories of the present" that treat historical change as contingent, marked by ruptures and discontinuities. The basis of official knowledge (*connaissance*) must be explored in the power relations (*pouvoir*) of practical activity (*savoir*). The third aspect is an *ethics* that focuses on the techniques for formation and care of the self. If Kant is the villain of this book, Foucault is its hero. His work pulls together the threads of our various concerns. He explored the linkages between formal knowledge and embodied social know-how. He emphasized the constitution of both objects and knowing subjects in practical relations of power. He emphasized history without reducing it to logic or progress. He practiced a form of inquiry intended to be emancipatory without being authoritarian. He searched for local truths – ways in which an aspect of life is problematized – rather than universal, objective truth with a capital T. For Foucault, inquiry needs to problematize problematization.

At the end of his life, Foucault articulated the three central questions that he had tried to answer and that defined a broad program of research he called a "historical ontology of ourselves." The questions were: How are we constituted as subjects of our own knowledge? How are we constituted as subjects who exercise or submit to power relations? How are we constituted as moral subjects of our own actions? This is not a "how to" book, so I do not end with a discussion of techniques for posing these questions or detailed specifications for the program of a *historical ontology of ourselves*. Instead, Chapter 15 returns to the larger question of what science is in the light of what we have learned about the critical investigation of constitution. I propose that we think of human science itself as a program of research with theoretical, practical, and ethical dimensions.

Such a program has three phases: an archaeological phase (fieldwork), a genealogical phase (the study of practice), and an ethical phase (ethology, in its original sense as the study of character). Researchers conducting fieldwork will acknowledge that they can rarely be members of the form of life they study. They are strangers, visitors from the academy, and their fieldnotes and ethnographic accounts – accounts of the *regional ontology* – need to have local accountability. Their detailed analysis of practical interactions will go beyond the kinds of critical discourse analysis currently available to focus on the pragmatics of interaction, how it is embedded in material settings, and the *ontological work* that is accomplished. Their analysis of interviews will attend to the way rhetorical devices are used to invite us to see the world in new ways and show the *ontological complicity* of the speaker with a form of life. Their research will be reported in texts that offer both a way of saying and a way of seeing because thinking is a social practice of seeing and saying that exploits the power of language. Scientific accounts can offer *phronesis*, practical /political relevance. Scientific inquiry, practiced this way, can open our eyes to fresh ways of being human. This is the excitement, and the importance, of qualitative research.

One final, parenthetical remark. I wear two hats, suffering from a divided professional identity with one foot in methodology and the other in child development. Much of what is discussed in this book on the former has relevance to the latter. There simply is no space to detail the brilliant work of Lev Vygotsky in Russia or the groundbreaking investigations of Michael Cole and his colleagues in the United States, although this work "studies that zone of proximal development where the cultural becomes individual and individuals create their culture" (Laboratory of Comparative Human Cognition, 1983, pp. 348–349). Constitution, in other words.

PART I

THE OBJECTIVE STUDY OF
SUBJECTIVITY

1

What Is Science?

[T]he logical empiricists sold us an extraordinary bill of goods.
Taylor, 1980, p. 26

Science is not hypothetico-deductive. It does have hypotheses, it does
make deductions, it does test conjectures, but none of these determine
the movement of theory.
Hacking, 1983, p. 144

In an article titled "What drives scientific research in education?" Shavelson and
Towne (2004) note that the debate over how to define social science has gone
on for more than 100 years. They try to calm what have become politicized
arguments by recommending that scientific inquiry should be defined not
by a particular methodology but by a way of posing and answering questions.
Summarizing the conclusions of a National Research Council (NRC) committee convened in 2001 by the National Educational Research Policies and
Priorities Board, they recommend (see Table 1.1) that all scientific research, in
both the natural and the social sciences, should pose significant questions that
can be investigated empirically, should be linked to relevant theory, should use
methods that permit direct investigation of the questions, should provide a
coherent and explicit chain of reasoning to rule out counterinterpretations,
should replicate and generalize findings across studies, and should disclose
research data and methods to enable and encourage professional scrutiny and
critique (see Feuer, Towne, & Shavelson, 2002; Shavelson & Towne, 2002).
Overall, "It's the question – not the method – that should drive the design of
education research or any other scientific research. That is, investigators ought
to design a study to answer the question that they think is important, not fit the
question to a convenient or popular design" (Shavelson & Towne, 2004).

These recommendations seem reasonable, and the effort to overcome
competition among polarized camps seems admirable. However, the questionable assumptions that underlie their recommendations start to become
evident when the NRC committee identifies three fundamental types of

TABLE 1.1. *Key Characteristics of Scientific Research*

Scientific Research Should
Pose significant questions that can be investigated empirically
Link research to relevant theory
Use methods that permit direct investigation of questions
Provide a coherent, explicit chain of reasoning to rule out counterinterpretations
Replicate and generalize findings across studies
Disclose research to encourage professional scrutiny and critique

TABLE 1.2. *NRC's List of Questions, Answers, and Methods*

Questions	Answers	Methods
1. What's happening?	Asks for a description	Case studies
2. Is there a systematic (causal) effect?	Asks for a causal connection: X caused Y	Randomized clinical trials. Quasi-experiments and correlational studies when necessary
3. What is the causal mechanism? *or* How does it work?	Asks for a causal model	Longitudinal studies Artifact constructions

questions and the methods they consider most appropriate to answer them (see Table 1.2).

The three questions are (1) What's happening? (2) Is there a systematic (causal) effect? and (3) What is the causal mechanism, or how does it work? The committee judged that the first type of question is asking for a description, and they recommended that this should be provided by a survey, ethnographic methods, or a case study. The second type of question is asking whether X caused Y. Here the most desirable method is a randomized clinical trial. Quasi-experimental, correlational, or time-series studies may by needed when random assignment is either impractical or unethical, but "logically randomized trials should be the preferred method if they are feasible and ethical to do." The third type of question – how does it work? – asks for identification of the causal mechanism that created a described effect. Here it seems that mixed methods could do the job. (The committee seemed a bit confused here, perhaps because they believed that causal mechanisms can never be directly observed.)

A significant problem with these recommendations, well intended though they undoubtedly are, is that they perpetuate a widely held but incorrect belief that qualitative research can answer only descriptive questions, whereas

TABLE 1.3. *The Clichéd View of Qualitative and Quantitative Research*

Quantitative Research	Qualitative Research
Provides explanations	Provides only descriptions
Is objective	Is subjective
Studies causes	Studies experiences
Can test hypotheses	Can only generate hypotheses

quantitative research is able to answer explanatory questions, and, in addition, that such questions are always answered by identifying a causal mechanism (see Table 1.3). If this were so, qualitative research would be adequate for *generating* hypotheses, but measurement and experimentation would be needed to *test* these hypotheses. This was indeed the committee's position. Experimentation, they asserted, "is still the single best methodological route to ferreting out systematic relations between actions and outcomes" (Feuer, Towne, & Shavelson, 2002, p. 8). Although they regretted that "the rhetoric of scientifically based research in education seems to denigrate the legitimate role of qualitative methods in elucidating the complexities of teaching, learning, and schooling," they saw this "legitimate role" as a limited one. "When a problem is poorly understood and plausible hypotheses are scant – as is the case in many areas of education – qualitative methods such as ethnographies ... are necessary to describe complex phenomena, generate models, and reframe questions" (Feuer, Towne, & Shavelson, 2002, p. 8). In other words, qualitative research can *invent* hypotheses but can never *test* them, so it can never provide explanations.

Perhaps it is true that the committee avoided the temptation of allowing method to drive their choice of research design, but their unexamined assumptions about the nature of science led them to a very short list of the types of questions that can be asked. The committee adopted without question the model of scientific research as a process of "hypothesis testing," the application of a "hypothetico-deductive" logic. The basic idea in this model is that science proceeds by taking two steps. First is the speculative step of proposing a hypothesis. Second is the logical step of testing this hypothesis to see whether its predictions hold up. Science builds knowledge, on this account, by systematically testing hypotheses and eliminating those that are found to be false.

The randomized clinical trial has in recent years been called the "gold standard" of research in social science. For example, the U.S. Department of Education considers use of this design the main sign that a study is supported by "strong evidence." In the department's view, "All evidence is NOT created equal" (U.S. Department of Education, n.d.), and the evidence from a randomized clinical trial is much stronger than evidence from other kinds of investigations.

TABLE 1.4. *The Elements of a Randomized Clinical Trial*

It evaluates a *treatment* (a medicine, an intervention) intended to change people (improve health, foster learning).
By *comparing* two or more groups, only one of which receives the treatment
The participants are *assigned* randomly to these groups in order to control for unknown variables.
The outcomes are *measured* with appropriate tests and instruments.
The *null hypothesis* is that the treatment has no effect.
No effect means that any differences among the groups in measures of the *dependent variables* are caused by chance alone.
Analysis takes the form of statistical tests to decide how probable it is that the differences are caused by chance alone and thus how *significant* these differences are.

A randomized clinical trial (see Table 1.4) is a comparison of two or more groups to which participants have been randomly assigned. Its purpose is to test the hypothesis that some kind of treatment – a drug, a method of teaching, an intervention – makes a measurably significant difference to some characteristic of these groups, the dependent variable. Random assignment means that the groups are most likely to be similar not just on factors that the researcher knows but also on unknown factors. Ideally neither the participants nor the researchers know who is in the treatment group: this is a "double-blind" trial. Independent (treatment) and dependent (outcome) variables are given "operational definitions": each variable is defined in terms of the operations with which it will be manipulated and/or measured, with appropriate tests or measurement instruments. The results are analyzed statistically to decide how likely the measured differences between the groups are to be caused by chance alone.

These elements of the randomized clinical trial follow from the assumption that scientific research is a "hypothetico-deductive" process that builds knowledge by systematically testing hypotheses and eliminating those that are found false. The clinical trial is designed to test a specific hypothesis about a treatment. The emphasis on assignment to groups, on operational definition of measurable variables, and statistical testing reflects assumptions about science and scientific knowledge that have become second nature but should not go unquestioned. These assumptions have a history, though it is one many people have forgotten.

THE LOGICAL POSITIVISM OF THE VIENNA CIRCLE

For there is but one science, and wherever there is scientific investigation it proceeds ultimately according to the same methods; only we see everything with the greatest clarity in the case of physics, most scientific of all the sciences. (Hahn, 1933, 1959, p. 147)

The roots of the hypothesis-testing model of science can be traced back to the start of the 20th century, when a group of scientists, mathematicians, and philosophers formed what they called the Vienna Circle. The group met informally in Vienna during the early 1920s to discuss science and philosophy. Led by Moritz Schlick, they coined the term "logical positivism" (see Table 1.5) to reflect their agreement with the "positive philosophy" of French thinker Auguste Comte (1798–1857). The addition of the word "logical" was intended to signal the importance of formal logic in scientific investigation. As one of them put it, empirical work and logical construction "have now become synthesized for the first time in history" (Neurath, 1938, p. 1). Empiricists like John Locke and David Hume had seen experience as the basis for knowledge. Rationalists like Rene Descartes had instead based knowledge in the human capacity for reason. These disagreements were finally to come to an end, the Vienna Circle believed, because logical positivism defined the roles in science for *both* experience (in the form of measurements) *and* reason (in the form of logic). The logical positivists had been influenced by Immanuel Kant, who, as we shall see in Chapter 7, had proposed that human knowledge of the world draws its content from sensory perception and its form from innate cognitive categories. In logical positivism, logic and mathematics would provide the form, while observations would provide the content. We can see why this reconstruction of science was also called "empirical rationalism" and "logical empiricism" (Hanfling, 1981).

The Reaction to Einstein's Revolution

One of the main reasons the logical positivists believed that a new model of science was needed was the revolutionary new physics of Albert Einstein. At the end of the 19th century, physicists had begun to think their work was finished and that every physical phenomenon had received an adequate scientific explanation. Newtonian physics, whose laws of motion applied with equal accuracy to the flight of an arrow, the rotation of the planets, and the movement of atoms, seemed flawless and complete. His optics, with studies of color, refraction, and reflection, was equally powerful. Although Newton had lectured in the 1670s and 1680s and published his *Principia* (*The Mathematical Principles of Natural Philosophy*) in 1687, in the late 19th century his work was still unsurpassed. It combined, in powerful and compelling ways, detailed empirical observation with mathematical analysis – especially the calculus, which Newton invented (independently of Leibniz). Newton's famous laws of motion and theory of gravity were considered the model for scientific reasoning 200 years after he formulated them. At the start of the 20th century, physicists were confident that they were near the end of their task; the laws of physical nature were almost complete and perfect. New observations of electromagnetic radiation needed to be fitted in, but this was generally thought to present no real problem.

But then Einstein knocked over the Newtonian applecart. His theory of relativity not only predicted empirical phenomena that had not previously been observed but also explained known phenomena that had proved troubling to Newtonian physics. The most famous example was the precession of the perihelion of the planet Mercury: the way the point on the planet's orbit at which it is closest to the sun shifts slightly with each revolution. But much more important than this, Einstein's physics contradicted most of the basic tenets of the Newtonian worldview. Einstein insisted that motion is relative, so that a body moving in one frame of reference may be stationary in another. He proposed, outrageously, that mass changes with velocity (which means that mass is relative, too). The very concept of a frame of reference, fundamental in Einstein's physics, is simply missing from Newtonian physics; Newton's laws were written as though things are observed from nowhere, or perhaps from everywhere. The Newtonian physicist had, without noticing it, adopted a God's-eye view. Such a position, Einstein declared, is impossible.

It is clear why Einstein's new physics shocked the scientific establishment. In the meetings of the Vienna Circle, the chief topic of debate was what had gone wrong with Newtonian science. How could a system of explanation that had seemed so compelling, so complete, and so consistent turn out to have been so wrong in so many ways? One of their conclusions was that Newtonian physics had, despite appearances, included assumptions that were "metaphysical" rather than truly scientific. Central among these was the concept of gravity. In Einstein's physics, gravity is a local phenomenon – it is the way a body follows the path of least energy through space that has been curved by the presence of a mass. In Newton's physics, gravity was something that now seemed mysterious – a force that one body somehow exerted on another across empty space. Action at a distance, with no intermediary: how could that be? This, surely, was metaphysics (see Box 1.1) rather than genuine science!

THE RECONSTRUCTION OF SCIENCE

In their effort to put science on a firm footing, the Vienna Circle began to define principles that would prevent this kind of embarrassment in the future: principles for any genuine science. Einstein's theory of relativity seemed to offer important guidelines for the way science ought to be done. Bridgman – who developed the notion of an "operational definition" (Bridgman, 1945) – put it as follows:

> The Relativity Theory of Einstein is the result of, and is resulting in, an increased criticalness with regard to the fundamental concepts of physics. ... The general goal of criticism should be to make impossible a repetition of the thing that Einstein has done; never again should a discovery of new experimental facts lead to a revision of physical

BOX 1.1. *Metaphysics*

Metaphysics is the investigation of the fundamental principles of reality and the nature of being. The term comes from the Greek words μετά (metà) (meaning "beyond" or "after") and φυσικά (physikà) (meaning "physical"). Its first use was based on the order of the texts in published editions of the writing of the Greek philosopher Aristotle (384 BC – 322 BC). The chapters on "physics" were followed by chapters on what Aristotle himself called "first philosophy" (Aristotle, 1988).

Ontology is one of the central branches of metaphysics and is the investigation of the types of entities that exist and the relations these entities have. Epistemology is not usually considered part of metaphysics. It is the investigation of the ways people know the world and questions such as what distinguishes knowledge (*episteme*) from mere opinion (*doxa*). Ontology (or metaphysics) and epistemology are together the central pursuits of philosophy.

Science was originally considered to be part of metaphysics, "natural philosophy." Modern science considers the scientific method to be empirical and philosophy to be speculative or purely theoretical, so that many people now consider metaphysics to be something distinct from, and even opposed to, empirical science.

concepts simply because the old concepts had been too naïve. (Bridgman, unpublished manuscript, 1923, cited in Miller, 1962/1983, p. ix)

The reconstruction of science depended on a number of key tasks. The first was to distinguish between statements that were properly formed, or "meaningful," and those that were nonsensical because they referred to metaphysical notions. The criterion here was whether or not a statement could be tested using either empirical or rational procedures. For example, there are clearly ways to test empirically whether "the moon is round," so such a statement is meaningful. But a statement like "all life is a void" cannot be tested, so it is unscientific and meaningless. References to "absolute" space or time, independent of all observers, were now judged to be meaningless. More surprisingly, causality also was now seen as a metaphysical notion. A theory may make causal claims, but all we can observe empirically are associations of events, so a meaningful hypothesis must refer only to these. Value judgments, both ethical ("killing is wrong") and aesthetic ("coffee tastes good"), were also considered meaningless unless they could be transformed into factual statements about people's preferences: "more people say they prefer cheese."

The next task was to clarify the role of observation in science. The positivists proposed that reports of observable phenomena provided the "protocol sentences" on which all scientific knowledge is based. These simple, basic

TABLE 1.5. *The Logical Positivist View of Science*

Scientific knowledge involves matters of fact, not of value.
(Values are merely personal preferences or subjective opinions.)
The goal of science is a network of knowledge statements
(a unified theory of everything).
Research is based on measurement and logical inference.
Measurement is the objective application of an instrument.
Observation and theory are distinct.
Scientific method is the same everywhere.
Scientific knowledge accumulates.
Meaningful scientific statements include no metaphysics.
They include only logical propositions and statements of empirical regularities.
Observation provides the elementary statements.
Propositional logic combines these to build theoretical statements.

statements of experience would provide the empirical basis for any science. They could be combined to form more complex statements, but for this to be successful the truth value of the elementary statements needed to be unambiguous. Observations should be self-evidently true, incorrigible; that is to say, requiring no interpretation or prior knowledge:

> The analysis of knowledge leads to the search for the simplest basic sentences upon which further development can rest. We find them in the so-called "protocol sentences," i.e., short linguistic indications of the immediately observed present. (Von Mises, 1939/1956, p. 368)

A clear and unambiguous language was needed for observation so that empirical data would be as free as possible from personal bias and theoretical contamination. The way to prevent concepts from being naive or metaphysical was to define them "operationally," in terms of operations of observation and measurement. Einstein's physics built from very simple and straightforward observations: the reading of clocks, the observation that two bodies coincide, the application of measuring sticks. This was the "empiricist" side of logical empiricism. The other side was the "logical" component. In a genuine science, the basic data must be built inductively into coherent theoretical statements using the laws of formal logic. Logic would provide the language in which "theory becomes a logical short-hand for expressing facts and organizing thoughts about what can be observed" (Hacking, 1983, pp. 169–170). Scientists deal with linguistic statements, not mental states. But natural language is misleading; it is filled with unscientific, metaphysical notions. "It is the oldest experience and the primitive theories derived from them that are preserved in the traditional stock of language" (Von Mises, 1939/1956, p. 368). Formal logic offered a scientific language that would avoid these problems and preserve the truth value of scientific observations.

The rules of formal logic could be used to combine protocol sentences and produce increasingly complicated statements, culminating in the general theoretical statements that are scientific laws. "Starting from single observations, general propositions are set up in a constructive manner as conjectures (the so-called inductive inference)" (Von Mises, 1939/1956, p. 368). The members of the Vienna Circle were impressed by the theory of language that Ludwig Wittgenstein had introduced in his *Tractatus-Logico-Philosophicus* (1922). Wittgenstein wrote of "atomic propositions" that "mirror" or "picture" the world. Such propositions – each of them either true or false – can be systematically combined in "truth tables." The truth or falsity of more complex propositions follows logically (i.e., without interpretation) from the truth values of the atomic propositions. All this seemed exactly the kind of combination of the empirical and the rational that the Vienna Circle wanted. But when in 1927 they finally convinced Wittgenstein to meet with them, they realized his point of view was very different. As we will see in Chapter 3, Wittgenstein himself came to repudiate this view of language in his later works, such as his *Philosophical Investigations* (1953). But for the Vienna Circle the central task of philosophy was to analyze the formal language of logic in science, which they considered the only valid way to talk about knowledge:

> *Philosophy is to be replaced by the philosophy of science* – that is to say by the logical analysis of the concepts and sentences of the sciences, *for the logic of science is nothing other than the logical syntax of the language of science.* (Carnap, 1937/2002, p. xiii, emphasis original)

However, the logical positivists' apparently straightforward proposals immediately led them into disagreements and conflict. Defining meaningful statements and linking observation and theory turned out to be difficult. First, statements about empirical regularities could clearly be tested empirically, but science also contains logical statements, tautologies such as "a bachelor is an unmarried man," which are true by definition. Obviously one wants to consider these meaningful, too, but they cannot be given operational definitions.

The logical positivists originally proposed that the meaning of a scientific statement *is* its method of verification. But this ran into immediate problems: how can a "meaning" be a "method"? What if there is more than one method? So they proposed instead that a meaningful statement is one that *has* a method of verification. But this also had problems. It excluded universal claims: "All swans are white" is impossible to verify but hardly meaningless. It excluded historical theoretical statements, such as: "The universe once had a diameter of 1 meter." It excluded speculative counterfactuals, such as: "If the moon were hit by a sufficiently large asteroid, it would collide with the earth." And it excluded hypothetical counterfactuals: "If the moon were plastic, its orbit would be...." And what if we don't yet have a method to test a statement? Amusingly the verification criterion was itself meaningless by its own definition because it could not be verified (see Hempel, 1935; Hempel & Oppenheim, 1948; Schlick, 1935, 1936)!

The proposal that elementary observational statements are infallible was also unexpectedly controversial. How can science be based on experiences when these are presumably private? Does the scientist describe basic sensations, physical objects, public events, or measures of properties? Can an observation be directly confirmed – if, for example, I claim that a cathedral has two spires, can others judge whether my statement corresponds with the facts? Or is the truth of a statement a matter of its coherence with other statements?

It soon became clear that, as far as the physical and natural sciences were concerned, the logical empiricist model was highly inaccurate. It just didn't describe what physicists, biologists, or chemists actually do. But, despite this, logical positivism became highly influential. The social sciences were differentiating themselves from philosophy and seeking their own identity, especially in the United States but also in the United Kingdom and other countries. They adopted the logical positivist program as a blueprint for truly "scientific" inquiry. In a real sense, the logical positivist model of science was *made* a reality by investigators of psychological, social, and cultural phenomena who were keen to have the status of scientists.

The Project for a Unified Science

The logical positivists' reconstruction of science to define more clearly the appropriate roles of empirical observation and logical reasoning was part of an ambitious project to create a "unified science." The "unity of science movement" was intent on "bringing together scientists in different fields and in different countries" (Neurath, 1938, p. 1) in a spirit of collaboration and tolerance, adding individual elements together to complete the mosaic of scientific activities. The *Encyclopedia of Unified Science* (planned as 26 volumes) was to be both product and record of this collaboration. If every scientist, in every scientific discipline, followed the proposals the logical positivists had defined, then it seemed that scientific progress and human well-being would be inevitable results: "We expect from the future that to an ever-increasing extent scientific knowledge, i.e., knowledge formulated in a connectible manner, will control life and the conduct of men" (Von Mises, 1939/1956, p. 368).

Modern Empirical-Analytical Inquiry

Today not many researchers accept the label "positivist." "The term has become one of opprobrium, and has been used so broadly and vaguely as a weapon of

---

BOX 1.2. *Popper's Modification*

Karl Popper raised an important objection to the logical positivist model of science, though in other respects his view of science was like theirs. He pointed out that reasoning by induction can never lead to certainty. A general statement – one with the form "All X are Y" – can never be finally demonstrated to be true. If I see one white swan after another, inductive reasoning will lead me to conclude that all swans are white. I may feel increasingly confident of this conclusion as more white swans appear, but I can never be certain because there may still be a black swan lurking out of sight. But we can, Popper pointed out, be certain when such a statement is *false*. A single black swan will "falsify" the statement "All swans are white."

Popper argued that certainty, not probability, is the goal of scientific research, and that science requires a "critical attitude" that seeks to test and falsify laws, not a "dogmatic attitude" that seeks to verify and confirm them (Popper, 1959, 1963). The criterion for a meaningful scientific statement is not that it can be verified but that it can be falsified. In place of inductive reasoning, Popper proposed a logic of falsification, of the *testing* of hypotheses.

Popper insisted on a distinction between "the context of discovery" and "the logic of validation." There are all sorts of ways for a scientist to invent or guess at a hypothesis, but imagination, speculation, and invention are not part of the logic of science. Science, as a rational activity, is about hypothesis testing, not hypothesis generation. Science, Popper insisted, follows a "hypothetico-deductive" logic (Popper, 1959).

---

critical attack ... that it has lost any claim to an accepted and standard meaning" (Giddens, 1974, p. i, cited in Carr & Kemmis, 1986, p. 61). But they continue to use the approach to investigation that was the logical positivists' reconstruction of the scientific method. They continue to accept without question the Vienna Circle model of science, despite its many problems: "[D]espite a dozen subsequent qualifications and changes of name, the same basic dichotomies (e.g., between the factual and the logical, the cognitive and the emotive) still preserve a central place in the 'logical empiricism' of the present day" (Janik & Toulmin, 1973, pp. 212–214).

One clear sign of this is the emphasis on randomized clinical trials. The influence of the Vienna Circle, and Karl Popper (see Box 1.2), is clear: a clinical trial is designed precisely to test a hypothesis. The outcome of such a trial is a decision about how confident we may be in rejecting the "null hypothesis" that a treatment makes no difference. The operational definition of variables is an effort the positivists would have applauded to avoid constructs with hidden metaphysical connotations. The mantra that correlation is not causation reflects the Vienna Circle's insistence that causality is a metaphysical notion.

A meaningful hypothesis must be framed in terms of the association of events, even when it is deduced from a causal theory.

Similarly, the emphasis on measurement continues the logical positivists' efforts to ground scientific knowledge in simple, objective empirical observations, free from personal bias or theoretical preconceptions. The logical positivists themselves were not especially committed to quantification or measurement in science (Michell, 2003). Carnap, in particular, granted that numbers are just one among a variety of possible ways to describe observed phenomena (Carnap, 1937/2002). But quantification offers apparent solutions to several of the problems that the logical positivist reconstruction of science ran into. First, measurements are publicly available, thus avoiding the objection that observations are private and unverifiable. Second, quantitative variables allow ready application of statistical procedures to assess reliability and make decisions about the outcome of a trial. The reliance on statistical tests reflects the logical positivists' emphasis on formal logic to ensure the validity of theoretical formulations. The logic of statistical inference provides a link between measurement and the acceptance or rejection of a hypothesis, though this logic is not inductive but deductive, showing Popper's influence. In addition, the use of inferential statistics to generalize from a sample to a population reflects the positivists' belief that a theory makes general claims.

Finally, randomized clinical trials show the distinction the logical positivists drew between fact and value. Hypothesis testing has no place for ethical or aesthetic norms and values. The "gold standard" for research must be a purely objective, factual enterprise.

A DIFFERENT MODEL OF SCIENCE

The logical positivists of the Vienna Circle tried to reconstruct science as they believed it *ought* to operate, as a process of interpretation-free observations and logical inferences. They wanted to build the foundations for a universal and unified science that would be free from metaphysics and so would never again suffer the embarrassment that Einstein's new physics had caused. This was an extremely ambitious program, and with time they became considerably more humble:

> To me it now seems incomprehensible that I should have ever have thought it within my powers, or within the powers of the human race for that matter, to analyze so thoroughly the functioning of our thinking apparatus that I could confidently expect to exhaust the subject and eliminate the possibility of a bright new idea against which I would be defenseless. (Bridgman, 1959, cited in Miller, 1962/1983, p. xi)

A "bright new idea" did indeed soon appear, in the form of a completely different interpretation of the relationship between Newtonian and Einsteinian science and a fresh view of science in general. It was offered by the historian of science Thomas Kuhn (see Table 1.6), whose book *The Structure of Scientific Revolutions* (1962) was, ironically, originally published in the grand encyclopedic *Foundations of the Unity of Science*, edited by the positivists Otto Neurath, Rudolf Carnap, and Charles Morris (Neurath, Carnap, & Morris, 1938).

Kuhn proposed that the structure of science is one of repeated revolutions. A revolution involves "the pieces suddenly sorting themselves out and coming together in a new way," and a science may go through hundreds of such revolutions in its history. Kuhn drew a distinction between "normal" and "revolutionary" science. In normal science, there is a widely held and accepted understanding of the central phenomena, and an agreed list of questions and problems to investigate. Teachers, textbooks, peer review panels, and editors all know what matters and what counts as a solution to a problem. Normal science operates within what Kuhn famously called a "paradigm."

A crisis begins to develop when a number of unsolvable problems or "anomalies" arise. Anomalies, such as the shift of Mercury's orbit, consistently violate scientists' expectations and consequently attract growing attention. They persist despite the best attempts to make them conform to existing theories or to explain them away as the products of experimental error or poor equipment. Eventually scientific research goes into crisis. A growing part of the scientific community feels that existing theories and concepts are inadequate and that things have gone badly wrong. The only solution seems to require a new way of working, a change in "the way in which one conceives of the phenomena, a break that may involve junking much former work, not as false but as part of a quickly forgotten example of how not to think" (Hacking, 1983). This is a "revolution." As Kuhn put it, scientific revolutions are "those non-cumulative developmental episodes in which an older paradigm is replaced in whole or part by an incompatible new one" (Kuhn, 1970, p. 92).

Scientific Paradigms

Kuhn emphasized that every scientist conducts their daily work in circumstances of continual uncertainty and ambiguity. In any area of scientific investigation, much more is unknown than is known. This would be paralyzing if it were not the case that basic "conceptual boxes" are provided by the accepted scientific paradigm. A paradigm is necessary, Kuhn argued, for research to proceed. Furthermore, paradigms are needed if young scientists are to be trained effectively. The education of a new generation of scientific researchers would be impossible without general agreement on what they need to know and do. It follows that normal science is a conservative enterprise that generally resists change. It is "a strenuous and devoted attempt to force nature into the conceptual boxes supplied by professional education. Simultaneously, we shall wonder whether research could proceed without such boxes, whatever the element of arbitrariness in their historical origins and, occasionally, in their subsequent development" (Kuhn, 1962, p. 5).

Why Does a Science Have Revolutions?

The occurrence of revolutions is a consequence of the existence of paradigms. If scientists must have a paradigm in order not to be paralyzed by uncertainty, and if normal science is a matter of extending the application of the conceptual boxes of the accepted paradigm rather than challenging them, change in the boxes cannot come easily. Kuhn proposed that change can only be a sudden, dramatic transformation. Paradigms must be defended, often at great cost, and novelties must be suppressed, so once problems with a paradigm become apparent, a battle will be necessary. It is rare for scientists to abandon the paradigm in which they were trained and adopt a new one, so the history of scientific progress is that younger, fresher scientists explore the possibilities of a new paradigm, whereas the established scientists defend the old one.

A scientific revolution involves several kinds of changes. It involves, of course, a rejection of the dominant conceptual framework and acceptance of a new, incompatible one. It also involves a change in the problems considered important, and in the standards for these problems and their solutions. It involves changes in the rules governing practice, the reconstruction of old theories, and the reevaluation of accepted facts. And it involves what we must understand, Kuhn argued, as a *transformation of the world* in which science is done. This point is of central importance to my argument in this book, which will culminate in Chapter 15 with Foucault's notion of *problematization*, which I have already mentioned. To understand Kuhn's point more clearly, we must consider paradigms more closely.

Kuhn's idea of scientific paradigms immediately became influential, but it was not without its critics. Several commentators noted ambiguity in Kuhn's use of the term "paradigm." Kuhn responded by acknowledging that it indeed had two different senses (see Kuhn, 1974/1977):

> On the one hand, it stands for the entire constellation of beliefs, values, techniques, and so on shared by the members of a given community. On the other, it denotes one sort of element in that constellation, the concrete puzzle-solutions which, employed as models or examples, can replace explicit rules as a basis for the solution of the remaining puzzles of normal science. (Kuhn, 1970, p. 175)

Paradigm as Exemplar

Kuhn emphasized that exemplars, or concrete problem solutions, are core elements of a paradigm. "Concrete scientific achievement is prior to the various concepts, laws ... that may be abstracted from it" (Kuhn, 1970, p. 11). Cases of exemplary scientific research define what is considered a well-formed problem in a scientific paradigm and what is a good solution to such a problem. These are "[u]niversally recognized scientific achievements that for a time provide model problems and solutions to a community of practitioners" (p. viii). They are models of scientific practice that provide the basis for a coherent program of research.

Paradigm as Disciplinary Matrix

The second sense of paradigm is as a "disciplinary matrix": "'disciplinary' because it is the common possession of the practitioners of a professional discipline and 'matrix' because it is composed of ordered elements of various sorts, each requiring further specification" (Kuhn, 1974/1977, p. 297). These elements include "symbolic generalizations" ("the formal expressions of a science"), particular models that provide preferred metaphors and analogies, values, and shared examples. It is important that "models" – of electricity as a fluid, gas as tiny billiard balls, fields, and forces – "provide [researchers] with preferred analogies or, when deeply held, an ontology":

> Effective research scarcely begins before a scientific community thinks it has acquired firm answers to questions like the following: What are the fundamental entities of which the universe is composed? How do these interact with each other and with the senses? What questions may legitimately be asked about such entities and what techniques employed in seeking solutions? At least in the mature sciences, answers (or full substitutes for answers) to questions like these are firmly embedded in the educational initiation that prepares and licenses the student for professional practice. Because that education is both rigorous and rigid, these answers come to exert a deep hold on the scientific mind. That they can do so does much to account both for the peculiar efficiency of the normal research activity and for the direction in which it proceeds at any given time. (Kuhn, 1970, pp. 4–5)

Kuhn's proposal that assumptions about the kinds of entities that exist (*onto-logical* assumptions) and how we can know them (*epistemological* assumptions) are in some sense "*embedded*" in the practices and institutions of a science was radical. The proposal that practical activity involves a kind of knowledge that shapes how we understand the world, and ourselves, will run throughout this book. It is a point of fundamental difference from the positivist reconstruction of science, which as we have seen aimed for the complete *elimination* of such assumptions.

Kuhn's interpretation of the relationship between Newtonian and Einsteinian physics was also very different from that of the logical positivists. In Kuhn's account, they were two distinct paradigms. The passage from one to the other was not gradual and progressive, a matter of winnowing out the unscientific, metaphysical concepts and replacing them with ones that had been appropriately defined operationally. Rather, the change was one of "crisis" and "revolution" in which the fundamental entities of physics, and their properties of position, time, and mass, became completely different. Einsteinian physics was a decisive transformation. Physical objects and concepts were completely reinterpreted, along with the "fundamental structural elements of which the universe to which they apply is composed" (Kuhn, 1970, p. 102). This was the "revolutionary impact" of Einstein's theory.

The difference was not that Newtonian physics contained metaphysical notions whereas Einsteinian physics did not. *Both* paradigms involved untestable metaphysical assumptions – assumptions about fundamental entities, their modes of interaction, the questions to be asked about them, and the ways to find answers. It is intrinsic to a paradigm that it tells us "about the entities that nature does and does not contain and about the ways in which those entities behave" (Kuhn, 1970, p. 109). "Successive paradigms tell us different things about the population of this universe and about this population's behavior" (p. 103). Far from it being the case that "[p]rogress of research leads in every sphere away from metaphysics, toward the realm of connectible, scientific theories" (Von Mises, 1939/1956, p. 368), Kuhn pointed out that even what *counts* as metaphysics (in the bad sense) changes: "the reception of a new paradigm often necessitates a redefinition of the corresponding science" and in doing so it changes "the standard that distinguishes a real scientific solution from a mere metaphysical speculation" (Kuhn, 1970, p. 103).

"The World Itself Changes"

This brings us to another important aspect of Kuhn's analysis of the structure of scientific paradigms and scientific revolutions, his insistence that paradigms are "incommensurate": they have no common measure. It is often said that Einstein's equations of motion approximate very closely Newton's equations when velocities are small compared with the speed of light, so Newtonian and Einsteinian physics are compatible: the former is simply a special case of the latter. Kuhn argued, on the contrary, that "Einstein's theory can be accepted only with the recognition that Newton's was wrong" (Kuhn, 1970, p. 98), though he acknowledged that "[t]oday this remains a minority view."

The incommensurability of paradigms follows from that fact that, as Kuhn described it, a change in paradigm is a transformation of the world in which science is done:

> [T]he historian of science may be tempted to exclaim that when paradigms change, the world itself changes with them. Led by a new paradigm, scientists adopt new instruments and look in new places. Even

more important, during revolutions scientists see new and different things when looking with familiar instruments in places they have looked before. It is rather as if the professional community had been suddenly transported to another planet where familiar objects are seen in a different light and joined by unfamiliar ones as well. Of course, nothing of quite that sort does occur: there is no geographical transplantation; outside the laboratory everyday affairs usually continue as before. Nevertheless, paradigm changes do cause scientists to see the world of their research-engagement differently. In so far as their only recourse to that world is through what they see and do, we may want to say that after a revolution scientists are responding to a different world. (Kuhn, 1970, p. 111)

How exactly should this claim be understood? Was Kuhn merely saying that after a scientific revolution scientists observe the world differently, or did he truly mean that they are "transported" to a new world? It is hard to make sense of this proposal and much easier to think of a paradigm as providing each scientist with a pair of spectacles through which to see the world we all share. But through the course of this book I hope to make sense of Kuhn's suggestion that a new paradigm really does provide a *new* world. There is a tendency to think of a scientific paradigm as the *individual* scientist's "conceptual model" (e.g., Sluka & Robben, 2007, p. 4), but Kuhn described paradigms as shared. He was also adamant that they involve a tacit and practical kind of knowledge. He recognized that scientific knowledge is more than explicit hypotheses and formal theories, and he insisted that paradigms provide a form of knowing that cannot be captured in rules, laws, and formal criteria. Both the disciplinary matrix and the exemplars involve "tacit knowledge":

> When I speak of knowledge embedded in shared exemplars, I am not referring to a mode of knowing that is less systematic or less analyzable than knowledge embedded in rules, laws, or criteria of identification. Instead I have in mind a manner of knowing which is misconstrued if reconstructed in terms of rules that are first abstracted from exemplars and thereafter function in their stead. (Kuhn, 1970, p. 192)

Tacit Know-How and Seeing the World

Kuhn proposed that this tacit mode of knowing makes possible a *way of seeing* the world. It is "embodied in a way of viewing physical situations rather than in rules or laws" (Kuhn, 1970, pp. 190–191). The scientist looking into a cloud chamber sees "not droplets but the tracks of electrons, alpha particles, and so on" (p. 197). Kuhn was aware that he was challenging both a long-standing view of scientific knowledge and the accepted view of scientific vision. "What I am opposing in this book is therefore the attempt, traditional since

TABLE 1.6. *Kuhn and Logical Positivism Compared*

	The Logical Positivists and Karl Popper	Thomas Kuhn
Relation of Theory and Observation	Theoretical statements and observational statements are linked by one-way "correspondence rules." Observational statements are independent of, and epistemologically prior to, theory statements.	Fact and theory are "not categorically separable." What counts as an "observation" is determined by the paradigm and by the theory. At the same time, observations count for (and occasionally against) the paradigm.
Character of Theory	A theory has a tight, logical deductive structure. A change in theory is a logical procedure of hypothesis testing.	Theory involves evaluative judgments, and a change in paradigms is not deductive; it is a choice based on evaluative consideration, not procedural decision.
Nature of Scientific Discovery	The "context of discovery" is distinct from the "context of justification." The former is psychological; the latter is logical. Hypotheses are like guesses; testing these guesses is where logic operates.	Theory is rejected through the intransigence of anomalies. We need to study discovery to understand justification.
Definition of Scientific Terms	Terminology is precise. Theoretical terms are grounded in observational statements by operational definition. Scientific progress is in part the elimination of metaphysical terminology.	Analogy and loose definition abound. In a paradigm shift, there will be a semantic change in terminology (e.g., the meaning of "mass" is quite different in Newtonian and Einsteinian physics). There is terminological incommensurability between paradigms.
Relations among the Sciences	Unity of science: there is a general "scientific method." There is a hierarchy of reduction: sociology to psychology to biology to chemistry to physics.	No unity over time; no single method. Different sciences operate in very different worlds, so no reduction from one to another is possible.
The Structure of Science	The structure of science is distinct from its history; it is this logical structure that interests the philosopher of science.	Science is essentially historical; its structure has changed over time. And its structure is a temporal one, of paradigms and revolutionary change.
Scientific Progress	Science progresses cumulatively by piecemeal accumulation of individual discoveries and inventions. Science is the progressive correction of error	Change is revolutionary and nonprogressive. Change is sudden and thorough. Criteria of progress are generally internal to each paradigm, so

(continued)

TABLE 1.6 (*continued*)

The Logical Positivists and Karl Popper	Thomas Kuhn
and superstition, the elimination of metaphysics.	they change, too. Every paradigm has its own metaphysics. Old views are not just superstition; we need to try to understand them in their own terms.

Descartes but not before, to analyze perception as an interpretive process, as an unconscious version of what we do after we have perceived" (p. 195). What Kuhn calls "interpretation" here is the kind of reasoning where we apply explicit criteria and rules: "a deliberative process by which we choose among alternatives" (p. 194). This, he argues, is *not* what happens when we perceive. On the contrary, scientific vision is shaped by forms of tacit knowledge that are not individual but "are the tested and shared possessions of the members of a successful group" (p. 191). Members of a scientific community "learn to see the same things . . . by being shown examples" (p. 193).

Kuhn stressed that his position was not one of epistemological relativism because "very few ways of seeing will do" and "the ones that have withstood the tests of group use are worth transmitting from generation to generation" (Kuhn, 1970, p. 196). What a scientist learns is "a time-tested and group-licensed way of seeing" (p. 189). Kuhn added that this kind of learning "comes as one is given words together with concrete examples of how they function in use; *nature and words are learned together*" (p. 191, emphasis added). We learn how to see as we learn how to talk about what we see. The order of words orders the things we talk about. This is a fundamental point that we shall return to; in subsequent chapters I will develop Kuhn's criticism of the model that individuals know the world by forming mental representations and applying explicit rules and criteria. I will trace this model to Immanuel Kant rather than to René Descartes, but Kuhn was correct to say that it has been traditional for more than three centuries. It turns out to be the case that *both* empirical-analytic research and much qualitative research continue to use this inadequate model of knowledge.

THE METAPHYSICS OF RANDOMIZED CLINICAL TRIALS

Now we can return to the importance that the NRC committee found in randomized clinical trails, knowing that it reflects their assumption that research is a logical process of testing hypotheses. We now know that this assumption is part of a larger reconstruction of science as an objective and neutral activity in which knowledge is built from elementary observations using formal logic, in which explanations have the form of statements about causal mechanisms that are visible as associations of events (correlations) and can be identified by manipulation. Science, in this model, can be and ought to be free from metaphysics.

But if Kuhn is correct, and I think he is, the practice of randomized clinical trials itself presupposes a particular way of seeing the world, and in particular a way of seeing human beings. This way of seeing is *prior* to experimentation, measurement, and analysis, yet it generally goes unexamined and unquestioned. And it amounts to a metaphysics: assumptions about what exists that are embedded in the practices themselves.

The epistemological and ontological commitments embedded in randomized clinical trials, and in contemporary empirical-analytic research in general, are not difficult to discern once we look for them. Randomized clinical trials assume that the world is made up of entities with isolable, independent properties. Some of these properties may be hidden from us, for the moment or ultimately unknowable, but enough can be observed and measured that we can learn about them and formulate theories. These entities can be identified, observed, categorized, and measured without reference to the settings or circumstances in which they happen to be found.

People are just one kind of entity. They exist independently of one another, and each has isolable and independent properties. Observing, categorizing, and measuring these properties is how we obtain knowledge about people. Many of the properties are objective, such as height, weight, age, and even psychological traits such as personality, cognitive style, and intelligence. Some characteristics require that we burrow into the brain to observe them. Other properties are subjective: they are characteristics of what and how people experience the world around them. These also must be categorized and measured. Knowledge about human subjectivity arises from objective study of people's perceptions, beliefs, constructs, memories, and conceptions. What exists are matters of fact: if values exist, it is only as the subjective opinions, beliefs, and evaluations of individual people, and these are themselves matters of fact.

Problems with the Gold Standard

We can see now some of the problems with the claim that randomized clinical trials are the "gold standard" of scientific research. First, the focus on clinical trials – and on hypothesis-testing research generally – completely ignores the revolutionary aspect of science. Clinical trials are one way that research is carried out during times of (some forms of) *normal* science. To view scientific research only in terms of such activity is to look only at the comfortable side of science. It is to ignore those aspects of science that make revolutions necessary – the strong ontological and epistemological commitments that scientists adopt when they work within an accepted paradigm.

When researchers in the social sciences are told that they *must* perform research this way, a misleading description has become an oppressive prescription. Scientists who are engaged in normal science do not test the paradigm in which they work. Such a researcher "is a solver of puzzles, not a tester of paradigms" (Kuhn, 1970, p. 144). Research in normal science is like playing chess – trying out various moves but not challenging the rules of the game. To *force* scientists to conduct only hypothesis-testing research is to prevent them from challenging the rules of their game, from questioning or even examining the assumptions of the prevalent scientific paradigm.

But more importantly, the gold standard also prevents researchers from studying, let alone questioning, the forms of life in which people find themselves and in which things are found. People are *not* in fact independently existing entities. We exist *together*, in shared forms of life. Princeton political theorist Sheldon Wolin has gone so far as to call these the "moral paradigms" in which we live:

> From this viewpoint society would be envisaged as a coherent whole in the sense of its customary political practices, institutions, laws, structure of authority and citizenship, and operative beliefs being organized and interrelated. . . . This *ensemble* of practices and beliefs may be said to form a paradigm in the sense that the society tries to carry on its political life in accordance with them. (Wolin, 1968, p. 149)

I shall adopt Wolin's term for the moment, though we shall see it needs refinement and the proposal that society forms a *coherent whole* needs examination.

The philosopher Richard Bernstein has explored the implications for social science of this proposal that a society should be considered a moral paradigm – moral in that it defines "what is correct, appropriate, or 'rational' behavior" (Bernstein, 1976, p. 104) and "the acceptance of norms of behavior and action by political agents" (p. 106). Bernstein points out the important difference between the social sciences and the natural sciences is that "the very reality with which we are concerned in the human sciences is itself value-constituted, not an indifferent value-neutral brute reality" (p. 104). The world that humans perceive, think about, and act in is also defined by their form of life.

There is "no necessity for the practicing [natural] scientist . . . to study the history of his discipline" (Bernstein, 1976, pp. 101–102), but for human scientists it is crucial to become aware of the dominant moral paradigm of their society and question it. Hypothesis-testing research in the social sciences cannot study moral paradigms. It takes for granted a particular moral paradigm and then tests claims about the identifiable entities, distinguishable events, and recognizable institutions *defined* by that paradigm. It grants the *subjective* reality of people's thoughts, beliefs, and judgments about these entities, events, and institutions but avoids value judgments of its own. Studies of voting behavior and political opinions, of buying patterns and brand preferences, and so on, take for granted the reality of a particular "ensemble of practice and beliefs." In doing so, they assume that this moral paradigm is how things *must* be rather than how things are for the moment, something factual rather than something normative. But this is "mistaking historically conditioned social and political patterns for an unchangeable brute reality which is simply 'out there' to be confronted":

> In the eagerness to build a new natural science of human beings, there has been a tendency to generalize from regularities of a regnant moral paradigm, and to claim we are discovering universal laws that govern human beings. The most serious defect in this endeavor is not simply unwarranted generalizations, but the hidden ideological bases. There has been a lack of critical self-consciousness among mainstream social scientists that the admonition to be "realistic," to study the way things are, is not so much a scientific imperative as a dubious moral imperative that has pernicious consequences in limiting human imagination and political and social possibilities. Scientism in social and political studies has become a powerful albeit disguised ideology. (Bernstein, 1976, p. 106)

Bernstein anticipates that hypothesis-testing social science will fail to find regular and predictable patterns of behavior because the people it studies live in a variety of different moral paradigms and these are continually changing. When researchers find regularities and correlations, these merely reflect the stable normative assumptions of a particular moral paradigm, not general laws of human nature. Seemingly universal laws can be found only to the extent that people do not challenge and change the assumptions of their form of life, its constraints on what is acceptable and rational behavior. Bernstein suggests that "[t]he success ... in explaining and predicting human behavior may result from men's acceptance of rigid normative constraints about what is rational and acceptable behavior" (p. 105).

When empirical-analytic inquiry takes for granted the normative assumptions of a dominant moral paradigm, we are prevented from criticizing these assumptions and even from noticing them. Unexamined, they become ossified, even harder to change. Yet, as I suggested in the introduction, it is precisely our current moral paradigm – our attitude toward the planet, our attitude toward other cultures – that needs to be changed. We need to conduct research that is sensitive to the culturally diverse ways we live in a world of natural complexity if we are to learn how to transform these ways. To do this, we need to move beyond the testing of hypotheses.

CONCLUSIONS

In this chapter, we have examined two views of science. The first presumes that all scientific investigation employs the methodology of logical procedures for testing hypotheses. The belief that randomized clinical trials are the "gold standard" for research in the social sciences follows from this view. All "metaphysics" – untestable presumptions about the kinds of entities that exist – should be avoided, for science is at its core a logical process.

The second view, far from seeking to eliminate metaphysics, sees it as central and essential to all scientific activity. Thomas Kuhn proposed that every science operates within a "paradigm": shared ontological and epistemological commitments that are embedded in the practices of that science. Kuhn argued that the successful practice of science requires a kind of tacit knowing that makes possible a particular way of seeing the world.

The point of criticizing logical positivism and randomized clinical trials is not to argue that an empirical-analytical social science is *impossible*. This is clearly not the case. The point is to question "present and prevailing emphases, concerns, and problems" (Bernstein, 1976, p. 41) in order to explore alternatives, and to consider the costs we pay when we restrict our conception of what counts as scientific research: "The most important and interesting challenges to any dominant orientation are those which force us to question the implicit and explicit emphases that make us self-conscious not only of what is included in the foreground, but excluded or relegated to the background as unimportant, illegitimate, or impractical" (Bernstein, 1976, p. 41).

The randomized clinical trial is one small part of the practice of modern scientific inquiry. Testing hypotheses is just one thing that scientists do. The efforts today to convince social scientists that we ought to ask questions of the form "did X cause Y?" and that to answer these questions we should conduct clinical trials prevent us from asking other important questions. Bernstein noted that questioning a moral paradigm can have a broad range of consequences, including new ways of thinking about "the essential aims of social inquiry, the type of education appropriate to social scientists, the role of the theorist, the relation of theory and action, as well as that of fact and value" (Bernstein, 1976, p. 43). These are some of the topics that we will explore in the chapters that follow.

Of course, discussion of the differences between qualitative and quantitative research has frequently been framed in terms of paradigms. Talk of "competing paradigms" (Guba, 1990; Guba & Lincoln, 1994) played an important part in the development of qualitative research from the 1970s through the 1990s. The debate was described as "paradigm wars" (Gage, 1989), and the war was then declared over (Anderson & Herr, 1999; Rizo, 1991). The declaration of peace was decried as closing down conversation (Smith & Heshusius, 1986), and this was followed both by celebration and dismay over "paradigm proliferation" (Donmoyer, 1996). The number of different paradigms identified in qualitative inquiry has ranged from two to ten. Now "mixed-method" research is announced as a new paradigm (Johnson & Onwuegbuzie, 2004).

Some people have asked whether all this talk of paradigms has been useful (Morgan, 2007). Often the appeal to paradigms has been unnecessarily prescriptive. Guba and Lincoln advised that "no inquirer, we maintain, ought to go about the business of inquiry without being clear about just what paradigm informs and guides his or her approach" (Guba & Lincoln, 1994, p. 116), even though Kuhn noted that paradigms usually operate unnoticed, tacitly accepted until a crisis develops. There has been a lack of clarity about the definition of a paradigm, which is perhaps forgivable given Kuhn's ambiguous use of the term. Guba and Lincoln defined it as "the basic belief system or worldview that guides the investigator" (p. 105), a worldview that takes a stand on ontology, epistemology, and methodology. Morgan (2007) has suggested that this is a "metaphysical" conception of paradigms that imposes external criteria drawn from the philosophy of knowledge. Lincoln and Guba indeed write of "basic foundational philosophical dimensions" (Lincoln & Duba, 2000, p. 169). Morgan recommends that, rather than paying attention to ontological and epistemological assumptions, we should look at the practices actually engaged in by communities of researchers.

But an important point of Kuhn's analysis was that practices and ontological and epistemological commitments are not separate. His view was that scientific practices *require* and *embody* such commitments: they are commitments made *in* practice. A paradigm provides shared exemplary practical activities that offer – in a manner we need to explore further and that Kuhn himself wasn't entirely sure how to describe – a way of seeing to which particular kinds of entities become visible. Being scientific is not a matter of *avoiding* metaphysical commitments, as the logical positivists believed, for this would be to have no way of seeing the world. Science is continually trying *new* ways of seeing, offered by new practices of investigation, in order to solve as best it can the problems that we face.

In the chapters that follow, I will recommend that to redefine qualitative inquiry we need to question our paradigmatic commitments and those of the forms of life in which we live. We need a form of research that is able to explore and investigate moral paradigms and give us the tools to change them. As Bernstein pointed out, great developments in the human sciences have been responses not to theoretical crises but to crises in the world. The planetary crises we face today should spur us to take up the radical potential of qualitative inquiry rather than allowing it to be absorbed into the status quo.

We can begin by examining the ontological and epistemological commitments embedded in the practices of qualitative researchers today, to explore the ways of seeing these practices offer and ask how we could modify research practice to see the world, and ourselves, differently. The way to an alternative to empirical-analytic research is not simply to replace "quantitative" data with "qualitative" data, or even to offer a new set of procedures to replace those of hypothesis-testing research design. What is necessary is a new methodology in the sense of a *program of inquiry* that replaces the inappropriate metaphysics that underlies empirical-analytic research at the start of the 21st century as it underlied the logical positivism of the early 20th century. The result will not be *freedom* from metaphysics but a *better* metaphysics, one that is more adequate to the complexities of human life and human being. I fully agree with Shavelson and his colleagues (Shavelson & Towne, 2004) that the design of research should follow from the question thought important. But as I explained in the introduction, I believe the most important questions concern not causation but constitution. We need to figure out how to design research that can ask and answer "how?" questions about constitution. Clinical trials cannot do this.

In the remaining chapters of Part I, I will examine the practices in which interviews are conducted, transcripts are coded, and the results are reported as summaries written in formal language. This kind of qualitative research is supported by many textbooks and "how to" manuals, and it cuts across a range of disciplines in the social sciences. In Chapter 2, I examine the conduct of the qualitative research interview and then in Chapter 3 the practices of analysis. In Chapter 4, I will draw on the history of hermeneutics – the theory of interpretation – to propose a new way of thinking about the purpose of interviews, and in Chapter 5 I explore a new approach to the analysis of interview material.

2

The Qualitative Research Interview

Without any doubt, the most popular style of doing qualitative social research, is to *interview* a number of individuals in a way that is less restrictive and standardized than the one used for quantitative research.
ten Have, 2004, p. 5, emphasis original

A standard practice for qualitative research has become accepted in which interviews are conducted, the data are coded, and the results reported in the form of summaries written in formal language. In this chapter, I examine the practice of the qualitative research interview and explore the assumptions about language – and about people – that are embedded in its practices.

Interviews are a ubiquitous way of collecting data throughout the social sciences. Although qualitative researchers today work with a wide variety of perspectives and study an enormous range of phenomena – from rituals to the unconscious – the technique used to obtain data is overwhelmingly the interview (Potter & Hepburn, 2005). Verbal data are sometimes elicited in the form of solitary words or phrases, or brief statements or explanations, but interviewing has become a familiar, taken for granted component of qualitative research, as well as research in cognitive studies, developmental psychology, the learning sciences, and elsewhere. Qualitative approaches that are otherwise very different – grounded theory, thematic analysis, empirical phenomenology, and interpretative phenomenological analysis – all agree that interviews are the method of choice to obtain qualitative data (Table 2.1). Many qualitative research projects use *only* interviews as their source of empirical data.

We live today in what has been called "an interview society" (Atkinson & Silverman, 1997). On television, on the radio, and in newspapers and magazines, we watch, hear, and read interviews every day. Bogdan and Biklen suggest that "Most of us have conducted interviews. The process is so familiar we do it without thinking" (Bogdan & Biklen, 1992, p. 96). Everyone knows

TABLE 2.1. *The Ubiquity of Interviews*

Research Perspective	Object of Study	Technique of Data Generation
Ethnography	Cultures, rituals, groupings	Interviews
Phenomenology	Experience, consciousness	Interviews
Psychoanalysis	The unconscious	Interviews
Narrative psychology	People's life stories	Interviews
Grounded theory	Highly varied	Interviews
Discourse analysis and discursive psychology	Talk and texts	Interviews and naturalistic data

Source: Modified from Potter & Hepburn, 2005.

what an interview is – or at least we think we do. Perhaps the very ubiquity of interviews prevents us from looking at them carefully.

Typically the qualitative research interview is "semistructured." Other possibilities exist – focus group interviews, interviews that are completely unstructured, life histories, and so on – but the semistructured interview is the workhorse of qualitative research today. In a semistructured interview, the researcher has a general plan for the topic to be discussed but does not follow a fixed order of questions or word these questions in a specific way. Interviewees are allowed a great deal of latitude in the way they answer, the length of their responses, and even the topics that they discuss. The aim of such an interview is to encourage the person to speak "in their own words" to obtain a first-person account. The interview is generally audio-recorded and transcribed, and – as we shall see in the next chapter – analysis typically involves comparison, coding, and summarization.

This all seems clear and straightforward, the only difficulties technical ones that can be resolved with practical advice (e.g., "[w]rite memos to yourself about what you are learning"; Bogdan & Biklen, 1992, p. 159). But we will follow Kuhn's lead and interrogate the steps of this process to discover its embedded ontological and epistemological commitments.

CHARACTERIZATIONS OF THE QUALITATIVE INTERVIEW

Researchers have defined and justified the qualitative research interview in two principal ways. The first is to emphasize how this kind of interview differs from those conducted in conventional survey research. The second is to emphasize how it is similar to everyday conversation. Each of these definitions has merit, but when we put them side by side we find a surprising *similarity* between the qualitative research interview and the conventional survey interview.

Contrasted with Conventional Survey Interviews

Qualitative research interviews are often contrasted with the interviews that have been conducted for many years in traditional survey research – for example, in opinion polls, voting exit polls, and consumer satisfaction surveys. The "standard practice" (Mishler, 1986) for survey research interviewing has been to predesign the questions, follow a script, and strive to be unbiased. The conventional survey interview is thoroughly standardized. The wording of the preliminary instructions and the questions is determined in advance, as is the sequence of questions. Different interviewees will be asked exactly the same questions, except when some alternative branching is allowed on the basis of earlier responses. Interviewers strive to adopt an attitude of neutrality and avoid any evaluation of the replies to their questions. They are not allowed to clarify, elaborate, or modify questions, and they are looking for specific categories of responses, so they will tend to encourage brief replies that fit predetermined categories that can then be analyzed statistically.

The aim in conventional survey research is to make the interview a neutral and reliable social scientific instrument, analogous to a measuring stick or a clock. The assumption is that to obtain information about people's beliefs and attitudes one must present everyone with the same standardized stimuli and elicit their responses. [See Gorden (1980) for a detailed and sophisticated presentation of this standard practice.] Survey researchers have discovered that responses in a conventional interview are influenced by the wording of questions, by their grammatical form and order, and by characteristics of the interviewer and the data-collection situation. These influences are "unsystematic," which means that there is no easy way to eliminate them. The reaction has been to try even harder to impose external constraints on the way questions are phrased, the sequence in which they are posed, and the categories of answers that are acceptable.

A number of people have argued that, ironically, these efforts to make the interview a reliable scientific instrument have undermined its ability to provide valid information, let alone genuine understanding. Lucy Suchman and Brigitte Jordan point out how "[i]n the interest of turning the interview into an instrument, much of the interactional resources of ordinary conversation are disallowed" (Suchman & Jordan, 1990, p. 232). The result is various kinds of "interactional troubles" that cannot be avoided without breaking the rules of the traditional interview. "[T]he standardized interview question has become such a fragile, technical object that it is no longer viable in the real world of interaction" (p. 241).

Because the interviewer is following a script with fixed questions, the resources that in ordinary conversation enable people to establish and maintain understanding cannot be used. The interviewer is not allowed to respond to the way the interview is unfolding or react to the particular circumstances, and this means that inappropriate or even nonsensical questions get asked. Requests for clarification by the interviewee must be ignored. The interviewer is looking for responses that fit fixed categories, and this leads to contradictory problems: on the one hand, detailed answers are discouraged, which prevents interviewers from exploring, or often even noticing, any equivocations or qualifications in the interviewee's responses. On the other hand, short and elliptical answers that would be perfectly acceptable in ordinary conversation are questioned as the interviewer tries to fit them into the appropriate category.

These restrictions mean that the conventional interview runs into trouble whenever interviewer and respondent don't share assumptions about the world. One of the illustrations offered by Suchman and Jordan begins as follows:

I: Generally speaking, do you usually think of yourself as a Republican, Democrat, Independent, or what?
R: As a person.

When understanding breaks down like this, neither interviewer nor respondent is allowed to do the work necessary to repair the problem.

Ambiguous questions and responses are not caused by poor interview technique but rather are consequences of the effort to standardize. They can't be eliminated by more control of wording and sequence. As Suchman and Jordan argue:

> The meaning of an utterance is not inherent in the language, but is a product of interaction between speakers and hearers. In ordinary conversation, utterances can be elaborated to whatever level of detail participants require. So if the response to a question is cursory, vague, or ambiguous, or if the questioner simply wants to hear more, he or she can ask the respondent to say more. The producer of a question is taken to have the license and authority to clarify what it is that the question is designed to discover. In the interview, however, the person who asks the question is simply its administrator, trained to resist the respondent's appeals to elaborate on its intent. (Suchman & Jordan, 1990, pp. 237–238)

Everyday conversation involves continuous work to detect and repair ambiguity. Because the traditional interview prohibits this work, the likelihood that the interviewer and interviewee understand each other, and that valid data are obtained, is low. The interviewee is responding to questions whose meaning cannot be clarified, and the meaning of their answers is often unclear. Suchman and Jordan recommend "rethinking the relations between participants" (p. 240) in the interview. This might involve small changes such as using a questionnaire that both interviewer and respondent can see or, more radically, it might involve allowing the interviewer to engage in interaction with the interviewee to clarify and elaborate questions and negotiate their meanings. Clarification of questions, they point out, is not the same thing as bias.

I said earlier that "interactional troubles" cannot be resolved without breaking the rules of the conventional interview. Given the extent of these troubles, it is no surprise that in practice survey interviewers do indeed often break their own rules. Interviewers deviate from their scripts, and they ignore the prohibition against using the resources of everyday conversation. Eliot Mishler has argued that there is a "gap" between the way the conventional research interview is supposed to be conducted and what actually happens. (Mishler focused on the analysis of interviews, but the same points can be made about conducting the interview.) Interviewers smuggle in the very characteristics of ordinary conversation that they claim to avoid. Interviewers must make sense of "an isolated response to an isolated question" (Mishler, 1986, p. 3). In the absence of "the shared assumptions, contextual understandings, common knowledge, and reciprocal aims of speakers in everyday life" (p. 1), they must make their best guesses by imagining what the context might be and assuming that words mean what they "usually" mean. As Mishler puts it, "[b]y suppressing the discourse and by assuming shared and standard meanings, this approach short-circuits the problem of meaning" (p. 65). Interviewers make the best of a situation in which genuine communication is virtually impossible, but nonetheless the data they collect are flawed and the products of such data analysis are questionable.

Mishler as well as Suchman and Jordan emphasize that everyday conversation is always a "joint production." The meaning of questions and answers is negotiated, topics are introduced collaboratively, and ambiguity is resolved together. The conventional survey interview prohibits these characteristics of conversation at great cost.

If conventional interviews are filled with interactional troubles and interviewers are forced to break the rules that are intended to ensure the interview's reliability, then one might argue, as Mishler does, that qualitative interviews, far from being "unscientific," are actually more adequate for social scientific inquiry than survey interviews. After all, qualitative interviewers explicitly recognize "the nature of interviewing as a form of discourse between speakers" (Mishler, 1986, p. 7) and the "ordinary language competence" (p. 7) that conducting an interview always draws on.

There is a *flexibility* to the qualitative research interview that is entirely at odds with the emphasis on *control* in the conventional interview. Whereas the wording of questions is scripted in the conventional interview, in the qualitative research interview the interviewer can phrase questions on the fly. The order of questions can vary; indeed, control over at least the subtopics of the interview is likely to be given to the respondent. Elaborated responses are often encouraged; extended narratives may be solicited in reply to a single short question. And for at least some qualitative researchers, the interviewer does not strive to be neutral and unbiased: sometimes "he lets his personal feelings influence him ... thus he deviates from the 'ideal' of a cool, distant, and rational interviewer" (Fontana & Frey, 2000, p. 653), answering the interviewee's questions, offering personal information. All these character-istics make for a very different interview, one that, as Suchman and Jordan recommended, allows flexible use of the resources of everyday conversation.

Compared with Everyday Conversation

So far, so good. Given the problems that follow from the emphasis on control in the conventional survey interview, it would seem that we are putting our finger on a strength of the qualitative research interview, not a weakness, when we say that it resembles everyday conversation. And this has been the second way in which qualitative research interviews are often defined and justified. "Qualitative interviewing is based in conversation," writes Warren (2002, p. 83), and Kvale (1996) suggests that "[a]n interview is a conversation that has a structure and a purpose" (p. 6). Blaxter, Hughes, and Tight (2001, p. 171) propose that "the informal interview is modeled on the conversation and, like the conversation, is a social event with, in this instance, two participants." Similarly, Bogdan and Biklen (1992) write that "[a]n interview is a purposeful conversation, usually between two people but sometimes involving more."

This way of characterizing the qualitative research interview doesn't contradict the approach of contrasting it with conventional interviews. If the interview is an everyday conversation, then it avoids the artificiality of the conventional interview, with its emphasis on control and standardization, and surely is superior in its sensitivity to the unique experiences and sub-jectivity of the interviewee. But the emphasis on similarities to everyday conversation exposes some ways in which the qualitative interview has its own artificialities, for it must be considered a *special* kind of conversation. Bogdan and Biklen, for example, add that "[i]n the hands of the qualitative researcher, the interview takes on a shape of its own" (Bogdan & Biklen, 1992, p. 96). Kvale acknowledges that: "The interview is a specific form of con-versation. ... It goes beyond the spontaneous exchange of views as in everyday conversation, and becomes a careful questioning and listening approach with the purpose of obtaining thoroughly tested knowledge" (Kvale, 1996, pp. 6, 19).

It would be misleading, then, to say that the qualitative research interview is *just* an ordinary conversation. First, there are of course ordinary conversa-tions of various kinds. A conversation between a parent and a child will be quite different from one between a teacher and a student or between a shopper and a storekeeper. To ask whether a qualitative interview is "ordinary" would be pointless, but it is illuminating to try to characterize the kind of conversation that qualitative interviews typically involve. Although it is certainly true that many of the resources of everyday conversation are allowed in qualitative research interviews, when we compare it with the conversation that might occur between two neighbors who bump into one another on the street we find, rather surprisingly, that the qualitative interview has its own peculiar characteristics, and these create a "gap" between theory and practice that is analogous to the gap in the conventional survey interview.

First, the semistructured qualitative research interview is a scheduled, not spontaneous, event. The very fact that an interaction is identified as an interview suggests that it has been distinguished from spontaneous conversation. Qualitative researchers resemble conventional researchers insofar as they still schedule a place and time to talk with their respondent rather than recording conversations that occur spontaneously.

Second, it often takes place between strangers. Qualitative interviews vary in the degree of prior familiarity between interviewer and interviewee. Some interviews occur during long-term fieldwork, with people whom the researcher gets to know well. But frequently a qualitative interview takes places between two people who are relative strangers, with minimal or no prior relationship.

Third, it is not an interaction between equals. Qualitative interviews resemble conventional survey interviews in their presumption that it is the interviewer who chooses the topic of the conversation and asks the questions, while the interviewee answers. There is what Kvale calls an "asymmetry of power" to the qualitative research interview that makes it unlike a conversation between equals. "It is the interviewer who has the power to ask questions, to stop and start the exchange, and so on" (Honey, 1987, p. 80).

Indeed, some have questioned the ethics of the qualitative research interview on this basis, seeing it as an interaction in which the interviewer tries to gain trust and confidence but does not reciprocate. "What seems to be a conversation is really a one-way pseudoconversation" (Fontana & Frey, 2000, p. 658). Oakley (1981) has argued that the preoccupation with rapport (working to be trusted) and neutrality serves to obscure the hierarchical and exploitative character of the interviewer–interviewee relationship.

Fourth, it is conducted for a third party. The qualitative research interview – just like the conventional interview – is conducted for people who are not present. In this sense, "it is not a conversation in the usual sense of the word; rather, the interview is a spoken text" (Honey, 1987, p. 80). Even when interviewers will analyze the data themselves, the tape recorder "symbolizes the potential audience and, in a certain sense, insures that the interview is already staged" (p. 80).

Fifth, qualitative interviewers are often told they need to adopt a special attitude distinct from that of everyday interaction. The interview is said to require a special "methodological awareness of question forms, a focus on the dynamics of interaction between interviewer and interviewee, and a critical attention to what is said" (Kvale, 1996, p. 20). Interviewers are told they must cultivate a "double attitude" that combines faith and skepticism: "a willingness to listen coupled with an attempt to illuminate what is said" (Honey, 1987, p. 80). The interviewer is advised to "consider with the utmost seriousness everything that the [interviewee] says, while *simultaneously* casting a critical eye on the interviewee's discourse" (p. 81, emphasis original). Kvale (1996) has traced the origins of semistructured interviewing back to Carl Rogers's (1945) recommendations for clinical interviewing. In Rogers's view, the interviewer should follow the general approach of nondirective therapy, repeating the respondent's words with unconditional positive regard and an attitude of acceptance, avoiding anything that might trigger defenses. With such a conception of the interviewer as unbiased and avoiding evaluations or making suggestions, we are approaching something very similar to the conventional interview. Yet Oakley has argued convincingly that "personal involvement is more than dangerous bias – it is the condition under which people come to know each other and to admit others into their lives" (Oakley, 1981, p. 58).

Sixth, the qualitative research interview generally is not about the here and now. The topic of the conversation is frequently a past event or an abstraction (e.g., "beauty") rather than something present, so that "event and meaning do not coincide in the here and now" (Honey, 1987, p. 81). In contrast, everyday conversations frequently focus on the here and now.

Finally, the qualitative research interview is an occasion to obtain accounts or descriptions from an interviewee. Bogdan and Biklen propose that "the interview is used to gather descriptive data in the subject's own words so that the researcher can develop insights on how subjects interpret some piece of the world" (Bogdan & Biklen, 1992, p. 96). "Good interviews," they suggest, "are those in which the subjects are at ease and talk freely about their points of view. . . . Good interviews produce rich data filled with words that reveal the respondents' perspectives" (p. 97). Kvale agrees that the qualitative interview's "purpose is to obtain descriptions of the life world of the interviewee with respect to interpreting the meaning of the described phenomena" (Kvale, 1996, pp. 5–6). In his view, "it is in fact a strength of the interview conversation to capture the multitude of subjects' views of a theme and to picture a manifold and controversial human world" (p. 7). But "to gather descriptive data" is not the aim of the typical everyday conversation. Most conversations have a motive that is pragmatic, not informational. When information is solicited, it is not as an end in itself but to further practical purposes. And if a conversation is defined as "an oral exchange of sentiments, observations, opinions, ideas" (*Webster's New Collegiate Dictionary*), then the qualitative interview is not a conversation because generally it involves not an *exchange* but a one-way delivery, in which the interviewee's opinions and sentiments are solicited but the interviewer does not reciprocate.

FLEXIBILITY IN SERVICE OF ASYMMETRY

In summary, the qualitative interview is an unusual kind of conversation. It is usually an interaction between relative strangers. One participant adopts a professional role and the other does not. The qualitative interview is a scheduled event rather than one occurring spontaneously. Frequently the topic of the interview is not something in the here and now but an abstraction. The interviewee is encouraged to elaborate to an unusual degree. The conversation is tape-recorded, introducing an anonymous third party. One of the participants asks the questions, while the other provides the answers. Although the qualitative interview does allow use of the resources of everyday conversation, such as clarification and elaboration, to negotiate the meaning of questions and answers, there is a central asymmetry in the use of these resources.

The consequence is that, as Wiesenfeld (2000) has noted, there is a "gap between the theory and the practice of qualitative inquiries." Although many of the proposals for qualitative research have insisted that the researcher should have a relationship of "symmetry, dialog, cooperation, and mutual respect" with participants, the practice of qualitative research interviews rarely shows these characteristics:

> Considering that a) the researcher is the person who generally chooses and approaches a context with which he/she becomes familiar, but that is not the case for the informants in that context, b) that the researcher is motivated by certain purposes which – though they may be negotiated or modified in the course of the interaction – define him/her as the promoter of a process, c) that the researcher has access to the informants' subjectivity and intimacy as a result of his/her training to inquire, argue, and report, and d) that in general the researcher asks and the informants answer but not the reverse, we are forced to admit that this is not a symmetrical relationship. (Wiesenfeld, 2000)

Just as Mishler identified a gap between what traditional interviewers claim to be doing and their practice (they draw on their understanding of ordinary language to get through the interview successfully, although claiming the contrary), there is a gap between what qualitative interviewers claim to be doing and their practice. In theory, the qualitative research interview manifests respect for the interviewee in a more equitable kind of interaction, but in practice it shows an asymmetry of power and control. Wiesenfeld is concerned with the ethical problems this gap creates. But the gap also has epistemological and ontological implications.

I have already mentioned a central point, made by Mischler and by Suchman and Jordan, that the interview – whether it is a conventional survey interview or a qualitative research interview – is always a "joint production." The meaning of questions and answers is negotiated, successful questions are tailored to suit the circumstances, and the interviewer and interviewee work collaboratively. I am not suggesting that these elements are missing from the qualitative research interview; far from it. But they are put to work to achieve a particular end. The asymmetry of question-and-answer places the burden (and pleasure) of explication on the interviewee. The focus is shifted away from the interviewer and falls chiefly on the interviewee. The topic is often an event that only the interviewee has experienced. The presence of the tape recorder has the effect of deemphasizing the listener who is actually present, namely the interviewer, and introducing a listener who does not interact with the interviewee. In sum, interviewers in large part control the interaction, yet in doing so assign themselves a supporting role, playing only second fiddle.

The overall effect of these characteristics of the semistructured interview is to give the impression that the words recorded on the tape recorder, transcribed onto the computer screen and subsequently analyzed, are the product of only one person. Yet if Mishler and Suchman and Jordan are correct, the spoken words are the result of a joint production. The asymmetry of power in the qualitative research interview is employed to create also an asymmetry of *visibility*. The interviewer becomes invisible and the interviewee is the center of attention.

This asymmetry doesn't stop when the interview ends; it continues in the analysis and reporting. When qualitative interviews are reported, almost everything that characterizes them as interactions is left out. Published reports rarely tell us about the interviewer, the setting of the interview, or how the study was described to the participants. The interviewer's comments and responses are omitted from the transcript of the interview, they are excluded from the analysis, and they are expunged from the reports of qualitative research. When excerpts from the interview transcript are published, the questions to which they were replies are often omitted. But if the meaning of both questions and answers is *negotiated* in the qualitative interview, then showing only half of the negotiation is unacceptable. If it is the case that a qualitative interviewer "may paraphrase, rephrase, or elaborate ... when a respondent asks for clarification or does not respond in an adequate way" (Mishler, 1986, p. 46), then it is crucial to be able to read the question that was actually asked. A first-person account has to be given *to another person*, in this case the interviewer. The way interviewers present and introduce themselves, the way they explain the purpose of the interview, and the setting in which the interview takes place will all influence the account that is given. What if a child is interviewed by an adult? A working-class man by an upper middle-class woman? A worker by their boss? A housewife by a graduate student? These examples quickly show the importance of the relationship between interviewer and interviewee, yet this relationship is rendered invisible.

At first glance, this seems entirely appropriate. After all, it is the interviewee whose experiences, subjectivity, or understanding are of interest. For many qualitative social scientists, the principal motivation for exploring an alternative to conventional empirical-analytic research has been to add a missing *subjective* component. From this point of view, empirical-analytic research can tell us about objective events but not about how they are experienced and what they mean to people.

For example, an empirical-analytic study of poverty might explore the association between level of income and measures of physical and mental well-being, but such a study seems to neglect the *experience* of poverty. The task of adding this subjective element is often described as the study of "meaning," where "[m]eaning is taken primarily as a condition under which a person's life, or significant events in it, 'make sense' (i.e., have worth and relate to the subject's feelings of integrity, wholeness, and self-mastery)" (Wuthnow, 1987, p. 35). So we find talk of "meaning-making," the subjective process of con-ceptualizing the world around us (e.g., Bruner, 1986). From this point of view, qualitative research *is* the study of meaning, and this requires access to the subjective interpretations people attach to their objective circumstances. As Seidman says, "[a]t the root of in-depth interviewing is an interest in under-standing the experience of other people and the meaning they make of that experience" (Seidman, 1998, p. 3). Qualitative research, from this perspective, is *the objective study of subjectivity:* "What qualitative researchers attempt to do . . . is to objectively study the subjective states of their subjects" (Bogdan & Biklen, 1992, pp. 46–47).

There is an ethical dimension to this move. Empirical-analytic research is viewed as objectifying and dehumanizing. Qualitative research is motivated by respect for the unique and creative ways that individuals understand the world. Researchers pride themselves on conducting interviews that are flexible and responsive, in which we act toward the interviewee as a fellow human being with something important to say and interact with them with sensitivity and reciprocity. Research is a process of cooperation based on dialogue and mutuality.

MIXED METAPHORS

It seems self-evident that the access to a person's subjectivity, to the personal meaning of some event or phenomenon, that is needed for this approach to qualitative research will be provided by interviewing. Kvale put the point well:

> If you want to know how people understand their world and their life, why not talk with them? In an interview conversation, the researcher listens to what people themselves tell about their lived world, hears them express their views and opinions in their own words, learns about their views on their work situation and family life, their dreams and hopes. The qualitative research interview attempts to understand the world from the subjects' points of view, to unfold the meaning of peoples'

experiences, to uncover their lived world prior to scientific explanation. (Kvale, 1996, p. 1)

But precisely how does an interview provide access to subjectivity? Is it "a window into human experience" (Ryan & Bernard, 2000, p. 769)? Is it an "entranceway" that the researcher can "go through" (McCracken, 1988, p. 44)? These metaphors presume that experience is hidden inside a person and that the researcher needs to create an opening to see and examine it. They imply that a person is closed to other people, their experience visible only to themselves; that subjective experience is contained within the individual mind. They imply, too, that it is nonetheless possible to create an opening so that another person can see what is usually hidden. But visual metaphors ignore the verbal character of the interview. If "the best way to learn about people's subjective experience is to ask them about it, and then listen carefully to what they say" (Auerbach & Silverstein, 2003, p. 23), how is what is said related to what is experienced?

This usual view is that the interviewee's words are an "expression" of their experience, of what is hidden inside them. The interviewee "expresses in their own words" their opinions and beliefs, "putting into words" what they have experienced or the meaning of these experiences.

The Conduit Metaphor

This is a common metaphor for the way language functions; Michael Reddy (1979) called it the "conduit metaphor." We can see this metaphor at work in many phrases, such as "he got his idea across well," "it's difficult to put my ideas into words," or "her words carry little meaning." Reddy described it in these terms:

> (1) [L]anguage functions like a conduit, transferring thoughts bodily from one person to another; (2) in writing and speaking, people insert their thoughts and feelings in the words; (3) words accomplish the transfer by containing the thoughts or feelings and conveying them to others; and (4) in listening or reading, people extract the thoughts and feelings once again from the words. (Reddy, 1979, p. 290)

George Lakoff and Mark Johnson (Lakoff & Johnson, 1980) identify three components to the conduit metaphor:

IDEAS (or MEANINGS) ARE OBJECTS
LINGUISTIC EXPRESSIONS ARE CONTAINERS
COMMUNICATION IS SENDING

The conduit metaphor presents language as a channel through which containers travel into which thoughts and feelings have been placed in order to be transferred to another person. "The speaker puts ideas (objects) into words (containers) and sends them (along a conduit) to a hearer who takes the ideas/ objects out of the words/containers" (Lakoff & Johnson, 1980, p. 10). The image of language as a conduit and words as containers encourages us to suppose that ideas, opinions, beliefs, and experiences are entirely separate from the words they are "put into." We think that ideas are like physical objects and that the purpose of language is to package them up for transfer between minds.

Lakoff and Johnson believe that all our understanding is metaphorical (Moser, 2000), so one might ask why the conduit metaphor is a problem. Their view is that it is a "subtle case of how a metaphorical concept can hide an aspect of our experience" (Lakoff & Johnson, 1980, p. 10) because:

> The LINGUISTIC EXPRESSIONS ARE CONTAINERS FOR MEANINGS aspect of the CONDUIT metaphor entails that words and sentences have meanings in themselves, independent of any context or speaker. The MEANINGS ARE OBJECTS part of the metaphor, for example, entails that meanings have an existence independent of people and contexts. The part of the metaphor that says LINGUISTIC EXPRESSIONS ARE CONTAINERS FOR MEANINGS entails that words (and sentences) have meanings, independent of contexts and speakers. These metaphors are appropriate in many situations – those where context differences don't matter and where all the participants in the conversation understand the sentences in the same way. (Lakoff & Johnson, 1980, pp. 11–12)

The notion that "the meaning is right there in the words" trivializes the process of communication, and its difficulties. In this image of communication: "[I]t is possible to objectively say what you mean, and communication failures are matters of subjective errors: since the meanings are objectively right there in the words, either you didn't use the right words to say what you meant or you were misunderstood" (Lakoff & Johnson, 1980, p. 206). Reddy also saw a danger in the conduit metaphor: "This model of communication objectifies meaning in a misleading and dehumanizing fashion. It influences us to talk and think about thoughts as if they had the same kind of external, intersubjective reality as lamps and tables" (Reddy, 1979, p. 308).

The conduit metaphor is so ubiquitous – Reddy estimated that it operates in at least 70% of the expressions we use to talk about language – that we don't notice it. It is hard not to draw on this metaphor when we conduct qualitative research interviews. If language is a conduit and talk is a container, we have no need to consider the *interaction* between the two people in the interview because the meaning of what was said is independent of this interaction. What was said by the interviewer can be omitted from the transcript and from subsequent analysis because it is irrelevant: the interviewer's words were containers sent *to* the interviewee, and it is the containers we receive *from* the interviewee that are important.

A metaphor for language is a model of how humans interact, and implicitly an ontology of the kind of beings that humans are. The conduit metaphor encourages us to see humans as separate individuals, each with their own private experiences and subjective mental states, and to see communication as sending messages: these mental states are packaged into a linguistic container that is transmitted across the airwaves to another individual. It is easy to see the appeal of this model; it is, after all, how telephone calls are made, how television images are broadcast, and how email is sent. In each case, the form – electronic impulses traveling along a wire or as radio waves, packages of digital data – is irrelevant to, and distinct from, the content.

But what if we change our point of view or, better, try looking at things differently, guided by a different metaphor, that of language as a joint production? Mishler recommended that we recognize that interviews are "speech events or speech activities," and it is "allowed, even required" of a qualitative researcher that we consider "the joint construction of the discourse by interviewers and interviewees, the prerequisite of an explicit theory of discourse and meaning for interpretation, and the contextual basis of meaning" (Mishler, 1986, p. 137).

Viewed this way, it becomes apparent that the interviewing practices we have identified – the elicitation of descriptions, the abstract topic, the special attitude, the recording for a third party – are a specific use of the resources of everyday conversation. The qualitative research interview places the interviewee in a reflective, contemplative mode, talking about a past event or an abstraction to an attentive listener with seemingly inexhaustible patience and a tape recorder that will store their words for posterity, for science. Who could resist this opportunity to wax eloquent, to formulate strong opinions, to tell a good story? The scheduling defines the event as something personal; the interviewer's attitude of polite skepticism invites the interviewee to make an effort to be convincing; the asymmetry of power lends a tone of confession.

There is a fundamental contradiction here. In practice, talk is treated as a collaborative activity requiring negotiation, repair, and other kinds of cooperative work. But in theory, talk is viewed as the expression of subjective opinions and personal experiences. This contradiction raises ethical issues, as Wiesenfeld points out. Equally important, it amounts to an "anomaly" in this approach to qualitative research, suggesting that its metaphysics, its ontological commitments, are fundamentally unsound. The belief that the qualitative research interview is like a conversation doesn't jibe with the metaphor that language is a conduit. The proposal that the interview is adding the "subjective component" otherwise missing from social science doesn't fit with the metaphor of the interview as a "speech event" or as "'speech activity." A "joint production" cannot be equated with "one person's subjectivity."

The research interviewer uses the resources of everyday conversation to fashion an interaction in which their involvement is rendered invisible, while the interviewee is encouraged to talk of personal tastes, feelings, and opinions in a manner detached from the here and now. One starts to wonder whether it is not the case that these practices *form* what we call "subjectivity," that subjectivity is an *effect* of the semistructured qualitative interview.

CONCLUSIONS

In this chapter, I have explored the practices of conducting the qualitative research interview. Perhaps because interviews are so familiar, we don't stop to look at them closely, so I have compared two ways in which the qualitative research interview has been described. The first highlights its differences from the conventional survey research interview, and the second points to its similarities with everyday conversation. Each of these characterizations has merit, but when we put them side by side, a *similarity* between the qualitative research interview and the conventional survey interview becomes evident. Despite its greater flexibility and rejection of the view that validity requires control and standardization, the qualitative research interview involves an *asymmetry* between interviewer and interviewee. Indeed, the flexibility of the qualitative interview seems to be *in service* of the asymmetry: the interviewer is flexible in order to foster disclosure by the interviewee. The qualitative research interviewer uses the resources of everyday conversation that are ruled out in survey interviews, but uses them to create an asymmetrical interaction that operates to hide the interviewer's contribution.

One response to the recognition of limitations in the qualitative research interview has been to propose that the interviewer should be more *active* (Fontana, 2002; Holstein & Gubrium, 1995). But insofar as this is merely a strategy for better access to subjective experience, it cannot resolve the contradictions that have become apparent here. The implication of my argument is that interviewers are *always* active, even when this activity is directed toward their own invisibility and apparent inactivity. What is needed is not a better strategy but a rethinking of the aims of the qualitative research interview.

The assumption that what is said in the interview is an expression of the inner subjectivity of the interviewee clashes with the idea that the interview is an everyday conversation between two fellow human beings. The "active invisibility" on the part of the researcher reflects the enduring assumption that they are engaged in an objective study of this inner subjectivity. The clash between theory and practice exposes problems in the ontological commitments of this standard practice. We will need to rethink the purpose and the conduct of the qualitative research interview, and the kind of data that it provides, in the light of a consistent recognition that every interview is a joint production. Before we do this, however, we need to look at the ways in which interview material is analyzed. Will the conduit metaphor operate there, too? Will analysis be viewed as the "unwrapping" of packages of speech? This is the topic of the next chapter.

3

The Analysis of Qualitative Interviews

Coding is analysis.

Miles & Huberman, 1984, p. 56

The declared aim of modern science is to establish a strictly detached, objective knowledge. Any falling short of this ideal is accepted only as a temporary imperfection, which we must aim at eliminating. But suppose that tacit thought forms an indispensable part of all knowledge; then the ideal of eliminating all personal elements of knowledge would, in effect, aim at the destruction of all knowledge. The ideal of exact science would turn out to be fundamentally misleading and possibly a source of devastating fallacies.

Polanyi, 1967, p. 20

In the standard qualitative research project, the step after conducting an interview is to transcribe it and analyze the material obtained. The analysis of qualitative material causes much anxiety and confusion for researchers, especially students conducting research for the first time. Yet remarkably little is said about analysis in many introductory qualitative research textbooks, and what is said is often unclear.

For example, Seidman's (1998) comprehensive book *Interviewing as Qualitative Research* includes only 14 or so pages on the topic of analysis out of a total of 124. Maxwell, in an otherwise excellent book titled *Qualitative Research Design*, writes in the chapter titled "Methods: What Will You Actually Do?" that his discussion "is not intended to explain how to *do* qualitative data analysis" (Maxwell, 2005, p. 95, emphasis original). This is odd coming from someone who offers an "integrative approach" to qualitative research design and insists, surely correctly, that all the elements of project design should interrelate. Maxwell talks only in general terms about analytic strategies of "categorizing" and "connecting," and it is not clear how these link to the research questions that orient a study or to the other components – goals of the study, theoretical framework, and others – whose interconnections he considers carefully.

These are not isolated cases. Generally only extremely brief characterizations of data analysis are offered; for example, that it is "a process of looking for significant statements, and comparing what was said in different interviews" (Blaxter, Hughes, & Tight, 2001). But what counts as "significant"? Is an interview really composed of "statements"? Why compare interviews, and how is the comparison made? What is the outcome of this comparison?

But despite the lack of detail, there is general agreement that analysis is a matter of "coding." Miles and Huberman state baldly that "[c]oding is analysis" (Miles & Huberman, 1994, p. 56). Ryan and Bernard write that "[c]oding is the heart and soul of whole-text analysis" (Ryan & Bernard, 2000, p. 780).

What precisely is coding, and why is it necessary? It is difficult to find clear answers to these simple questions. Seidman describes coding as the effort "to organize excerpts from the transcripts into categories" (Seidman, 1998, p. 107). He goes on: "The researcher then searches for connecting threads and patterns among the excerpts within those categories and for connections between the various categories that might be called themes" (p. 107). Ryan and Bernard explain that coding involves "finding themes," and "themes are abstract (and often fuzzy) constructs that investigators identify before, during, and after data-collection" (Ryan & Bernard, 2000, p. 780). Charmaz says that to code requires "defining actions or events" within the data (Charmaz, 2000, p. 515). Bogdan and Biklen have a similar view: "Developing a coding system involves several steps: You search through your data for regularities and patterns as well as for topics your data cover, and then you write down words and phrases to represent these topics and patterns. These words and phrases are *coding categories*" (Bogdan & Biklen, 1992, p. 166, emphasis original). Categories, threads, patterns, themes, regularities, topics, constructs, actions, events? The novice could be forgiven for becoming confused.

There seems to be a common approach to the analysis of interview transcripts. The transcripts are read and a coding system is developed, generally "inductively," seemingly on the basis of no prior theoretical framework or conceptual assumptions. Typically it focuses on what is common to several interviews. Frequently, researchers write that their coding categories "emerged" from the data. With the set of categories in hand, the researcher combs through the interview transcripts and applies a category (sometimes more than one) to everything that was said. The result is a collection of excerpts, each with one or more categories as a label. The results section of the paper lists these categories, quotes one or two excerpts for each, and adds a gloss to each excerpt.

Some approaches to qualitative research have spelled out this practice of analysis in enough detail for us to examine it. These approaches include thematic analysis (Boyatzis, 1998), empirical phenomenology (Giorgi, 1985), interpretative phenomenological analysis (Smith, 2004), and grounded theory

(Glaser & Strauss, 1967; Strauss & Corbin, 1990) . I will focus in this chapter on grounded theory because of its worldwide popularity (Mruck, 2000; Rennie, Watson, & Monteiro, 2000) and the large number of texts explaining it. I will propose that coding in grounded theory involves the twin practices of *abstraction* and *generalization* (Guignon, 1983). Abstraction is the practice of dividing a whole into elements that are distinct from one another and from their original context. Generalizing is the practice of finding what is common or repeated among these elements. In grounded theory, abstraction and generalization function together to (1) divide an interview transcript into separate units, (2) remove these units from their context, (3) identify abstract and general "categories," (4) extract the "content" from these categories, and (5) describe this content in formal terms. Once again, my interest is in the ontological and epistemological commitments embedded in these practices, and I will argue that, just like interviewing practices, coding practices embody contradictory notions about language and knowledge.

CODING AS ABSTRACTION AND GENERALIZATION

I begin with an example from a "how to" book, *Qualitative Data: An Introduction to Coding and Analysis* (Auerbach & Silverstein, 2003). To explain how analysis is done, the authors present material from their research on Haitian fathers. They illustrate coding with the following excerpt from a focus group interview:

> AG: Sometimes we cannot go to our father and say to our father, "You know something, I love you."
>
> F: Yeah, it is part of our culture. For me especially, even though I never heard such a word from my father's mouth such as "I love you." The way they act to us and the way they deal with us makes me feel that definitely this guy loves me.
>
> L: I tell you my father also never uttered the word[s] "I love you." ... But you knew he did. I make corrections in my own family. I must repeat to my children, I love them, I do not know, every several hours. Maybe every one hour. (Auerbach & Silverstein, 2003, p. 37, ellipsis in original)

Auerbach and Silverstein write, "We noticed that different research participants often used the same or similar words and phrases to express the same idea. These ideas are called *repeating ideas*, and they shed light on our research concerns" (p. 37, emphasis original):

> AG, F, and L all expressed the idea that their own fathers did not express love and affection. They said this in the language summarized as repeating idea number 4 . . . : "My father never said I love you." We assumed this idea was important because so many of the fathers in the study expressed it. (Auerbach & Silverstein, 2003, p. 38)

This example is fascinating because it illustrates very clearly the specialized way of seeing that coding requires. Although the words "I love you" were indeed spoken by each of the three participants, when one reads these words in their original contexts AG seems to be saying that *he* was unable to use them with his father, not that his father didn't use them. Perhaps his father didn't say the words either, but AG does not say so. F does attribute these words to his father, but he adds that his father expressed his love in other, nonverbal ways. And L, too, insists that he knew that his father loved him. So in *none* of these three utterances is it accurate to claim that the speaker states that "their own fathers did not express love and affection." To say that the speakers were "expressing the same idea" is very strange. To judge that the words "I love you" are "the same or similar" in the three cases requires seeing a transcript as made up of *units* – individual words and short phrases – that are read in isolation from their context. The words "I love you" appear in all three cases, but for this phrase to be identified as "repeated" it has to be *abstracted* from its setting. It has to become an abstract unit, decontextualized from the words that preceded and followed it, from the person who spoke it, and the moment in the interview when it was spoken, at the same time as it becomes seen as *general* to all three speakers.

The search for repeating words or phrases appears to accomplish two things: it identifies units that can be taken to be *common* to all cases, and at the same time it permits dividing the transcript into units whose context can now be ignored. Notice the assumption that what is common is *general*. Now the "meaning" of each unit needs to be defined. The common unit is rewritten in a way that avoids the concrete particularities of the original transcript. In the case we are considering here, the rewritten version is "My father never said I love you." Apparently this meaning must be both abstract and general. Finally, this rewritten version is attributed to the speakers: it is "the idea" that was "expressed" by each speaker.

This example clearly illustrates the practices of abstraction and generalization that are central to coding, as well as the peculiar kind of *vision* that coding requires. These practices can be found in the other most popular approaches to the analysis of qualitative material, such as empirical phenomenology, thematic analysis, and interpretative phenomenological analysis, though there is not space here to consider each of them. In the rest of this chapter, I will look in more detail at the approach Auerbach and Silverstein were trying to illustrate, "grounded theory." What we learn from examining this approach applies equally to the others.

GROUNDED THEORY

Grounded theory, one of the most popular and well-articulated approaches to analysis in qualitative research, was introduced by Barney Glaser and Anselm

Strauss in 1967. In *The Discovery of Grounded Theory*, Glaser and Strauss aimed to shift the focus of research in the social sciences away from testing theory (they called this testing "verification," but "falsification" would work equally well) to the *generation*, the *discovery*, of theory. As we saw in Chapter 1, the empirical-analytic model of research emphasizes hypothesis testing, and the construction of theories is left a mysterious and illogical process, a matter of guesses and hunches. Grounded theory was intended to fill this gap by offering a systematic way of producing theory from empirical data. Glaser and Strauss insisted that one doesn't have to be a "genius" to develop a theory that is appropriately "grounded" in data. The generation of theory is not "impressionistic," "exploratory," or "unsystematic" (Glaser & Strauss, 1967, p. 223); it proceeds by careful attention to empirical material, often interviews but also fieldnotes, documents, and archival library materials.

Glaser and Strauss viewed the generation of theory as "an inductive method" (Glaser & Strauss, 1967, p. 114) in which the researcher develops general concepts through abstraction from empirical data by "bring[ing] out underlying uniformities and diversities" (p. 114). Central to this is what they called a "comparative" approach to analysis – more technically, the "constant comparative method of qualitative analysis." They described four stages (see Table 3.1) to this method, emphasizing that in practice these stages are not completely distinct; indeed, one of the characteristics of grounded theory is that the collection of data, coding, and analysis go hand in hand (p. 43), so that questions raised in the analysis can be answered quickly by returning to the field for further investigation and to gather additional data, in a process called "theoretical sampling."

The first stage of comparative analysis is "explicit coding" of the data. Coding for Glaser and Strauss is a matter of associating one or more "categories" with each "incident" of data. The researcher reads through the material, "coding each incident in his data into as many categories of analysis as possible, as categories emerge or as data emerge that fit into an existing category" (p. 105). A "conceptual category" is said to be "generated" from the evidence. It is "a relevant theoretical abstraction" from the data. "In discovering theory, one generates conceptual categories or their properties from evidence; then the evidence from which the categories emerged is used to illustrate the concepts" (p. 23). For example, interviews with nurses who were caring for dying patients led to the abstraction of the category of "social loss" (Glaser & Strauss, 1964).

The "basic, defining rule for the constant comparative method" is "while coding an incident for a category, compare it with the previous incidents in the same and different groups coded in the same category" (Glaser & Strauss, 1964, p. 106). Glaser and Strauss insisted that it is this constant process of comparison that enables the analyst to identify a conceptual category and subsequently define its dimensions or properties.

Some categories are "abstracted" from terms used by interviewees and informants; these are "in vivo" concepts, and they tend to label the processes and behaviors to be explained (in other words, they are descriptive). Other categories are "constructed" by the researcher; these will tend to be the explanations. "For example, a nurse's perception of the social loss of a dying patient will affect (an explanation) how she maintains her composure (a behavior) in his presence" (Glaser & Strauss, 1964, p. 107). The in vivo category of "composure" is linked to, and explained by, the "constructed" category of "social loss." During analysis, the researcher writes notes – "memos" – on the categories and their application to the data. The second rule of the constant comparative method is to frequently "stop coding and record a memo on your ideas." These memos are described as being as important as attaching codes, if not more so.

The second stage involves "integrating categories and their properties." The "integration" of a theory – that it be "related in many different ways, resulting in a unified whole" (Glaser & Strauss, 1964, p. 109) – was an important goal for Glaser and Strauss. In this stage, comparison shifts to a more abstract level in order to establish relationships among the conceptual categories and their properties. For example, nurses calculated social loss on the basis of factors such as a patient's age and education. As they came to know a patient better, they did work to balance out the newly discovered social loss factors. This "calculus of social loss" was linked by the researchers to what they called a "social loss story" developed about each patient.

The third stage is "delimiting the theory." A theory "solidifies" as fewer modifications to it become necessary, and higher-level concepts are used to "delimit" the terminology. This may increase the theory's scope: Glaser and Strauss were able to generalize their theory of social loss "so that it pertained to the care of all patients (not just dying ones) by all staff (not just nurses)" (Glaser & Strauss, 1964, p. 110). The categories become "theoretically saturated": new incidents fit easily into existing categories and only need to be coded when they add something new to the theory. For example, a dying patient age 85 was viewed as a high social loss because her advanced age was balanced by her "wonderful personality," so "personality" became a new property of the category of "social loss."

The fourth stage is that of "writing theory." The theory is described in reports and publications. Memos for each category are collated and summarized. The memos contain "the content behind the categories, which become the major themes" of the written theory (Glaser & Strauss, 1964, p. 113).

What is a theory? For Glaser and Strauss, a theory is "a reasonably accurate statement of the matters studied" (Glaser & Strauss, 1964, p. 113), written either formally as a set of propositions or informally as a narrative. It may be "formal theory," dealing with a conceptual area of academic inquiry such as formal organization, or "substantive theory," dealing with a particular

TABLE 3.1. *Stages of Analysis in Grounded Theory*

Stage	From . . .	To . . .
1. Explicit Coding	Compare incidents *"Mr. Smith was dying."*	Conceptual categories *"Social Loss"*
2. Integrating Categories	Compare categories *Types of social loss: age, education, etc.*	Relationship among categories *"Calculus of social loss"*
3. Delimiting the Theory	Compare across cases *Nurses, patients*	Higher-level concepts *"Care for all patients by all staff"*
4. Writing the Theory	Collate and summarize memos *"The content behind the categories"*	Statements of generalized relations among categories *"A substantive theory of patient care"*

practical topic such as patient care. A theory predicts and explains, and it is intended to be generalizable. It involves statements about "generalized relations among the categories and their properties" that have been identified (p. 35). The aim of grounded theory, Glaser and Strauss wrote, is not that of generating a single, all-encompassing causal theory but "generating and plausibly suggesting (but not provisionally testing) many categories, properties, and hypotheses about general problems" (p. 104). These may include hypotheses about causes, but hypotheses need not be causal. They can also be about "conditions, consequences, dimensions, types, processes, etc." (p. 104).

The Practices of Analysis in Grounded Theory

It is clear that grounded theory involves the practices of abstraction and generalization that we saw in Auerbach and Silverstein's example. The very definition of a theory is based on these two practices: a theory is made up of statements about the "generalized relations" among "abstracted" conceptual categories and the properties of these categories. Comparing incidents leads to their coding into abstract categories, such as "social loss." Comparing categories leads to the identification of their properties and kinds, so that types of "social loss" such as "age" and "education" are related in a "calculus of social loss." Delimiting requires comparing different cases and leads to more general concepts, such as "care for all patients, by all staff." Finally, the memos for each category are collated and summarized, leading to a theory that has the form of statements about generalized relations among categories, such as a substantive theory of patient care. The emphasis on comparison runs throughout grounded theory: "constant comparative analysis" is central to the whole approach and combines abstraction and generalization.

BOX 3.1. *Grounded Theory and Quantitative Data*

It may surprise the reader to learn that a large portion of *The Discovery of Grounded Theory* deals with *quantitative* data. Glaser and Strauss held the view that "there is no fundamental clash between the purposes and capacities of qualitative and quantitative methods or data. . . . [T]he process of generating data is independent of the kind of data used" (Glaser & Strauss, 1964, pp. 17–18). They felt that "categories and properties emerge during the collecting and analyzing of quantitative data as readily as they do with qualitative" (p. 193). This leads one to suspect that the basic view of knowledge and reality in grounded theory is no different from that of traditional empirical-analytic science. Indeed, the very notion that the "discovery" of theory was a gap that needed to be filled reflects their acceptance of the logical positivists' definition of science, and in particular of the positivist distinction between the psychology of discovery and the logic of verification. Glaser and Strauss, just like the Vienna Circle logical positivists and like conventional empirical-analytic social scientists, continued to view the process of discovery as leading to a theory that consists of "hypotheses" that will subsequently be tested.

The basic unit of empirical material is what Glaser and Strauss call the "incident" (Glaser & Strauss, 1964, pp. 105ff), but although they emphasize that data must be "fractured" into these units, they never define the term. Glaser and Strauss wrote that: "The generation of theory requires that the analyst take apart the story within his data. Therefore when he rearranges his memos and field notes for writing up his theory, he sufficiently 'fractures' his story" (p. 108). Glaser and Strauss considered fracturing central to the first stage of analysis, and here we can start to see the assumptions embedded in grounded theory (see Box 3.1).

The term "incident" suggests that Glaser and Strauss drew no distinction between an event and its description. For example, they described how "the analyst codes an incident in which a nurse responds to the potential 'social loss' of a dying patient" (Glaser & Strauss, 1964, p. 106). But an interview is not made up of incidents. It is made up of verbal *descriptions* of incidents. Glaser and Strauss apparently saw their data – the text of an interview or a researcher's fieldnotes – as a *transparent* representation of what they called "the matters studied." They assumed that they could see "through" the interview to the incidents being described (Charmaz, 2000, p. 514). They assumed that what happens in the hospital, the nurses' accounts of these incidents, and the researcher's theory can all correspond because all three share a common "meaning." From this point of view – or with this type of vision – verbal descriptions of incidents can be studied with no attention to their context.

Interviews can be fragmented into separate incidents, and this is done by comparing them to identify what is *common*. Fragmentation is abstraction in the service of generalization.

Once incidents have been abstracted, they are coded into conceptual categories. For Glaser and Strauss, a "concept" is "a relevant theoretical *abstraction*" (emphasis added). Because it is based not on a single incident but on what is *common* to multiple incidents, it is assumed to be *general*. This process of categorization is then repeated at a higher level of abstraction, comparing not incidents but categories. This enables the analyst to define dimensions or properties shared across conceptual categories. Next, categories are generalized and abstracted across cases because the emphasis is on building a theory that is general because it is grounded in multiple cases. Comparing categories leads to "more abstract categories" and identification of "underlying uniformities and diversities," and ultimately to integration of the theory into a unified whole. Throughout, abstraction and generalization operate on empirical data to "bring out underlying uniformities and diversities" (Glaser & Strauss, 1964, p. 114). This move from the particular to the general is the ascent from the empirical "ground" to the theory.

In the final step, the researcher rewrites the memos in either formal or narrative terms. The memos contain, as we have seen, the researcher's interpretations of the "abstracted" categories. Terms such as "social loss" have replaced the nurses' original words. Even in vivo categories are "labels" invented by the researcher for everyday actions and events. Glaser and Strauss put the matter very clearly: in their view, the memos contain "the content behind the categories" (Glaser & Strauss, 1964, p. 113), just as the categories code "the content" in the data. By the end of the inquiry, they suggested, "the sociologist finds that he has 'a feeling for' the everyday realities of the situation, while the person in the situation finds he can master and manage the theory" (p. 241). But can a researcher's theory simply trade places with the nurses' situated know-how? Whereas theory is built from abstractions, everyday reality surely involves practical know-how, concrete and tacit knowledge. Theory may be "grounded" in the concrete particularities of everyday life, but as it "rises" through abstraction and generalization it becomes formal and theoretical. It should hardly be taken to be equivalent to the everyday reality of the people being studied, but this distinction was invisible to Glaser and Strauss.

Words, Things, and Concepts

In the 40 years since *The Discovery of Grounded Theory* was published, there have been modifications and refinements, and the two authors now differ in their views of how it should be practiced. Recent texts have sought to clarify the process of coding. But the drive to replace the particular and concrete with the abstract and general continues to motivate grounded theory today.

For example, Strauss and Corbin (1998; Corbin & Strauss, 2008) continue to emphasize the important function that comparison serves of "moving the researcher more quickly away from describing the specifics of a case . . . to thinking more abstractly about what . . . is share[d] in common and what is different" (Corbin & Strauss, 2008, p. 77). They explain that "[u]sing comparisons brings out properties, which in turn can be used to examine the incident or object in our data" (p. 75). "[W]e always work with concepts rather than specifics of data or cases. It is not the specific incident per se but rather what the incident symbolizes or represents" (Strauss & Corbin, p. 81). So when a nurse says "I prefer to work with another experienced nurse," Strauss and Corbin explain, "we turn to thinking comparatively about the terms 'experienced' and 'inexperienced' and not so much about the fact that this nurse does not like to work with some people. . . . It is the concepts 'inexperienced' and 'experienced' that interested us rather than the particulars" of the case (p. 81), so "we can draw our comparisons from any area or situation in which experience or inexperience might make a difference, such as with driving or house painting" (Corbin & Strauss, 2008, p. 76).

The notion that a researcher should be concerned not with particulars but with general and abstract concepts rests on several assumptions. It assumes that the words "experienced" and "inexperienced" stand for concepts and that the meaning of these concepts is available to the researcher by reflecting on the terms rather than by talking to the nurses (now "we are using incidents from our own experience," Corbin & Stauss, 2008, p. 76). It assumes that the meaning of the terms lies not in the specifics of their use but in some kind of shared conceptual space. Strauss and Corbin propose that a comparison (which they say might just as well be of "inexperienced seamstresses or drivers instead of nurses") will disclose the "list of properties" that characterizes each of these concepts. But who is to identify these "experienced" and "inexperienced" seamstresses and drivers? To do so would require that we already know what the words mean – or at least how to use them. The result of the comparison, they promise, will be "some idea of what it means to be inexperienced," perhaps "the properties of being cautious, apprehensive . . . prone to make errors, unsure of him or herself," and so on. This would then enable the researcher to "observe both experienced and inexperienced nurses, watching how they function and how they handle problems under various conditions" (Strauss & Corbin, 1990, p. 82).

But don't we already know that "inexperienced" means something like "lack of knowledge or proficiency"? We can look in the dictionary and discover that "inexperience" means "lack of practical experience" or "lack of knowledge or proficiency gained by experience." What is gained by a comparison, empirical or "theoretical" (whatever this means), of inexperienced drivers and nurses?

Surely a better question is, "How does the nurse use such words?" What is she accomplishing by calling someone "inexperienced"? Strauss and Corbin have no reply to Mishler's point that in traditional analysis "[t]he central problem for coding may be stated as follows: because meaning is contextually grounded, inherently and irremediably, coding depends on the competence of coders as ordinary language users" (Mishler, 1986, p. 3).

Does a focus on what is *common* to particular cases truly provide a way to grasp what is *general*, and therefore what they "mean"? The "ordinary language" philosopher Ludwig Wittgenstein (1953) pointed out that many of our everyday concepts are "polymorphous" and show what he called "family resemblances." Like members of a family, each particular exemplar of a concept can resemble another in one feature or another, but there may be no features shared by every case. A son may have his mother's mouth and his father's eyebrows. His sister has her brother's fair hair. But none of these features need be found in every member of the family. The search for a universal feature, common to all, would come up empty-handed. There are indeed resemblances, but they are "a complicated network of similarities overlapping and criss-crossing." To return to Auerbach and Silverstein's example, do we really believe that all Haitian men experience fatherhood the same way? Surely not. Do we believe that some aspects of the experience are common to all Haitian men? Perhaps. But does this mean the differences are unimportant? Not necessarily.

Wittgenstein's favorite example of an everyday concept was "game." It is worth quoting at length from his invitation to try to find a single feature that all games have in common:

> Consider for example the proceedings that we call "games." I mean board-games, card-games, ball-games, Olympic games, and so on. What is common to them all? – Don't say: "There *must* be something common, or they would not be called 'games'" – but *look and see* whether there is anything common to all. – For if you look at them you will not see something that is common to *all*, but similarities, relationships, and a whole series of them at that. To repeat: don't think, but look! – Look for example at board-games, with their multifarious relationships. Now pass to card-games; here you find many correspondences with the first group, but many common features drop out, and others appear. When we pass next to ball-games, much that is common is retained, but much is lost. – Are they all "amusing"? Compare chess with noughts and crosses. Or is there always winning and losing, or competition between players? Think of patience. In ball-games there is winning and losing; but when a child throws his ball at the wall and catches it again, this feature has disappeared. Look at the parts played by skill and luck; and at the difference between skill in chess and skill in tennis. Think now of games like ring-a-ring-a-roses; here is the element of amusement, but how many other characteristic features have disappeared! And we can go through the many, many other groups

of games in the same way; can see how similarities crop up and disappear.

And the result of this examination is: we see a complicated network of similarities overlapping and criss-crossing: sometimes overall similarities, sometime similarities of detail. (Wittgenstein, 1953/2001, p. 27)

Wittgenstein concluded, "I can think of no better expression to characterize these similarities than 'family resemblances'; for the various resemblances between members of a family: build, features, colour of eyes, gait, temperament, etc., etc. overlap and criss-cross in the same way. – And I shall say: 'games' form a family" (pp. 27, 29).

Wittgenstein was very skeptical about the approach of searching for common elements to find generalities. He wrote that "[t]he idea that in order to get clear about the meaning of a general term one had to find the common element in all its applications has shackled philosophical investigation" (Wittgenstein, 1958/1969, p. 19). But this is exactly how qualitative data are coded.

Wittgenstein's important insight was not merely that words are related in an untidy way, like members of a family. His point was that we are continually *changing* the relationships among words. The fact that we cannot identify common properties among the different activities we call "game" is not a failure on our part. Wittgenstein writes: "But this is not ignorance. We do not know the boundaries because none have been drawn" (Wittgenstein, 1953/2001, p. 69). When we call something a "game," we do so for a particular purpose, and in doing so we draw a boundary and define the concept of "game" for that moment. We might one day call something a game to emphasize that it is fun and the next day call something else a game to emphasize that it is a competition. What gives a word its meaning is its *use* in particular circumstances. The meaning of a word follows from the use it is put to. It is not some property or collection of features – or some concept – that it unchangeably names or labels. At first sight, we can all agree that the word "tree" refers to objects that are living and have a trunk and leaves. But what about a family tree? A decision tree? When we apply the word *tree* to a family, we do so to emphasize its relationships, but these are not *literally* branches.

To Strauss and Corbin, the relationship between things in the world and words in language is straightforward. A "concept," with its properties, is something both words and things have in common, so Strauss and Corbin can write of "properties inherent in the objects" (Strauss & Corbin, 1990, p. 105) but also claim that concepts "share certain properties" (p. 103). They write that: "[D]uring open coding, data are broken down into discrete parts, closely examined, and compared for similarities and differences. Events, happenings, objects and actions/interactions that are found to be conceptually similar in nature or related in meaning are grouped together under more abstract concepts termed 'categories'" (Strauss & Corbin, 1990, p. 102).

For example, a bird, a kite, and a plane all "have the specific property of being able to fly" and so can be categorized together as "having the ability to soar in the air." This is obviously a crucial part of the process of coding, in grounded theory and other approaches. But it is inherently circular. The outcome of such comparisons depends completely on what one chooses to compare. If we compared a bird not with a kite and a plane but with a rat and a camel, the common property would be quite different, something like "warm-blooded mammal."

Grounding Objectivity in Subjectivity

Abstraction and generalization work together to divide a transcript into decontextualized units, to abstract general concepts from these units, to extract the content of these concepts, and to rewrite it in new terms. This is very strange! We interview people to discover their personal experiences, but then we compare the interviews, divide what has been said into many pieces, search for common words and phrases, label each common phrase with a category that we have invented, and finally we replace the interviewees' words with our own!

Wiesenfeld, who noted the gap between the theory and practice of qualitative research interviews that I discussed in the last chapter, has also pointed out the contradictions in this kind of analysis. She writes that "the voice [of the research participant] is recovered only to be silenced again" (Wiesenfeld, 2000). When we compare interviews and search for commonalities, we fail to take account "of the context and the diversity of the participants," and this approach also "tends to conceal the researcher's reflexivity and underestimate his/her influence on the co-construction of the speech being analyzed."

Although qualitative research places emphasis on unique individual experience, there is evidently a strong motivation not to stop with an individual's subjectivity but to abstract, generalize, and formalize what each individual tells us. The character of this motivation is, I believe, stated clearly by Ericsson and Simon: "The problem with 'soft data' is that different inter-preters making different inferences will not agree in their encodings, and each interpreter is likely, unwittingly or not, to arrive at an interpretation that is favorable to his theoretical orientation" (Ericsson & Simon, 1984, p. 4).

The benefit of coding is thought to be that it "operationalizes one's subjective impression" (Chi, 1997, p. 282) and so enables a researcher to move from mere "inference" to truly objective knowledge. A formal language "seem[s] to offer the hope of intersubjective agreement free from interpretive dispute" (Taylor, 1980, p. 36). The motivation behind coding is to find *objective* knowledge in these expressions of *subjectivity*. We have seen that coding produces something abstract, general, and formal. The search for

---

BOX 3.2. *What Are Themes?*

Many qualitative researchers report their results as "themes" that "emerged" from their data. Themes are often summarized with simple phrases, such as "parental emotional abuse" or even simply (in a different study) "parents." Such phrases obviously refer to the *content* of the interview; that is, to the events or topics talked about. They are often not even framed as assertions (such as "parents are the cause of problems"), so they do not give the reader any information about the way that the content was presented.

Perhaps researchers find themes attractive because they have a long history and distinguished origins. Aristotle wrote in his *Poetics* (1967) of the "theme" or "thought" (Greek: *dianoia*) that accompanies the "fable" or "plot" (*mythos*) of a tragedy. By *dianoia* he meant the answer an audience might provide when they ask themselves the question "What is the point of this story?" But of course two members of the audience will often find a different point in the same story. A theme never simply "emerges"; it is the product of interpretation.

In his detailed discussion of interpretation, Heidegger suggested that "[i]f, when one is engaged in a particular concrete kind of interpretation ... one likes to appeal to what 'stands there,' then one finds that what 'stands there' in the first instance is nothing other than the obvious undiscussed assumption of the person who does the interpreting" (Heidegger, 1927/1962, p. 192). If this is the case, then the themes that "stand out" tell us more about the researcher than about the interviewee, and they should not be the starting point for analysis.

"regularities and patterns" (as Bogdan and Biklen put it) assumes that general, decontextualized units – whether they are categories, themes (see Box 3.2), or "essential" structures – are more objective than the particular, context-laden words that "contain" the interviewee's subjectivity.

Seeing Features and Categories

Grounded theory, like coding approaches in general, sees words, objects, events, and people as *exemplars* of more general categories, or *concepts*. This object in front of me, for example, is a chair. Researchers using grounded theory want to know not about this specific chair but about chairs in general. They are interested in the *category* of chair rather than any of its particular instantiations. To them this way of seeing the category is more real than the object itself. How can one gain access to these categories? In this way of seeing, things – objects, events, and people – have features or properties, and categories of things share features in common. The category *chair*, for example, amounts to having the properties of four legs, a back, and a seat. Some particular chairs are black and others are brown. Color, then, is not a common feature, so it is not part of the category of chair. The terminology of grounded theory – fragmentation, incidents, coding, concepts – fosters a way of seeing the world in which things have independent properties and are categorized on the basis of the properties they have in common.

The same is assumed to be true of people. The researcher interviews someone because that person is an example of the category "nurse," for instance. Researchers are not interested in the particular individual but in what one can learn about nurses in general. To obtain this kind of knowledge, they must compare different nurses to discover what features they have in common.

In this way of seeing, differences are far less important than similarities because similarities provide access to general concepts. The researcher is looking for the general that "underlies" the particular. Individual cases are seen not as wholes or unities but as collections of features or properties. It is as though when I look at a chair I see not a whole object but a collection of legs, surfaces, and colors that just happen to be together.

Where are these categories? They are not material, so they cannot exist in the world. But they can't exist in people's minds either, at least not directly. Certainly, in this way of seeing, people's minds contain things, *mental* things: experiences, thoughts, and feelings. These experiences also have features, and they, too, can be categorized: thoughts and feelings with common features are examples of the same *concept*.

But the concepts or categories can't exist in the individual mind either. If they did, they would be subjective. It seems that they can exist only in our *language*. Behind this approach to analysis once again lie unexamined assumptions about the nature of language – how one understands what has been said; how verbal descriptions relate to the events described; how thinking and speaking are related. Language is assumed to be a simple, direct, and universal way of relating subjective experiences to objective things. Once again, the conduit metaphor is operating. In this way of seeing, language is a collection of words, and each word is both the "expression" of something in the mind and the "name" of something in the world. Experiences are "expressed" in words, but words also "refer" to things and events in the world.

"Concepts" provide the bridge between person and person, and between person and world. The "meaning" of each word is a concept. Individuals are able to communicate because although each one has personal and subjective experiences, they share common concepts.

Seen this way, the researcher's job is to identify concepts. The researcher compares what has been said by different people in order to identify what is common to their words and therefore in their experiences. The common words and phrases in their interviews are unpacked to spell out the concepts they contain. This unpacking is done by the researcher, who draws on his or her "common sense" understanding of words, but the results are attributed to the speaker. It is assumed that the concepts that the researcher "unpacks" from the language were "put there" by the speaker, who "expressed" them in words. This is why grounded theorists are less interested in "the specifics of a case" than in "concepts" like "inexperienced." This is why they turn to words rather than to people to figure out the meaning of such a concept.

LANGUAGE AS NAMING

Strauss and Corbin, like Glaser and Strauss before them, have adopted a model of language that sees words (or phrases or sentences) as labels for concepts, which in turn name objects. The coding process involves spelling out the "concept" the interviewee used to express their experience, their subjective meaning, of some event. Strauss and Corbin (1990), as we have seen, describe open coding as "taking apart an observation, a sentence, a paragraph, and giving each discrete incident, idea, or event a name, something that stands for or represents a phenomenon" (Strauss & Corbin, 1990, p. 63). In their view, coding articulates the concepts that the interviewee uses to refer to things and events, and then the properties of these concepts.

This is, once again, the conduit model of language. It is precisely the model of language that Wittgenstein rejected in his *Philosophical Investigations* (1953). There is a close link between Wittgenstein's criticism of the strategy of comparison and his criticism of what he called the "naming" model of language, which is as old as Augustine's *Confessions*, written in 397 AD. Saint Augustine described how he first learned to speak: "When they called some thing by name and pointed it out while they spoke, I saw it and realized that the thing they wished to indicate was called by the name they then uttered. . . . Thus I exchanged with those about me the verbal signs by which we express our wishes" (Augustine, 397/2002, p. 8).

In this model of language, as Wittgenstein notes, "Every word has a meaning. The meaning is correlated with the word. It is the object for which the word stands." Wittgenstein points out that although this model might seem fine for concrete nouns, it is hard to apply it to actions or properties. What does "boiling" stand for? What does "green" stand for? Or "ugly"? This way of thinking about meaning "has its place in a primitive idea of the way language functions. But one can also say that it is the idea of a language more primitive than ours" (Wittgenstein, 1953, paragraph 2). To say that "all words in language signify something" is like saying "all tools in a toolbox modify something." We have really said nothing at all. We do sometimes use a name in order to attach a label to an object, but this is only one of the many ways words are used. Wittgenstein recommends that we consider not merely the words of language but also the forms of activity within which these words are used. "I shall ... call the whole, consisting of language and the actions into which it is woven, the 'language game'" (paragraph 37).

In the "naming" model of language, words label concepts, which stand for objects. In the "conduit" metaphor, thoughts and ideas are "put into" words, which can then be transmitted from speaker to hearer. To connect these two models of language, all one needs to add is the assumption that thoughts and ideas are made up of concepts. In the typical approach to the analysis of qualitative material, coding is assumed to reverse the process the interviewee used to put meanings "into" words. It "opens up" the words to "take out" the "meanings" they contain. The researcher's task is to identify the "content" contained in the verbal "form." The interview transcript is assumed to contain the meaning that the interviewee placed in it, and analysis is a matter of describing the contents of the container. Once this is done, the verbal forms – the interviewee's original words – can be discarded or simply used to "illustrate" the results of the analysis. Strauss and Corbin show their use of this model very clearly when they write that in coding "we open up the text and expose the thoughts, ideas, and meanings contained within" (Strauss & Corbin, 1990, p. 102). Speech, once again, is assumed to be a conduit; words are assumed to be containers.

Seeing the concepts "inside" words is not so easy, and this is one reason why learning to code is so frustrating. For example, it is not easy to see a transcript as being made up of individual units. Where does one unit end and the next begin? Is a unit a word, a phrase, or an utterance? Sometimes this problem is solved by recommending "a line-by-line analysis, asking, What is this sentence about? ... How is it similar or different from the preceding or following statements?" (Ryan & Bernard, 2003). Borgatti (n.d.) recommends that, in coding, "Each line of the transcript is examined, [and one asks] 'What is this about? What is being referenced here?'" But to equate the units of analysis with the lines of the transcript makes no sense because the division of a transcript into lines is completely arbitrary; lines don't exist in spoken language.

As Ryan and Bernard say, "Coding forces the researcher to make judgments about the meaning of contiguous blocks of text" (Ryan & Bernard, 2000, p. 780). But is being forced to do this such a good idea? What happens is that researchers draw on their personal knowledge of language, of "what words mean," and assume that this knowledge is shared by the speaker.

Replacing the Interviewee's Words

The final step in coding qualitative material is to replace the interviewee's original words with a formal description. The researcher's summary is both more abstract and more general that the original words. It claims to capture what is common to all cases, using abstract categories. In grounded theory, an interviewee's words are replaced first by the researcher's coding categories and then by the memos the researcher writes about these categories. Harold Garfinkel, a brilliant sociologist whose work we shall explore in more detail in Chapter 8, has pointed out how virtually all analysis in the social sciences attempts to replace what are called "indexical" expressions with nonindexical equivalents (Garfinkel, 1967). The term indexicality refers to the "pointing" that we accomplish with language: think of your index finger, the one you use to point. Everyday language has many links to its context, to the circumstances in which we speak. This kind of indexical reference to the landmarks that surround speakers can be achieved with pointing and other gestures or with verb tenses and adverbs (such as here, now). This is obvious with terms like "here," "there," "I," and "you": the place or person being referred to depends on the place and the person who is speaking. Consider the following utterance:

> "*I will put this here now.*"

To understand this, one needs to know who the speaker was and where and when they spoke. It seems natural to substitute what was "actually meant":

> "*I, Martin Packer, will put the cup on the table at 6:45 p.m.*"

Indexicals immediately challenge any suggestion that meaning is "in" words. In everyday interaction the indexicality of ordinary talk is not a problem because speakers and hearers know much more than what is evident in the words spoken. They know who and where they are, for a start. But this indexicality seems troublesome if one believes that social scientists are seeking knowledge that is not tied to particular times or places. Even if we are simply trying, as researchers, to report what someone is saying, indexicality seems a bother. Someone might say "I am here right now," but surely what they really mean is that "John is at the library on Saturday afternoon." It would seem that we must try to find nonindexical language that will spell out all the additional, implicit knowledge that is used to understand everyday speech. The typical analysis of qualitative material is attempting to do exactly this. The text from an interview – what the interviewee *said* – is replaced by the researcher with a description, a summary, a "gloss," which is supposed to capture what they really *meant*.

The typical view is that indexicals are useful but awkward (Garfinkel, 1967, p. 243), and most social scientists believe that genuine science, like logic, uses only objective (nonindexical) expressions. They assume that it is easy to distinguish between the two kinds of language and that only practical difficulties – limited time, limited money – prevent the substitution of one for the other.

But Garfinkel pointed out that this effort in countless areas in the social sciences to substitute nonindexical expressions for indexicals, this

TABLE 3.2. *Glossing Everyday Conversation*

Excerpt from Transcript of Everyday Conversation	Student's Gloss of "What they were actually talking about"
"Dana succeeded in putting a penny in a parking meter today without being picked up."	This afternoon, as I was bringing Dana, our four-year-old son, home from the nursery school, he succeeded in reaching high enough to get a penny in a parking meter when we parked in a meter zone, whereas before he had always had to be picked up to reach that high.

"programatical substitution," is never completed. He argued that it will always be "unsatisfied" in practice. In the preceding example, what day are we talking about? Which table? Which Martin Packer? (Googling turns up three!) There never seems to be enough time or money for researchers to complete the task of replacing all the indexical expressions in their data with objective expressions, and this ought to make us suspicious. Garfinkel offered two demonstrations that such a substitution is not merely difficult but impossible.

The Glossing Study

In the first demonstration, Garfinkel gave his students the task of transcribing an everyday conversation and then writing a "gloss." "Students were asked to report common conversations by writing on the left side of a sheet what the parties actually said, and on the right hand side what they and their partners understood they were talking about." For example, Table 3.2 shows a small part of an exchange between husband and wife. On the left is the excerpt from their conversation; on the right is a student's gloss. But of course the words in the right-hand column still show much indexicality. Who is the "we" who parked? What point in time does "before" refer to? Is the "he" who had to be picked up the same "he" who put in the penny? And so on, endlessly. Garfinkel reports that, "As I progressively imposed accuracy, clarity, and distinctness, the task became increasingly laborious. Finally, when I required that they [the students] assume I would know what they had actually talked about from reading literally what they wrote literally, they gave up with the complaint that the task was impossible" (Garfinkel, 1967, p. 254).

Garfinkel's students complained that their task was not just difficult but impossible. It became clear that they would never be able to spell out completely all of the ways in which the speakers drew on their own and others' knowledge of the situation in their talk. They came to recognize that "the very way of accomplishing the task multiplied its features" without end. The task of replacing all indexical expressions with objective expressions is an endless one.

TABLE 3.3. *An Incorrect Model of Language*

What Was Said	What Was Talked About
Words are *signs*	Objects are *referents*
The words spoken	What was "in mind," "beliefs," what was "intended," what was "meant"
The form	The contents
"The facts": what a tape recorder picks up	"More": what needs to be added
Indexical, sketchy, elliptical, ambiguous, incomplete	Elaborated, complete, nonindexical

Note: All based on a common background knowledge shared by speakers and the researcher (which provides the grounds needed to argue for the correspondence).

Yet this is exactly the task qualitative researchers are undertaking when they write their description of what was "really" meant by the people they interviewed.

In Garfinkel's view, the belief that everyday, indexical language needs to be replaced with a decontextualized, scientific language shows that people are using an inappropriate model of language. Researchers draw a distinction between what is *said* and what is *meant* and assume that the relation between the two is that of *signs* to their *referents* (see Table 3.3). They conclude that their task as researchers is to reconstruct, from what was said (the "facts" of speech that our tape recorders pick up), what was meant (what the speaker had "in mind," what they "intended"). They are given the *form*, and they must reconstruct the *contents* on the basis of a common background knowledge shared by the speakers and the researcher. But if language operated this way, there would be no reason for Garfinkel's students to find their task not merely difficult but impossible. This model of language fails to capture its impossibility. We need to drop the elements in italics in Table 3.3: the *theory of signs* and the *shared agreement* (i.e., the assumption in this model that everyone shares the same concepts). Garfinkel recommends that language should instead be "conceived, not simply as a set of symbols or signs, as a mode of representing things, but as a 'medium of practical activity,' a mode of doing things" (Giddens, 1977). Garfinkel argues that "the task of describing a person's method of speaking is not the same as showing that what he said accords with a rule for demonstrating consistency, compatibility, and coherence of meaning" (Garfinkel, 1967, p. 257). We shall return to this in Chapter 8.

The Coding Study

Garfinkel's second demonstration of the impossibility of substituting objective expressions for indexical expressions was a study of a psychiatric

BOX 3.3. *Coding as Nominal Measurement*

Part of the attraction of coding is undoubtedly that it amounts to a form of nominal measurement. Empirical-analytic researchers distinguish four levels of measurement: nominal, ordinal, interval, and ratio. These are listed in terms of increasing power: nominal measurement is no more than naming or categorizing, a very simple and unsophisticated kind of treatment. Nominal measures, such as male and female, left and right, or green, blue, and red, have no organization other than being a list. Ordinal measures have an intrinsic order, such as short, medium, and tall, or little, some, and much. Interval measures have in addition the characteristic that there is an equal interval between each pair of items. Temperature in degrees Celsius is a common example. But the zero point on this scale is arbitrary (the temperature at which ice melts), and ratio measures have in addition a nonarbitrary zero point on the scale, which means that a measurement can be read as a ratio against a unit quantity. Measures of mass, length, and time are ratio measures.

Nominal measures can be analyzed with nonparametric statistics such as chi-square analysis, which deals with frequency counts. We often find qualitative researchers counting data once they have coded it; there is comfort in familiarity, and measurement contributes to the sense that the results of the analysis are objective.

outpatient clinic. The researchers were interested in the criteria by which applicants were selected for treatment. The main research strategy was to code the data available in clinic records, patient files, and other paperwork. Clinic folders were examined, and a coding sheet was completed by research assistants who had been trained in the detailed operational definitions of each of the codes relating to clinic procedures and events.

In this case, then, we have a project in which the researchers were coding the messy, disorganized things that people wrote in the clinic folders in order to obtain an objective record using an organized, explicit, and unambiguous set of coding categories. Garfinkel, however, found that the coding was not what it appeared to be (see Box 3.3). The coders engaged in an unexamined level of "practical reasoning." Garfinkel observed that, although the coding manual *could* be strictly followed, when the coders wanted to be sure that they had correctly coded "what *really* happened" in the clinic, they invoked what he called "ad hocing" considerations – *interpreting* the coding instructions to assure their relevance on the basis of what they already knew about the operation of the clinic. When the coding rules generated what they knew to be the "wrong" code, the coders corrected it. As Garfinkel put it, "in order to accomplish the coding, coders were assuming knowledge of the very organized ways of the clinic that their coding procedures were intended to produce descriptions of."

This use of tacit understanding to guide coding can be found throughout qualitative coding. For example, Harry, Sturges, and Klingner (2005) illustrate their use of grounded theory:

> [I]n our data, a teacher, distressed about the large number of children in her class, exclaimed, "Oh, no! So many kids!" We compared the properties of the situation to which she was referring with a statement by another teacher: "There are 23 [exceptional education] kids lined up at my door." Noting that both teachers were complaining about the number of children they were expected to teach, we assigned both statements the code Class Size. (Harry, Sturges, & Klingner, 2005, p. 5)

Just as in Garfinkel's clinic, the coding of the data here is based on the researchers' *prior* understanding of what the teachers meant. These researchers apparently knew enough about the organized ways of the school to recognize that the two statements were "complaints." The coding as "Class Size" does not emerge from the data but from the researchers' understanding of the "situation" and of the teachers' "distress." The codes capture not regularities in the data but regularities in the organized ways of the school, which the researchers already understood. This understanding was a crucial *resource* for the coding, but it never became an explicit *topic*.

CONCLUSIONS

For many qualitative researchers, analysis is coding, and in this chapter we have seen that coding is accomplished through the practices of abstraction and generalization. Coding divides an interview transcript into separate units, removes these units from their context, identifies abstract and general "categories" among these units, extracts the "content" of these categories, and describes this content in formal terms. Grounded theory provides a clear example of how coding is supposed to work, but the same practices are employed in empirical phenomenology, interpretative analysis, and thematic analysis.

Coding supposedly avoids interpretation and so provides objective knowledge. The motivation for coding is that it "seem[s] to offer the hope of intersubjective agreement free from interpretive dispute" (Taylor, 1980, p. 36). The words "interpretation" and "subjective" are conjoined so frequently that it is clear that for most social scientists – and for many qualitative researchers – interpretation is assumed to inevitably lead to subjectivism and relativism. But in fact each step of coding introduces new problems (see Table 3.4).

TABLE 3.4. *The Problems Associated with Coding*

Step in Coding	Problems
1. Look for commonalities Find common (repeated) units 　(words, phrases, themes). Constant comparison.	Loses the voice of each individual. Assumes that words define concepts. Assumes that what is common is general.
2. Divide into parts Remove these units from their 　context. Fragmentation.	Destroys the interconnections in a transcript. Assumes that abstraction replaces subjective 　experience with objective knowledge. Assumes that words are merely exemplars of 　concepts.
3. Extract the content from each 　unit State what each part means. Categorize each unit.	Assumes that language is a conduit and that words 　are containers. Assumes that words are names. Assumes there is a single, fixed meaning. Conceals the researcher's influence.
4. Rewrite in objective language Remove all indexicality. Writing the theory.	Undertakes an impossible task. Treats indexicality as a problem to be eliminated. Depopulates the participants.

The emphasis on coding betrays the anxiety and confusion many qualitative researchers feel about the relationships between researcher and participant and between subjectivity and objectivity. When coding, are researchers active or passive? Are they merely observing patterns or meanings that "emerge" from the data, or are they actively inventing codes?

The confusion over coding and the anxiety about validity, objectivity, and method have their origin in the notion of subjectivity. Once we assume that all experience and knowledge are personal and subjective, we have to ask: Doesn't the researcher have a "subjectivity" too? Aren't they actively "making meaning"? But if so, how can they claim that their analysis is objective? How can a finite individual obtain knowledge that truly is valid?

The empirical-analytic sciences attempt to avoid the paradox by breaking out of the circle of interpretation and building on the seemingly interpretation-free foundations of measurement and logic (Taylor, 1971). Qualitative researchers cannot eliminate subjectivity in this way, so they need to find a way to resolve the paradox. The most common approach is to claim that although the research participant is an individual with personal, subjective experiences, researchers can overcome their subjectivity to obtain objective knowledge by following special rules or procedures. These procedures guarantee the objectivity of the knowledge obtained by a fallible individual researcher.

We see this strategy at work in the frequent proposal that qualitative investigators must "bracket" their presuppositions, as though in doing so they will become detached and neutral. (We will explore the origins of bracketing in Chapter 7.) We see it, too, in the efforts to bend over backward in search of "reflexivity," as though this reflection on subjectivity can compensate for its operation.

The mantra that "analysis is coding" is another form of the same strategy. Coding promises objectivity through its twin practices of abstraction and generalization. Coding works on the concrete and particular narratives obtained in an interview to produce abstract generalizations, presented in formal terms. This, surely, is objective knowledge! This is what scientists do! As Mruck and Breuer write, "the demand to exclude the researcher's subjectivity (and to include only what seems to be methodically controllable as a treatment) is one of the most important imperatives of the modern science" (Mruck & Breuer, 2003, paragraph 5). There is a powerful irony here: qualitative researchers struggle to gain access to the subjectivity of the people they study and then struggle to escape from their own subjectivity. It has been said of grounded researchers – though it is true of all the other efforts to be "objective about subjectivity" – that they "attempt to be subjective, interpretive, and scientific at the same time. It is not at all clear that this is either possible or desirable" (Denzin, 1988, p. 432).

But the strongest objection to coding as a way to analyze qualitative research interviews is not philosophical but the fact that it does not and cannot work. It is impossible in practice. As I noted in Chapter 2, Mishler pointed out the core problem with the coding of survey interviews:

> The central problem for coding may be stated as follows: because meaning is contextually grounded, inherently and irremediably, coding depends on the competence of coders as ordinary language users. Their task is to determine the "meaning" of an isolated response to an isolated question, that is, to code a response that has been stripped of its natural social context. Their competence consists in their being able to restore the missing context. ... In this act of interpretation, coders rely on the varied assumptions and presuppositions they employ as ordinary language users. (Mishler, 1986, pp. 3–4)

Exactly the same problem arises when coding qualitative interviews. With abstraction, each unit – a short phrase or a transcript line – is "stripped of its natural social context." Coders then have to restore this missing context, and to do this they draw on their skills as "ordinary language users." As Garfinkel pointed out, this coding of data can be done convincingly only by people who *already* understand the social processes that have produced the data. Coders unavoidably draw on their own common sense to interpret the basic units that coding produces. The whole enterprise of eliminating context and personal knowledge is undone, though these crucial "resources" themselves go unexamined. There is an inescapable "tacit dimension" to coding.

Michael Polanyi is well known for his exploration of this tacit dimension of all scientific practice. If researchers' tacit understanding of what has been said in an interview is always drawn on when they code, we have to take seriously Polanyi's proposal, cited at the beginning of this chapter, that "the ideal of eliminating all personal elements of knowledge would, in effect, aim at the destruction of all knowledge. The ideal of exact science would turn out to be fundamentally misleading and possibly a source of devastating fallacies" (Polanyi, 1967, p. 20).

This suggests that we should not try to eliminate the personal. Do we have to abandon objectivity as a consequence? It is becoming clear that the interview involves a complex set of relationships among the interviewer, interviewee, transcriber, and the person who does the analysis, as well as – once excerpts have been selected and published – the academic context, readers of the published report, and so on. To conduct, fix, analyze, and report an interview requires considerable interpretive work, interpretation that usually is entirely unexamined. But what exactly does it mean to say that "interpretation" is involved? How are we to understand these moments of interpretation? If a first-person account is not a "conduit" directly to the speaker's subjectivity, what is it? In the next chapter, we will explore how "hermeneutics," the theory of interpretation, can provide us with a new way of thinking about how a researcher understands what is said in an interview.

4

Hermeneutics and the Project for a Human Science

> The hallmark of the "linguistic revolution" of the twentieth century, from Saussure and Wittgenstein to contemporary literary theory, is the recognition that meaning is not simply something "expressed" or "reflected" in language: it is actually *produced* by it.
> Eagleton, 1983, p. 60

If we are to rethink the use of the interview as a tool in qualitative research, we must ask some fundamental questions. What does it mean to understand what someone says? What does it mean to understand a text? What is the "meaning" of a text? What is the relationship between a text and its author's subjective experience? The coding approach to analysis assumes that the answers to these questions can be found in the conduit metaphor for language. This metaphor implies that words or short phrases "represent" objects and events, that experience is "put into" words, that to understand is to "unpack" this content from the form, and that this "meaning" can be repackaged in language that avoids indexicality. We have seen how unsatisfactory these answers are.

But these answers are not the only ones possible. These questions have been asked for hundreds of years, and a variety of answers have been proposed. For 200 years, they have been topics of scholarly debate in philosophy, literary theory, and religion, in the field known as hermeneutics. Ironically, research with qualitative materials today more closely resembles the way people thought about these matters in the 18th century than it does contemporary views. The objective study of subjectivity has much in common with what is known as "Romanticist" hermeneutics. Yet the Romanticist view of interpretation, although very influential in the 1700s and for a considerable period afterward, searched for something unreachable. It required an endless circle of interpretation, or empathic leaps of identification with an author, or the appeal to a metaphysical notion of a universal life-force unfolding toward an objective end point.

The belief that there can be an objective *science* of subjectivity can be traced back to Wilhelm Dilthey, who in the 19th century made one of the earliest attempts to define a human science distinct from the natural sciences. His

Geisteswissenschaften (science of the mind or soul) had the two central elements that we can see today in the standard practice of qualitative research. It was to be a reconstruction of the subjectivity underlying or expressed in a text or other artifact, and it was to accomplish this reconstruction in an objective manner. To see both how this project arose and how it crashed, we need to learn a little about the history of hermeneutics and in particular the work not only of Dilthey but of Schleiermacher before him and Gadamer after him.

Hermeneutics is the theory of interpretation, named for Hermes, messenger of the Greek gods and interpreter of their messages for confused mortals. The Romans called him Mercury. Aristotle titled one of his books *De hermenuia* [*On Interpretation*], but explicit reflection on the character of interpretation blossomed in the 17th century when problems were encountered interpreting the Bible and other ancient texts. As the medieval epoch ended, these texts were now far from the circumstances of their original production, and understanding "the word of God" was becoming increasingly difficult. Hermeneutics became the term for the systematic study and interpretation of these religious texts. But in time it expanded to deal also with secular texts, cultural phenomena, and indeed all forms of human action (Ferraris, 1988/1996; Palmer, 1969). From its starting point in biblical exegesis, hermeneutics has become seen by some as a strong candidate to be the unifying basis for the social sciences and even for a new conceptualization of what it is to be human. There have also been attacks on hermeneutics from both the traditional empirical-analytic side and postmodernists (as we will see in Part III). But we can't understand their objections without first examining what they are objecting to (see Ormiston & Schrift, 1990a, 1990b). Hermeneutics has certainly been at the forefront of a 20th-century revolution in our understanding of language. Literary critic Terry Eagleton pointed to the key element in this revolution: "the recognition that meaning is not simply something 'expressed' or 'reflected' in language: it is actually *produced* by it" (Eagleton, 1983, p. 60).

RECONSTRUCTING THE AUTHOR'S INTENTION: FRIEDRICH SCHLEIERMACHER

We begin with Friedrich Schleiermacher (1768–1834), who, although he was a Protestant theologian, was one of the first people to appreciate the need for a secular "general" or "universalized" hermeneutics, a methodology that would deal with the interpretation of all types of discourse – religious, legal, and literary – both written and spoken. Schleiermacher envisioned this general hermeneutics as a systematic, lawful approach to textual interpretation, based on an analysis of the processes of human understanding. After all, he reasoned, the kinds of questions and issues that arise when we are reading a text also exist in everyday conversation or when we are listening to someone making a speech. He wrote that "the artfully correct exposition has no other goal than that which we have in hearing every common spoken discourse" (Schleiermacher, 1819/1990, p. 92). Hermeneutics, as Schleiermacher intended to define it, would systematically employ the skills of interpretation that operate within *all* occasions of understanding. This was the first attempt to define hermeneutics as the study of interpretation in general, outside the bounds of specific disciplines such as law, theology, or aesthetics. It was the first time hermeneutics was seen as providing a methodological grounding for the humanities and social inquiry. "Hermeneutics as the art of understanding does not yet exist in general, only various specialized hermeneutics exist.... [Hermeneutics is] the art of relating discourse [*Reden*] and understanding [*Verstehen*]" (Schleiermacher, 1819/1990, pp. 85–86).

What was needed to build such a hermeneutics was an exploration of the conditions for understanding in general. Just like many qualitative researchers today, Schleiermacher considered speech and writing to be an expression of the individual's thoughts and feelings – the external, objective manifestation of something private, inner, and subjective. He saw discourse as an activity in which the author's talent is a creative force that gives shape to the plastic, flexible medium of language. Just as a potter has the form of the pot in mind when she places her hands on a piece of clay, a speaker shapes language to give expression to his original thoughts. "The author sets a verbal object in motion as communication" (p. 96). This vision of the author as an inspired genius, the source of personal acts of creativity, was typical of 18th-century Romanticism: artists were believed to have godlike powers. Schleiermacher was unusual for his time in seeing this creative production at work everywhere, in everyday conversation and not only in classical texts.

Discourse, then, has twin origins: the author's personal creativity and the public medium of language. Thought without language would lack clear form; language without thought would have nothing to say. Understanding, in Schleiermacher's view, is the attempt to recapture or reconstruct the inner creative process, to grasp the author's thoughts. "[E]very act of understanding is the obverse of an act of discourse, in that one must come to grasp the thought which was at the base of the discourse" (Schleiermacher, 1819/1990, p. 87). Hermeneutics is the art of interpretation that will accomplish this reconstruction.

These two components of discourse – language and thought – provide its "objective" and "subjective" elements. The former is the relationship between discourse and the language as a whole, the latter its relation to the mind of the author. "[E]very discourse has a two-part reference, to the whole language and to the entire thought of its creator" (Schleiermacher, 1810/1990, p. 64). Consequently, although the goal of understanding is to reconstruct the author's inner creative processes, this cannot be achieved without understanding the language. Understanding involves "two elements – understanding the speech as

TABLE 4.1. *Schleiermacher's Grammatical and Psychological Interpretations*

Grammatical Interpretation	Psychological Interpretation
Builds on our general knowledge of the language	Studies what was personally intended and uniquely expressed
Assumes that language shapes all our thinking	Assumes that language is a means whereby the individual communicates his thoughts
The objective element of hermeneutics	The subjective element of hermeneutics
Objective reconstruction	Subjective reconstruction
One needs to know the vocabulary and history of the period; all the author's works	One needs to know the author's life

it derives from the language and as it derives from the mind of the speaker" (Schleiermacher, 1819/1990, p. 86). "One cannot understand something spoken without having the most general knowledge of the language, and at the same time, an understanding of what is personally intended and uniquely expressed" (Schleiermacher, 1810/1990, p. 64).

And so the art of hermeneutics involves two dimensions, corresponding to these two components. Schleiermacher (see Table 4.1) called them "grammatical interpretation" and "psychological interpretation." They interrelate and are of equal importance:

> Both stand completely equal, and one could only with injustice claim that the grammatical interpretation is the inferior and the psychological superior. (1) The psychological is the superior only if one views language as the means by which the individual communicates his thoughts; the grammatical is then merely a cleaning away of temporary difficulties. (2) The grammatical is the superior if one views language as stipulating the thinking of all individuals and the individual's discourse only as a locus at which the language manifests itself. (3) Only by means of such a reciprocity could one find both to be completely similar. (Schleiermacher, 1819/1990, p. 87)

Grammatical interpretation begins with the fact that in order to understand discourse one must know the language, with its grammar, components and rules, metaphors, and turns of phrase. When we express our personal thoughts and feelings in verbal form, the product is both enabled by and constrained by language. The "objective reconstruction" by means of grammatical interpretation "considers how the discourse behaves in the totality of the language, and considers a text's self-contained knowledge as a product of the language" (p. 93). Here "one seeks to understand a work as a characteristic type, viewing the work, in other words, in light of others like it" (p. 98). Schleiermacher believed that the whole of literature has significance for a single work. He recommended that "[o]ne must attempt to become the general reader for whom the work was intended, in order to understand allusions and to catch the precise drift of similes" (Schleiermacher, 1810/1990, p. 59). One needs to know not just a single text but all of the author's writing. One needs to know the vocabulary and syntax of the author's time and place.

The process of "psychological interpretation," on the other hand, studies the links between discourse and its author's subjectivity, experience, and life. It tries to explicate what has been uniquely expressed. Psychological interpretation involves what Schleiermacher called "subjective reconstruction," which seeks "to understand the discourse just as well and even better than its creator. Since we have no unmediated knowledge of that which is within him, we must first seek to become conscious of much which he could have remained unconscious of, unless he had become self-reflectingly his own reader" (Schleiermacher, 1819/1990, p. 93).

Schleiermacher did not think that we can achieve understanding either automatically or effortlessly. Previous theorists of hermeneutics had considered misunderstanding to be merely an occasional problem, so that explicit work at interpretation is needed only rarely. Schleiermacher considered misunderstanding to be the norm, and he believed that we must work constantly to overcome it. "[M]isunderstanding follows automatically and understanding must be desired and sought at every point" (cited in Gadamer, 1960/1986, p. 163). Indeed, he defined hermeneutics as "the act of avoiding misunderstanding." He was well aware, for example, that we encounter texts in circumstances very different from those in which they were produced, and we are not the people to whom they were originally addressed. To overcome these barriers to understanding, the interpreter needs to discover how the author related to the original audience: "One must keep in mind that what was written was often written in a different day and age from the one in which the interpreter lives; it is the primary task of interpretation not to understand an ancient text in view of modern thinking, but to rediscover the original relationship between the writer and his audience" (Schleiermacher, 1819/1990, pp. 89–90).

Consequently interpretation usually involves considerable work. Because one needs to know the author's inner and outer lives, "the vocabulary and the history of the period," and also read "all of an author's works" (Schleiermacher, 1819/1990, p. 94), the task of interpretation is certainly arduous. Schleiermacher acknowledged that there is "an apparent circle" to all this information, "so that every extraordinary thing can only be understood in the context of the general of which it is a part, and vice versa" (p. 94). Indeed, "Understanding appears to go in endless circles, for a preliminary understanding of even the individuals themselves comes from a general knowledge of the language" (p. 95) (see Box 4.1). Put this way, the task seems endless. But Schleiermacher dodged the question of how to bring interpretation to a successful conclusion:

BOX 4.1. *The Hermeneutic Circle*

Although Schleiermacher's conception of hermeneutics seemed to lead to an endless cycle of interpretation, it also involved a more benevolent circularity. Schleiermacher drew on a long-standing notion of the *hermeneutic circle*.

There are several related conceptions of the hermeneutic circle. One is that understanding a text as a whole requires considering its individual parts, but at the same time understanding each part requires a sense of the whole. This important notion of the mutual conditioning of part and whole is found in every formulation of hermeneutics. A circular relationship can also be said to exist between a text and its context, and more broadly between the text and the culture and historical tradition in which it is encountered. The relationship between reader and text is also a hermeneutic circle, a dialogue between present and past that never reaches a final conclusion.

Heidegger (1927/1962) proposed that a hermeneutic circle operates between understanding and interpretation. Understanding is the tacit, prereflective comprehension one has of a text or a situation. Interpretation is the "working out," that is to say, the articulation, of this understanding. In the process of articulating understanding, inconsistencies and confusion become evident, so interpretation can lead to a modified understanding. This hermeneutic circle is a dynamic relationship between the person and the world; humans are fundamentally embedded in the world and we can only understand ourselves in terms of our surroundings. But equally, the world only has sense in terms of our human concerns and cares.

Posed in this manner, the task is an infinite one, because there is an infinity of the past and the future that we wish to see in the moment of discourse. . . . However, the decision on how far one wishes to pursue an approach must be, in any case, determined practically, and actually is a question for a specialized hermeneutics and not for a general one. (Schleiermacher, 1819/1990, pp. 94–95)

The difficulty of actually achieving the goal of reconstructing an author's inner creativity becomes even more apparent when we dig deeper into what Schleiermacher believed was required. Both the objective and the subjective components of interpretation involve "two methods, the divinatory and the comparative" (Schleiermacher, 1819/1990, p. 98). The comparative moment is uncontroversial; it involves comparing the discourse with other examples from the same author and from others: "one seeks to understand a work as a characteristic type, viewing the work, in other words, in light of others like it" (p. 98). But because interpretation seeks to grasp the author's unique, individual spark of creativity, it also requires creativity on the part of the interpreter, and this is what Schleiermacher called the divinatory moment. Because all discourse is the expression of its author's thinking, and individual creative thought is a free construction, unconstrained by rules and norms, understanding and interpreting must be equally creative and unconstrained. "Using the divinatory [method], one seeks to understand the writer intimately to the point that one transforms oneself into the other" (p. 98). This is "the feminine force in the knowledge of human nature" (p. 98), and it requires an empathic act: "An important prerequisite for interpretation is that one must be willing to leave one's own consciousness [*Gesinnung*] and to enter the author's" (Schleiermacher, 1810/1990, p. 58). "By combining the objective and subjective elements one projects oneself into the author" (p. 81). For Schleiermacher, the goal of interpretation was not only to understand a text better than its author had but to know the author better than the author did. This projection into the mind of the author was possible, Schleiermacher believed, because there is a preexisting connection among all people: "Everyone carries a little bit of everyone else within himself, so that *divinatio* is stimulated by comparison with oneself" (cited in Gadamer, 1960/1986, p. 166).

In several ways, Schleiermacher's views resemble those of contemporary qualitative researchers. His view that discourse is an expression of the author's inner processes resembles the assumption that what is said in an interview is an expression of the interviewee's thoughts and feelings. His belief that interpretation reconstructs this subjectivity parallels the overall aim of much analysis of qualitative material.

But in other respects Schleiermacher's thinking was much more sophisticated. He recognized that understanding requires careful work of interpretation and, equally important, that interpretation requires detailed attention to language. He would never have studied a single sample of discourse. He would never have tried to understand an author by looking only at what is common to a selection of texts or studying only what is common to several authors. Schleiermacher's insistence that we must study the structure of the language in which a text is written is undoubtedly good advice. But the bottom line is that his goal for the interpretation of a text, to know the author "better than he knew himself," was impossible to achieve.

RECONSTRUCTING A SHARED FORM OF LIFE: WILHELM DILTHEY

Schleiermacher had a powerful influence on a better-known scholar, Wilhelm Dilthey (1833–1911). Like Schleiermacher (of whom he wrote a biography), Dilthey saw hermeneutics as a general methodology. But he broadened the scope of its application: in Dilthey's view, not only written texts and discourse but also cultural events and artifacts call for interpretation. Hermeneutics is the "methodology of the understanding of recorded expressions" (Dilthey,

TABLE 4.2. *Dilthey's Distinction between Natural Science and Human Science*

Natural Science (*Naturwissenschaften*)	Human Science (*Geisteswissenschaften*)
Provides explanation (*Eklaren*)	Provides understanding (*Verstehen*)
Fitting observed facts under general causal laws	Empathic attunement with another person's experiences
Study of nature	Study of mind
Requires study of the regularities of nature	Requires interpretation of the expressions of mind

1964/1990, p. 14). As a systematic theory of interpretation, it is "an essential component in the foundation of the human studies themselves" (p. 114). Dilthey has been an inspiration for many qualitative researchers.

Dilthey believed that the humanities and social sciences (*Geisteswissenschaften*) are equal in status to the natural sciences (*Naturwissenschaften*) but are distinct and autonomous (see Table 4.2). He was opposed to any program for a "unified science." Dilthey considered these human sciences to be founded on everyday understanding and what he called "lived experience" (*Erlebnis*). They have their own logic and method. Whereas the natural sciences seek explanation (*eklaren*), usually in terms of causal connections, the human sciences, as Dilthey conceived of them, offer something quite different: understanding (*verstehen*). The natural sciences can offer only reductionist and mechanistic analyses when they are applied to human phenomena, Dilthey argued. He pointed out that when we try to understand society, culture, history, art, and literature, we are *part of* these phenomena. We are already involved in the human world. This direct access should facilitate our inquiry, if we know how to make use of it.

Like Schleiermacher, Dilthey based his conception of hermeneutics on an analysis of understanding. "Action everywhere presupposes our understanding of other people," he wrote (Dilthey 1964/1990, p. 101). Understanding is a process we draw on not just with discourse and texts but with "the babblings of children ... stones and marble, musical notes, gestures, words and letters ... actions, economic decrees and constitutions" (p. 102). In all these, "the same human spirit addresses us and demands interpretation." And in every case, in Dilthey's view, we use the same process of understanding, "determined by common conditions and epistemological instruments ... unified in its essential features" (p. 102).

The key question was whether the basis of this understanding could be articulated systematically and whether its validity could be established to a point where we would feel comfortable and confident calling interpretation a science. How do we go from practical understanding to a scientific knowledge of discourse and artifacts? Dilthey agreed with Schleiermacher that: "All exegesis of written works is only the systematic working out of that general process of

Understanding which stretches throughout our lives and is exercised upon every type of speech or writing. The analysis of Understanding is therefore the groundwork for the codification of exegesis" (Dilthey, 1964/1990, p. 112).

Dilthey considered this process of understanding to be essentially historical. Our lived experience is always temporal, shaped by the context of the past and by the horizon of the future. Past and future form a structural unity, a temporal setting for our experience of the present. Dilthey proposed that our experience, always concrete and historical, is the basis for all understanding. As we shall see in Chapter 7, the philosopher Immanuel Kant had made the hugely influential proposal that the conditions for scientific knowledge are provided by innate and universal categories or concepts. Dilthey insisted that lived experience is made up not of static and atemporal cognitive categories, such as those Kant described, but of meaningful unities in which emotion and willing mingle with knowing: "That which in the stream of time forms a unity in the present because it has a unitary meaning is the smallest entity which we can designate as an experience" (Dilthey cited in Palmer, 1969, p. 107).

Lived experience is a direct, immediate, prereflective contact with life, an act of perceiving in which one is unified with the object one understands: "[T]he experience does not stand like an object over against its experiencer, but rather its very existence for me is undifferentiated from the *whatness* which is present for me in it" (Dilthey cited in Palmer, 1969, p. 109).

Lived experience, then, is a kind of understanding that is prior to any separation between subject and object, between person and world. It contains within it the temporality of living and of life itself. Dilthey considered the basis for a science of human phenomena to be reflection on lived experience and description of the structural relationships that are implicit in it. Such a human science would itself be historical, building on the fact that we understand the present in terms of the past and the future. It would grasp the ways in which discourse, texts, artifacts, and people are products of history. Our historicality means that we understand ourselves indirectly through "objectification," through cultural artifacts (both our own and those of others) that are scattered through time. Humans have, in Dilthey's view, no fixed and ahistorical essence. We make ourselves in history. "The totality of man's nature is only history" (Dilthey, 1914/1960, p. 166). "The type 'man' dissolves and changes in the process of history" (p. 6). We humans define ourselves in history, and we cannot escape from history. It is fair to say that "Dilthey gave the real impetus to the modern interest in historicality" (Palmer, 1969, p. 117), an interest that we will explore further in Part III: "[M]an is the 'hermeneutical animal,' who understands himself in terms of interpreting a heritage and shared world bequeathed him from the past, a heritage constantly present and active in all his actions and decisions. In historicality, modern hermeneutics finds its theoretical foundations" (p. 118).

Like Schleiermacher, Dilthey viewed the objects of the human sciences as "expressions of inner life." But he considered Schleiermacher's conception of understanding as a contact between subjectivities – the mind of the author and the mind of the interpreter – to be too limited. Understanding is not merely a contact between individual minds or a reconstruction of an author's mental state but a reconstruction of the historical process that has shaped a cultural product. To Dilthey there is an inner creativity to life itself. To interpret is to fulfill the highest needs of this creative process. This

> unified and creative power, unconscious of its own shaping force, is seen as receiving the first impulses towards the creation of the work and as forming them. . . . Such a power is individualized to the very fingertips, to the separate words themselves. Its highest expression is the outer and inner form of the literary work. And now this work carries an insatiable need to complete its own individuality through contemplation by other individualities. Understanding and interpretation are thus instinct and active in life itself, and they reach their fulfillment in the systematic exegesis of vital works interanimated in the spirit of their creator. (Dilthey, 1964/1990, pp. 110–111)

Interpretation is not simply penetrating an individual's mind but contact with a manifestation of the life process. Dilthey believed that what makes it possible for an interpreter to understand a cultural work is the fact that "both have been formed upon the substratum of a general human nature" (p. 112), a nature both formed by and informing the process of history, and indeed the unfolding of life itself.

What is expressed in a work of art, then, is not an idiosyncratic individual subjectivity but the life that is common to all of us. When individuals differ, Dilthey proposed, this is not a truly qualitative difference but a difference in their degree of development. When an interpreter "tentatively projects his own sense of life into another historical milieu," he is able "to strengthen and emphasize certain spiritual processes in himself and to minimize others, thus making possible within himself a re-experiencing of an alien life form" (Dilthey, 1964/1990, p. 112). Understanding is not merely a meeting of minds but a historical event. Understanding is

> a rediscovery of the I in the Thou: the mind rediscovers itself at ever higher levels of complex involvement: this identity of the mind in the I and the Thou, in every subject of a community, in every system of a culture and finally, in the totality of mind and universal history, makes successful cooperation between different processes in the human studies possible. (Dilthey cited in Ormiston & Schrift, 1990a, p. 15)

So hermeneutics, for Dilthey, is the theory of how life discloses and expresses itself in cultural works (Palmer, 1969, p. 114). Interpretation aims to go beyond subjectivity to the "thought-constituting work" of life itself. For Dilthey, understanding is not a purely cognitive matter but life grasping life in and through a full and rich contact that escapes rational theorizing.

How adequate is this romantic conception of "life" as something we all share, regardless of our cultural and historical differences? Although life is undoubtedly a universal and unitary process, Dilthey glossed over the fact that there are important differences in people's *ways* of life. The notion that there is a universal process of which all humans and all cultures are part was, for Dilthey, the only way to guarantee that an *objective* interpretation could ever be reached, and this was something he cared about deeply (Bauman, 1981). Although Dilthey insisted that the human sciences are distinct from the natural sciences, he believed that they have the same goal: objective knowledge about their respective domains. He insisted that the human sciences, too, must provide objective knowledge, knowledge that is and remains true for all times and places. He recognized that the objects of inquiry in the human sciences are historical phenomena, but he could not fully accept the implications of his own belief that the inquirer, the interpreter, is also always historically situated. It is ironic that someone who emphasized the historical character of our experience wanted to provide interpretations that would transcend history (but cf. Harrington, 1999). If we are thoroughly involved in history, it is difficult to see how we can achieve an objective viewpoint on human phenomena, yet this was the goal that Dilthey struggled all his life to achieve. He had accepted the dominant ideology of science as an activity that provides objective knowledge, but he could not identify a solid foundation for objective knowledge in the human sciences, whose legitimacy he sought to define.

APPLICATION AND MEANING AS AN EFFECT: HANS-GEORG GADAMER

The problems with Romanticist hermeneutics were clear to Hans-Georg Gadamer (1900–2002). His solution was not to discard hermeneutics but to rethink its aim (Gadamer, 1960/1976, 1960/1986) (see Table 4.3). Gadamer was critical of Schleiermacher and Dilthey for thinking that interpretation could reconstruct the creative act that originally produced a text, discourse, or cultural artifact. Such a view makes hermeneutics "a second creation, the reproduction of an original production," and this, he argued, is like thinking that if we take a work of art out of the museum and return it to its original setting we can reconstruct its original meaning. On the contrary, it just becomes a tourist attraction!

> Ultimately, this view of hermeneutics is as foolish as all restitution and restoration of past life. The reconstruction of the original circumstances, like all restoration, is a pointless undertaking in view of the historicity of our being. What is reconstructed, a life brought back from the lost past, is not the original. In its continuance in an estranged state it acquires only a secondary, cultural, existence. The recent tendency to take works of art out of a museum and put them back in the place for which they

TABLE 4.3. *Gadamer's Hermeneutics*

Understanding and interpretation are productive processes, a mediation between text and interpreter, a historical dialogue between past and present.

There is "a prejudice against prejudice" (which comes from the Enlightenment). "Prejudices are the biases of our openness to the world." We can correct them, but we cannot get rid of them.

Every interpreter has a "horizon" and participates in tradition.

The interpreter seeks a "fusion" between their own horizon and the horizon of the text.

The distance that separates us from the past is productive, not destructive – it enables us to filter the faddish from the classical.

were originally intended, or to restore architectural monuments to their original form, merely confirms this judgment. Even the painting taken from the museum and replaced in the church, or the building restored to its original condition, are not what they once were – they become simply tourist attractions. Similarly, a hermeneutics that regarded understanding as the reconstruction of the original would be no more than the recovery of a dead meaning. (Gadamer, 1960/1986, p. 149)

An interpreter can never get inside the mind – or the life – of the author. Understanding is not reproduction or reconstruction; it can never be a "repetition or duplication" of an experience "expressed" in the text. On the contrary, Gadamer (building on the work of Martin Heidegger, which we shall examine in Chapter 8) suggested that understanding is a *productive* process, a *mediation* between text and interpreter, a dialogue between past and present. Interpretation is an interaction in which neither interpreter nor text can step out of their historical context.

Like Schleiermacher and Dilthey, Gadamer viewed interpretation as grounded in understanding. His conception of understanding was based on Aristotle's notion of *phronesis* – a practical grasp of how to act well in specific concrete circumstances (Aristotle, 1980). Gadamer proposed that understanding is the skilled practical grasp of a specific situation, and interpretation, too, is always practical. When we read, understand, and interpret a text, we do it for its relevance to our present situation. Gadamer called this "application": to interpret a text is always to apply it to our contemporary circumstances and put what we learn from it to practical use. A text has relevance when it helps us better understand our situation and the challenges we face.

Interpretation, then, is a process of interrogating a text, asking it questions that arise from our own time. What we find in a text will depend on the questions we ask of it. For Gadamer, what a text means is not a matter of the author's thoughts or intentions but the *experience* that someone has when reading it, so that "understanding is an event" (Gadamer, 1960/1986, p. 441). Reading a text is a matter not of understanding the person who wrote it but understanding what they wrote about. As Gadamer put it, "understanding is not based on 'getting inside' another person, on the immediate fusing of one person in another. To understand what a person says is . . . to agree about the object [spoken of], not to get inside another person and relive his experiences" (p. 345).

It follows that there can be no single correct interpretation of a text, as Dilthey and Schleiermacher had believed. There is no final objective meaning that corresponds to the author's intention or to life's intrinsic purpose. The "openness" of every text makes it inevitable that each reader will find a different meaning. Gadamer wrote, "we understand in a different way, if we understand at all" (Gadamer, 1960/1986, p. 264).

He was surely correct. We all know, from our everyday experience, that a text will be read in different ways by different people. Who can claim to have found the one true interpretation? Why should we believe that some kind of technique of analysis – coding – will guarantee an objective reading? With deliberate irony, Gadamer titled his best-known book *Truth and Method* (1960) because in it he argued that the search for a foolproof method for interpretation can never lead to truth. But he did not believe that there is *no* truth. Instead, we need to think differently about truth: a true interpretation is one that points out something relevant in our present situation that we had not noticed. Meaning is not something placed inside the text to be extracted or uncovered by the interpreter. Because a text is, of course, something linguistic, its meaning depends on language, and language is first of all something cultural and historical, not private and subjective. Meaning is always an experience, an event, a moment of application: "to understand a text always means to apply it to ourselves and to know that, even if it must always be understood in different ways, it is still the same text presenting itself to us in these different ways" (Gadamer, 1960/1986, p. 359). The meaning of a text is changing and multiple, and an interpretation is true when it applies the text to successfully answer contemporary questions.

Gadamer argued that if being objective means being free from all preconceptions, then interpretation is never objective. Every interpreter has preconceptions or "prejudices" because he or she is the product of a particular time and place, and Gadamer insisted that these preconceptions play a *positive* role:

> Prejudices are not necessarily unjustified and erroneous, so that they inevitably distort the truth. The historicity of our existence entails that prejudices, in the literal sense of the word, constitute the initial directedness of our whole ability to experience. Prejudices are biases of our openness to the world. They are simply conditions whereby we experience something – whereby what we encounter says something to us. (Gadamer, 1960/1976, p. 9)

It is precisely our cultural-historical position that makes it possible for us to understand and interpret the human world and the cultural works that constitute human history. Gadamer was strongly opposed to the view that we should, or could, rid ourselves of all preconceptions. The receptivity to something unfamiliar "is not acquired with an objectivist 'neutrality': it is neither possible, necessary, nor desirable that we put ourselves within brackets" (p. 152). He challenged the belief (perhaps he would say the prejudice) that to be scientific we must be disinterested, neutral, and detached. This is impossible to achieve in practice, and it is a dangerous myth because those who believe they have achieved it are the most dogmatic. "What is necessary is a fundamental rehabilitation of the concept of prejudice and a recognition of the fact that there are legitimate prejudices, if we want to do justice to man's finite, historical, mode of being" (Gadamer, 1960/1986, p. 246).

Gadamer did not deny that there are prejudices that need to be changed. Some preconceptions may be inappropriate for a particular task of interpretation. There are forms of prejudice that we must have the courage to confront – such as racism or sexism. But our preconceptions are our involvement in history, our participation in traditional cultural practices. They "have a threefold temporal character: they are handed down to us through tradition; they are constitutive of what we are now (and are in the process of becoming); and they are anticipatory – always open to future testing and transformation" (Bernstein, 1983, pp. 140–141). Our preconceptions set limits beyond which we do not see, but they are not fixed and we are constantly testing them. One important way in which we test our preconceptions is through encounters with other people, such as occur in research.

Like Dilthey, then, Gadamer emphasized that an interpreter is always located in history: "In fact, history does not belong to us, but we belong to it" (Gadamer, 1960/1986, p. 245). Gadamer called this location in history a "horizon." Our horizon is all that we can see around us, and none of us can see everything. "The horizon is that range of vision that includes everything that can be seen from a particular vantage point.... To exist historically means that knowledge of oneself can never be complete" (p. 269). It is defined by our preconceptions: what "constitutes ... the horizon of a particular present [is] the prejudices that we bring with us" (p. 272).

Unlike Dilthey, Gadamer didn't shy away from the implications of historicity. He drew the obvious conclusions: we cannot step outside history, we cannot leap back in time, and we cannot walk in the shoes of a dead author. But this doesn't mean that we cannot understand a text written long ago. The distance that separates us from the past, which Dilthey and Schleiermacher had considered a problem to be overcome, is for Gadamer something productive. It enables us to filter the faddish from the classical. Distance in time "is not a yawning abyss, but is filled with the continuity of custom and tradition, in the light of which all that is handed down presents itself" (Gadamer, 1960/1986, pp. 264–266). For Gadamer, tradition is a living continuity that links the present with the past and enables us to span the distance between our time and that of a text. "[W]e stand always within tradition. . . . It is always part of us, a model or exemplar, a recognition of ourselves" (p. 250). This tradition makes possible a dialogue between present and past in what Gadamer called a "fusion of these horizons which we imagine to exist by themselves" (Gadamer, 1960/1986, p. 273). Fusion is not the interpreter going *back* in time, nor is it the text leaving the past and moving *forward* in time. There is always a "tension between the text and the present" (p. 273), and interpretation must "bring out" this tension rather than ignore it. But this dialogue, with both fusion and tension, is the way a tradition operates: "In a tradition this process of fusion is continually going on, for there old and new continually grow together to make something of living value" (p. 273).

So our location in history does not mean we cannot change and learn, for our horizon is flexible and mobile. "The historical movement of human life consists in the fact that it is never utterly bound to any one standpoint, and hence can never have a truly closed horizon. The horizon is, rather, something into which we move and that moves with us" (Gadamer, 1960/1986, p. 271). Understanding has an ongoing and open character, reflecting the historical character of human existence. Our preconceptions enable us to understand, and they change as our horizon changes and as the questions we ask change.

Serious questions have been raised about Gadamer's conception of tradition. We will see in Part III that Jurgen Habermas argued that Gadamer appealed to the continuity and universality of tradition a little too quickly and easily. We need to be able to question and criticize any tradition. After all, bearers of tradition can act irrationally, and traditions do impose inequities.

Gadamer's reply was that there is not an "antithesis between tradition and reason" (Gadamer, 1960/1986, p. 250). We cannot step completely outside our tradition to critique it, as Habermas seemed to propose. But nor do we need to. If our tradition completely destroyed our critical judgment, this would be a problem, but tradition doesn't often impose itself in this way. Often a tradition can foster openness and reflection. Sometimes a tradition is fragile and needs not to be criticized but actively preserved and cultivated. Tradition is the basis for our institutions and attitudes, the grounds for our ethical values. In Gadamer's view, we interpret texts in order to study tradition and to preserve and transmit what is best about it. He emphasized that we cannot achieve a methodological "distance" from our cultural and historical situation; there is no place to stand that would provide an "overview."

CONCLUSIONS

This brief exploration of the history of hermeneutics is enough to show that the attempt in much qualitative research today to objectively study the subjectivity of an interviewee faces insuperable difficulties. Like Dilthey and Schleiermacher, researchers believe they can escape from or overcome their *own* historical situation and achieve a timeless reconstruction of the meaning expressed in the words.

Gadamer insisted that interpreting a text has a completely different goal. He was able to see the problems of Romantic hermeneutics and identify the source of these problems, and even though his version of hermeneutics is not itself without problems, it contributes to our analysis of contemporary qualitative research. Gadamer argued, first, that understanding a text is always an active process. Readers – in our case researchers – are actively engaged in making sense of what they read. Second, we can never be free from preconceptions. Every reader encounters a text within a specific horizon, a place in history, and has expectations and interests that cannot be eliminated and according to Gadamer *shouldn't* be. The receptivity to something unfamiliar "is not acquired with an objectivist 'neutrality': it is neither possible, necessary, nor desirable that we put ourselves within brackets" (Gadamer, 1979, p. 152).

Third, understanding is always interested, not detached. Understanding a text always involves its "application" to our current situation. Our interpretation of a text is organized by its relevance to our current situation and our concerns, and guides our *phronesis*. Gadamer recognized that *both* social science (whether qualitative or quantitative) and natural science presuppose and depend on the prescientific activities of everyday life – on an unexamined moral paradigm.

Fourth, it follows that no text has a single correct interpretation. The meaning of a text is an *effect* of reading it, an effect that will be different for different people, in different situations, with different concerns, and with different preconceptions: "[T]he object of interpretation . . . is not a single meaning-in-itself but rather a source of possibilities of meaning which can be realized by future interpreters insofar as they investigate it from differing perspectives. In principle the object is continually open to new retrospections which depart from varied hermeneutical situations" (Mendelson, 1979, pp. 54–55).

Fifth, the encounter with a text can change the reader. It can enable us to become aware of at least some of our prejudgments, to test them, and to change them if necessary. "It is through the fusion of horizons that we risk and test our prejudices. In this sense, learning from other forms of life and horizons is at the very same time coming to an understanding of ourselves" (Bernstein, 1983, p. 144). We read, and we conduct research, to learn something.

These conclusions are deeply troubling for the coding approach to analysis that we considered in the last chapter. But if we take off the blinkers forced on us by a narrow conception of science and by the conduit metaphor of language, we will see that these conclusions open up important and exciting possibilities for qualitative research, our grasp of what an interview involves, and our theories of human understanding and even human being. Gadamer's hermeneutics implies that research should involve a dialogue between researcher and participant in which the researcher draws on their preconceptions rather than trying to be rid of them. The researcher is always a partner in dialogue. "The relation of observing subject and *object* is replaced here by that of participant subject and *partner*" (Habermas, 1968/1971, pp. 179ff, emphasis original).

We shall be returning in later chapters to the proposal, shared by Schleiermacher, Dilthey, and Gadamer, that interpretation is grounded in a more fundamental participation in a process of history and even of life itself, a process that is far larger than the individual. The suggestion that we humans are historical, with no fixed, natural essence, is an important one. But in the chapter that follows we will first explore the implications of this kind of hermeneutics for the analysis of interviews, for the practice of interviewing, and for our understanding of the character of qualitative inquiry.

Qualitative Analysis Reconsidered

[A]n interpreter can no longer claim to teach the reader the meaning of a text, for without a subjective contribution and a context there is no such thing. Far more instructive will be an analysis of what actually happens when one is reading a text, for that is when the text begins to unfold its potential; it is in the reader that the text comes to life. . . . In reading we are able to experience things that no longer exist and to understand things that are totally unfamiliar to us; and it is this astonishing process that now needs to be investigated.

Iser, 1980, p. 19

We have seen that standard practice in qualitative research – semistructured interviews followed by analysis through coding – embodies contradictory ontological commitments and notions of subject and object and of subjectivity and objectivity. On the one hand, each person is assumed to be a separate individual with personal and private experiences, beliefs, thoughts, and desires. On the other hand, scientific knowledge is assumed to be objective and general, impersonal and detached, with all personal elements eliminated. How, then, can scientific knowledge be derived from subjective experience?

The qualitative research interview also draws confusingly from two models of language: discourse as a joint construction and language as a conduit. In the conduit model, what someone says is an "expression" of their experience or subjectivity. In the joint practice model, what is said is a product of two people, interviewer and interviewee. The semistructured interview uses the collaborative resources of language asymmetrically to render the interviewer invisible and encourage "disclosure" by the interviewee. The interviewee is encouraged to contemplate and reminisce about a topic outside the here and now, to confess to a patient but skeptical listener. In this respect, the interviewee's subjectivity is an *effect* of the semistructured interview, not a preexisting, independent personal experience that is the content expressed in what is said.

Ironically, the ontological and epistemological assumptions embedded in the standard practices of qualitative research today turn out to be more or less the same as those of empirical-analytic inquiry. Individuals are assumed to be separate, with properties – opinions, experiences – that are internal and hidden. Rather than conduct surveys to measure these personal and private experiences, qualitative researchers try to gain access to them through the conduit of a language that they assume is composed of common concepts. Knowledge is taken to be scientific only when it is theoretical and general, decontextualized and abstract, and any suggestion (such as those by Wittgenstein and Polanyi) that knowledge can be practical and concrete, specific and local, is ignored.

What should we do? Some people have suggested that qualitative researchers should avoid interviews altogether because they provide artificial data, and instead study naturally occurring interactions. In Parts II and III of this book, I will explore naturalistic investigation in the form of ethnographic fieldwork, but the rejection of interviews fails to address the central problem, which is not that the interviewer influences what is said in an interview, so that we should try to collect data in such a way that this impact is avoided, but that we hold on to the notion that scientific knowledge can be achieved only when a researcher has no impact on the subject of research and no personal involvement (O'Rourke & Pitt, 2007).

We need to escape from the dualism of subjectivity–objectivity. This will be a central topic of Part II, where I will explore the origin of this dualism and efforts to escape it, efforts that will lead in Part III to a radically different project for qualitative investigation. By the end of Part III, we will be able to locate interviewing within a fresh program of research and make recommendations about how it should be used, along with other tools, to answer questions about constitution.

Although we are not at that point yet, we can explore in this chapter the basis for a different approach to the analysis of interviews. A powerful and informative approach to the analysis of interview material can be based on the hermeneutics of the last chapter. We saw there that understanding a text should not be considered a reconstruction of the author's intention. Meaning should be viewed as an *effect* of reading the text, one that will inevitably vary from one reader to another. In the present chapter I will develop two notions that are central to this kind of hermeneutics. First, researchers are always actively involved in the analysis of interview material, and their preconceptions are a key resource for understanding and articulating this material. Second, the language of an interview involves instructions, devices, techniques, and strategies that invite the interviewer to adopt a new way of seeing the world, including a way of seeing the speaker, the interviewee. A way of *saying* invites a way of *seeing*. Together these two notions provide the basis for a kind of analysis that returns to the fundamental fact that we want to *learn* from interviewees. Careful attention to the language of an interview transcript can enable us to understand how an interviewee is suggesting we see a new world or see our familiar world in a new way. If we design our research carefully, being clear about the question we are trying to answer, we will not try to replace an interviewee's words with a formal, general, and abstract substitute but will instead read them carefully for the answer they offer to our question.

This chapter, then, offers a kind of proof of concept of where we have come to so far. What would an analysis look like that doesn't try to eliminate the researcher's tacit understanding but tries instead to articulate and develop it, treating it as a resource but also as a topic? What would an analysis look like that explores the *interaction* between researcher and interviewee, as a joint production that extends beyond the event of the interview and continues in the analysis? In such an approach, "subjectivity" is not the object of inquiry, and the knowledge that is achieved is not an "objective" description produced through abstraction and generalization.

To articulate this approach to analysis, in this chapter I will draw on the work of several people. Literary theorist Wolfgang Iser has studied the *activity* of reading a text. Historian Hayden White has drawn attention to the onto-logical and epistemological choices that are made when a reader engages a text. Iser and White show us the kinds of textual elements that occupy a reader's attention and produce the effect of meaning, the linguistic devices that a reader must grapple with in order to understand a text.

We will also need to consider the fact that an interview elicits *oral* language, whereas both Iser and White were concerned with *written* texts. Often it is not appreciated how different the transcript of an interview is from the original conversation. But Paul Ricoeur explored how speech is trans-formed when it is "fixed" in writing. He has suggested that fixed speech escapes from the concrete setting of spoken language to have a general relevance and invite a reader to "project" a world.

THE POWER OF NARRATIVE

Lieblich, Tuval-Mashiach, and Zilber (1998) proposed that approaches to qualitative analysis can be distinguished in two dimensions. The first is whether analysis is conducted on units (a categorical approach) or on the text as a whole (a holistic approach):

> The first dimension refers to the unit of analysis, whether an utterance or section abstracted from a complete text, or the [text] as a whole. . . . In working from a categorical [approach] . . . the original story is dissected, and sections of single words belonging to a defined category are collected from the entire story or from several texts belonging to a number of narrators. . . . In contrast, in the holistic approach, the [text] is taken as a whole, and sections of the text are interpreted in the context of other parts of the narrative. (Lieblich, Tuval-Mashiach, & Zilber, 1998, p. 12)

TABLE 5.1. *Four Approaches to Narrative Analysis*

	Focus of Analysis	
Unit of Analysis	Content	Form
Part	Categorical analysis of content	Categorical analysis of form
	(e.g., content analysis; grounded theory)	(e.g., types of metaphor)
Whole	Holistic analysis of content	Holistic analysis of form
	(e.g., case studies; thematic analysis)	(e.g., types of plot)

Source: Lieblich, A., Tuval-Mashiach, R., & Zilber, T. (1998). *Narrative research: Reading, analysis, and interpretation* (Vol. 47). Thousand Oaks, CA: Sage.

The second dimension is whether the focus of analysis is on content or form. Lieblich et al. write that "[t]he second dimension, that is, the distinction between the content and the form ... refers to the traditional dichotomy made in literary reading of texts." We will see in this chapter that this dichotomy is not as straightforward as it might seem and that it really should not be called "traditional." We know now, for example, that the distinction between content and form is central to the conduit metaphor; in fact, a content–form distinction directly implies the metaphor of a container.

But let us accept this categorization for the moment. These two dimensions combine to define four possible types of analytic strategy: the categorical analysis of content, the holistic analysis of content, the categorical analysis of form, and the holistic analysis of form (Table 5.1). One interesting result of this categorization scheme is that the vast majority of analyses of qualitative material fall into just one of the four approaches: they are categorical analyses of content. They divide interview transcripts into decontextualized elements, search for commonalities, extract the meaning, and rewrite it in nonindexical language. Whether the analysis leads to "themes" or "codes" or "categories," the task is assumed to be one of capturing regularities and patterns in the *content* contained within each of these decontextualized units and then replacing the original *form* with decontextualized language. No attention is paid to the sequential organization of what was said. Such an analysis "excludes from consideration the course of the interview itself, that is, the internal history of the developing discourse, which is shaped by prior exchanges between interviewer and respondent" (Mishler, 1986, p. 53). When codes are attached to a fragment of the discourse – a sentence, a phrase, even a word – with no attention to where (or, really, when) in the account they appear, the analysis has no way of dealing with the fact that "terms take on specific and contextually grounded meanings within and through the discourse as it develops and is shaped by the speakers" (p. 64).

How should we pay attention to this shaping of discourse? One of the most common ways in which discourse is temporally organized is as a narrative. As

Honey puts it, "[I]nsofar as the interview has a beginning and an end it stands as a structured whole. The [semi-structured] research interview is certainly more than a sentence, it is an attempt to tell a story about a particular topic or issue" (Honey, 1987, p. 80). A narrative is "a recounting of one or more real events that do not logically presuppose or entail one another, with a continuant subject, constituting a whole, communicated to a narrator by a narratee" (Prince, 1987). It is "an account of something which develops and changes" (Paget, 1983, p. 75).

> By definition, narrative always recounts one or more events; but as etymology suggests (the term *narrative* is related to the Latin *gnarus*: "knowing," "expert," "acquainted with"), it also represents a particular mode of knowledge. It does not simply mirror what happens; it explores and devises what can happen. It does not merely recount changes of state; it constitutes and interprets them as signifying parts of signifying wholes (situations, practices, persons, societies). Narrative can thus shed light on individual fate or group destiny, the unity of the self or the nature of a collectivity. . . . In sum, narrative illuminates temporality and humans as temporal beings. (Prince, 1987, p. 60)

Interest in narrative has grown rapidly throughout the human sciences. The importance of narrative has been pointed out by anthropologists (Clifford Geertz), historians (Hayden White), psychiatrists (Donald Spence), psychologists (Jerry Bruner), and philosophers (Alisdair MacIntyre, Stephen Toulmin). Bruner has suggested that narrative is a natural cognitive form through which people try to order, organize, and communicate meaning. He has explored the notion that "logical thought is not the only or even the most ubiquitous mode of thought." An alternative mode of thinking involves "the constructing not of logical or inductive arguments but of stories or narratives" (Bruner, 1987). But Prince, in *A Dictionary of Narratology* (1987), describes the power of narrative as greater even than this. Narrative has the power to organize and explain, to decipher and illuminate; in short, the power to offer a way to see.

Narratives are generally considered to be organized on two levels, *plot* and *discourse*. Plot is the "what" of a narrative, the *narrated* situations and events in their causal sequence. Plot is what connects the incidents of the narrative, and "plot time" is the order in which these incidents are represented as having taken place. Discourse, in contrast, is the "how" of a narrative: the *narrating*. "Discourse time" is the order of presentation of events in the narrative, an order that will often differ from the order of their sequence in the plot.

There are many ways to analyze a narrative. A strong influence on the study of narratives has been structuralism, but there is a disadvantage to structuralist analyses because they draw a firm line between "synchronic" and "diachronic" study. Diachronic analysis is the study of change over time. Synchronic analysis takes a slice through time to study the organization of a phenomenon at a specific moment. Structuralist analyses of narrative undertake the latter.

This means that structuralists try to grasp the plot of a narrative "as a whole" rather than tracing the way it unfolds in time. In doing so they ignore, even destroy, the temporal flow of the interaction between reader and text. The result is an analysis that deemphasizes chronology, and temporality in general. The structure that supposedly "underlies" and "generates" a myth, for example, is the same no matter the temporal sequence in which the "mythemes" appear. In other words, structuralist approaches to narrative have what Ricoeur called "the general tendency of structural analysis to 'de-chronologise' the narrative, that is, to reduce its temporal aspects to underlying formal properties. . . . The tendency of structuralist literary criticism is to assign the chronological aspect of the narrative to the mere surface structure and to recognize nothing but 'achronic' features in the deep structure" (Ricoeur, 1981, pp. 281–282).

This is the danger of too narrow a focus on form. We need to find a way to pay attention to the "whole" interview, but not by trying to grasp this whole as though it were a fixed form, given all at once. The reader of a text is engaged by the plot as an *unfolding* sequence of acts and events and is moved to ask "*What next?*" An analysis that ignores or destroys the temporality of the process of reading will not show us how we understand a text or help us understand it better. The dynamic – goal-oriented and forward-moving – organization of narrative is responsible for its thematic interest and emotional effect, indeed for its very intelligibility. Rather than a holistic approach to analysis that looks only for a form that is "in" the text of the interview transcript, we need an approach to analysis that considers the interaction between text and reader and the *activity* of reading.

THE ACTIVITY OF READING: WOLFGANG ISER

Wolfgang Iser (b. 1926) was a student of Gadamer who has worked in an approach to literary criticism known as "reader-response theory." In his book *The Act of Reading* (1980), Iser took a close look at the activity of reading and explored the ways in which a text engages readers and guides the way they respond. Like Gadamer, Iser has criticized the notion that the meaning of a text is a "buried secret" that needs to be excavated. This idea, in Iser's opinion, is a remnant of the religious origins of hermeneutics, where interpretation was viewed as an effort to illuminate the lesson, the moral, the message, of a holy book. In the kind of interpretation that tries to reconstruct the author's original meaning, the "sole achievement is to extract the meaning and leave behind an empty shell" (1980, p. 5) – and we have seen that much the same can be said of coding. To view the text as a vehicle carrying the speaker's or writer's ideas, concepts, or experiences – the conduit metaphor – is to fail to recognize the active role of the reader.

Iser recommends that we pay attention to the *process* of reading, not just its *product*. The goal of interpreting a text should be to "reveal the conditions that bring about its various possible effects" (Iser, 1980, p. 18). We need to clarify "the *potential* of a text" and "no longer fall into the fatal trap of trying to impose one meaning on the reader, as if that were the right, or at least the best, interpretation" (p. 18). Iser insists that a text doesn't simply transmit a message that the reader decodes. If a reader can be said to "receive" a message, it is only because they *compose* it, in an interaction between reader and text. Like Gadamer, Iser proposed that meaning is an *effect* of this interaction: "As text and reader thus merge into a single situation, the division between subject and object no longer applies, and it thereby follows that meaning is no longer an object to be defined, but is an effect to be experienced" (Iser, 1980, pp. 9–10).

If the meaning of a text is an effect, an experience that requires the reader's active participation, any exegesis that detaches the reader from the text and looks for meaning in the text itself will fail. Any attempt to identify *the* meaning of a text is confronted by the fact that it can be read in many different ways. But this does not mean that every reading is merely a subjective interpretation. The effect of a text reflects its organization. Every text presents a reader with a collection of "intersubjective structures" that regulate the act of reading without determining the outcome. These structures are the same for every reader, though each reader will respond to them differently. Iser points out that "[h]owever individual may be the meaning realized in each case, the act of composing it will always have intersubjectively verifiable characteristics" (Iser, 1980, p. 22). The interaction between reader and text is neither idiosyncratic nor arbitrary.

Iser proposes that we think of a text as being made up of "a network of response-inviting structures." First, a text presents readers with a sequence of distinct perspectives. They experience a "wandering viewpoint" through the narrative, learning first of one character's experiences and then of another's, or learning first of one setting and then another, or shifting backward and forward in time. Each perspective provides a background to the one that follows, in a "constant reshuffling of perspectives" (Iser, 1980, p. 117). This continually changing vantage point is a consequence of the fact that the whole text can never be available at once: the text *implies* a totality without actually providing it. The reader must *construct* a sense of the whole – of a complete and consistent entity – from partial views. Iser argues that this mode of interaction is unique to narrative: it is unlike the relation between an observer and an object because the reader has "a moving viewpoint which travels along *inside* that which it has to apprehend" (p. 109). Readers are "entangled" in a text. They must assemble the totality, "and only then do the aspects take on their full significance" (p. 147). There will always be different ways to find consistency in a text, different ways to link the perspectives, and these will depend on the reader's preconceptions and their unique experiences as well as cultural norms of interpretation. "The structure of a work can be assembled in many different ways" (p. 17).

The reader's preconceptions are the basis for a "referential background against which the unfamiliar may be conceived and processed" (Iser, 1980, p. 38). We tend not to notice "all the experiences that we are constantly bringing into play as we read – experiences which are responsible for the many different ways in which people fulfill the reader's role set out by the text" (p. 37). As a reader works to bring different perspectives together, possible resolutions are projected, many of which turn out not to be fulfilled. The text gives rise to expectations on the part of the reader, but these must continually be modified. "The reader's communication with the text is a dynamic process of self-correction, as he formulates signifieds which he must then continually modify" (p. 67). In Iser's view, the basic inducement to every kind of communication is the *absence* of a shared situation and a *lack* of understanding. There is "an indeterminate, constitutive blank which underlies all processes of interaction" (p. 167). With a text, the lack of a common frame of reference is what initially motivates reading. A reader "is drawn into events and made to supply what is meant from what is *not* said" (p. 168, emphasis added). Until the very end, the reader's understanding is incomplete and this, together with the tension among changing perspectives, motivates a continual search for synthesis. What Iser calls the "basic hermeneutical structure of reading" is this dynamic activity of anticipation and fulfillment in which a reader's understanding is continually prestructured and restructured. This is why a reader's understanding "has the character of an event, which helps create the impression that we are involved in something real" (p. 67). The act of reading gives the reader the compelling sense of participating in a world that is real, not an illusion. Iser emphasizes that "as we read, we react to what we ourselves have produced, and it is this mode of reaction that, in fact, enables us to experience the text as an actual event" (pp. 128–129). When our expectations are disturbed, we experience surprise, frustration, even disappointment; when they are fulfilled, we are excited, satisfied, and gratified.

Every text contains what Iser calls "blanks" and "negations." The reader must fill these in; they provide "a kind of pivot on which the whole text–reader relationship revolves" (Iser, 1980, p. 169). Blanks and negations can perform a number of functions, but in general they "open up an increasing number of possibilities" that challenge and engage the reader (p. 184). Blanks are "empty spaces in textual structures" (p. 206). Iser calls them "the unseen joints of the text" (p. 183). Each blank is "a vacancy in the overall system of the text" that "induces and guides the reader's constitutive activity" (p. 202) and "maps out a path along which the wandering viewpoint is to travel" (p. 203). They are created by fragmentation of the narration, by "a suspension of connectability" between perspectives, by opaque allusions and references, and they disappear when the reader succeeds in linking perspectives. Blanks are not defects; on the contrary, they play a crucial role: "If one tries to ignore such breaks, or to condemn them as faults in accordance with classical norms, one is in fact attempting to rob them of their function" (p. 18). By impeding textual coherence, the blanks become challenges to a reader. By separating textual perspectives, they motivate a reader to create the "referential field" that regulates the connections among textual segments. It is when the pieces of the text don't seem to fit together smoothly that we start to see the complex, three-dimensional organization of the world of the text. Iser quotes art historian Rudolf Arnheim (1967): "It is one of the functions of the third dimension to come to the rescue when things get uncomfortable in the second" (Iser, 1980, p. 197).

Negations, in contrast, are gaps that throw into doubt a reader's knowledge of what is *outside* the text. They "invoke familiar or determinate elements only to cancel them out." They challenge a reader to question the familiar facts and norms of the situation in which they are reading. The resolution of these negations rests on conventions, assumptions of sincerity, and familiar types, but the reader has to figure out which particular conventions are relevant, guided by the text's narrative techniques and strategies. Once again, the "relative indeterminacy" of the text offers the reader the challenge and pleasure of making sense and allows for "a spectrum of actualization" (Iser, 1980, p. 24).

Different readers, with different interests and expectations and with different background knowledge, will fill these gaps in different ways, and each will produce a different meaning. Yet these various interpretations are set in motion by a single set of common structures. The effect is different for each reader, but the structures that challenge them to search for consistency and harmony are the same. But how a text is understood and what it means to a reader "is not *formulated* by the text; . . . [it] is the reader's projection rather than the hidden content" (Iser, 1980, p. 17, emphasis original). "[T]he object of interpretation . . . is not a single meaning-in-itself but rather a source of possibilities of meaning which can be realized by future interpreters insofar as they investigate it from differing perspectives" (Mendelson, 1979, pp. 54–55).

There is much more to Iser's analysis of what he calls the phenomenology of reading, but this is sufficient for us to see the possibility of a new approach to analysis of the material obtained in a qualitative research interview. The goal of such an analysis will not be to describe what a text means but to "reveal the conditions that bring about its various possible effects" (Iser, 1980, p. 18). And, for Iser, one of the key characteristics of literary language is that it "provides instruction for the building of a situation and so for the production of an imaginary object" (p. 64). Our analysis will focus on these instructions.

THE ONTOLOGICAL POWER OF NARRATIVE: HAYDEN WHITE

Historian Hayden White has also emphasized the capacity of a text to motivate a reader to see the world in a new way – that is to say, the relationship between a way of saying and a way of seeing. In his book *Metahistory* (1973a),

White studied the narrative structures, techniques, and devices used by 19th-century historians. White's work is relevant beyond the discipline of history, however. All narratives are historical in the sense that they are reporting events that have (or could have) occurred. When an interviewee tells us the story of their experiences, or even their life, the narrative they offer us is historical. And the social sciences are always studying events that occurred in the past, even when they are trying to predict the future. "Human science looks backward. It is inescapably historical" (Taylor, 1971, p. 57). "Although White believes that historical texts are an ideal place to study narrative realism because historians traditionally claim to represent reality itself rather than fictional simulacra, his inquiry into the ideology of narrative forms and his use of tropology extend to all narrative forms" (Kellner, 2005).

A narrative is not simply a matter of describing verbally the way that reality is already organized. White rejects the idea that people's lives already have a narrative organization, as some researchers have supposed (e.g., Connelly & Clandinin, 1990), and that the researcher's task is simply to collect these stories and retell them in a more objective narrative, "the truth of which would reside in the correspondence of the story told to the story lived by real people" (White, 1973a, p. x). White insists that "the notion that sequences of real events possess the formal attributes of the stories we tell about imaginary events could only have its origin in wishes, daydreams, reveries" (p. 24). The world just doesn't "present itself to perception in the form of well-made stories, with central subjects; proper beginnings, middles, and ends; and a coherence that permits us to see 'the end' in every beginning" (p. 24). White points out that a historian writing a narrative is not interested merely in describing what has occurred but is also trying to explain events. The aim of a history is "*explaining what* [past structures and processes] *were by representing* them" (p. 2, emphasis original).

Ricoeur made a similar point, that narratives not only describe events but also offer an explanation of them. The sequence of actions and events in a narrative plot gives us a sense of "directedness" (Ricoeur, 1981, p. 277). "We are pushed along by the development and . . . we respond to this thrust with expectations concerning the outcome and culmination of the process." This outcome cannot be *predicted*, but it must be *acceptable* once we get to the end. Ricoeur argued that narrative explanation builds on our ability to tell and follow a story and accept it as plausible or reject it as implausible. "Explanation must thus be woven into the narrative tissue" (p. 278). This means that narrative provides a form of explanation that is not formally equivalent to a prediction, as the logical positivists demanded all explanation should be and as a clinical trial is supposed to provide.

White also rejects the commonsense assumption that fiction and non-fiction are fundamentally different: that history refers to "real" objects, people, and events, whereas fiction refers only to "imaginary" ones. He proposes that both fictional and nonfictional narratives are "semiological apparatuses that produce meanings by the systematic substitution of signifieds (conceptual contents) for the extra-discursive entities that serve as their referents" (White, 1973a, p. x). Both fiction and nonfiction involve the real and the imaginary wrapped up together. Nonfiction narrative, just as much as fiction, invites readers to use their imagination:

> [N]arrative is revealed to be a particularly effective system of discursive meaning production by which individuals can be taught to live a distinctively "imaginary relation to their real conditions of existence," that is to say, an unreal but meaningful relation to the social formations in which they are indentured to live out their lives and realize their destinies as social subjects. (White, 1973a, p. x)

White argues that when we judge an account to be "objective" and "realistic" it is because we can easily find in it closure, coherence, and completeness. This is ironic, he suggests, because real life has none of these characteristics! A realistic narrative, White argues, is one that enables us to *bring* coherence and completion to the chaotic, unpredictable, and ultimately mysterious lives that each of us leads. White uses the term "narrative realism" to refer to the way a narrative creates the sense for a reader that what is narrated really occurred.

White proposes that a narrative is a *persuasive device* that plays on our reason, our emotions, and our aesthetics to invite us to see the world in a new way or even to see a new world. It uses various linguistics devices that, through indirect and figurative discourse, encourage the reader to imagine the *kind of world* that the narrative is about. White argues that "narrative is not merely a neutral discursive form that may or may not be used to represent real events in their aspect as developmental processes but rather entails *ontological and epistemic choices* with distinct ideological and even specifically political implications" (White, 1973a, p. ix, emphasis added). He explains:

> I treat the historical work as what it most manifestly is: a verbal structure in the form of a narrative prose discourse. Histories ... combine a certain amount of "data," theoretical concepts for "explaining" these data, and a narrative structure for their presentation as an icon of sets of events presumed to have occurred in times past. In addition, I maintain, they contain a deep structural content which is generally poetic, and specifically linguistic, in nature, and which serves as the precritically accepted paradigm for what a distinctively "historical" explanation should be. (White, 1969/1990, p. ix)

White focused on these "epistemological and ontological choices" inherent in a narrative. A "deep structural content" is introduced by an author's use of particular linguistic forms. It is deep not because it is hidden but because it precedes the levels of plot and discourse. At this level of deep content, the narrator "performs an essentially *poetic* act, in which he *pre*figures the ... field and constitutes it as a domain upon which to bring to bear the specific theories he will use to explain 'what was *really* happening' in it" (p. x, emphasis original). White uses the term "poetic" in the sense of the Greek word *poesis*, that is to say *making*. In subtle ways, of which a reader is usually not explicitly aware, a narrative achieves a *poetic* effect.

White proposes that this "poetic act" is "constitutive of the *concepts* [the narrator] will use to *identify the objects* that inhabit that domain and *to characterize the kinds of relationships* they can sustain with one another" (White, 1969/1990, p. 31, emphasis original). The narrator "both creates his object of analysis and predetermines the modality of the conceptual strategies he will use to explain it" (p. 31). At this deep level, "the field is made ready for interpretation as a domain of a particular kind" (p. 30).

Notice that in White's approach the distinction between form and content has disappeared. Linguistic forms *provide* content. In his book appropriately titled *The Content of the Form* (1969/1990), White argued that "narrative, far from being merely a form of discourse that can be filled with different contents, real or imaginary as the case may be, already possesses a content prior to any given actualization of it in speech or writing" (White, 1969/1990, p. xi).

Like Iser and Ricoeur, White suggests that reading a text can transform the reader, and he explains this in terms of constitution. Reading is "an experience which entails the reader constituting himself by constituting a reality hitherto unfamiliar to himself" (Iser, 1980, p. 151). "Ultimately, the whole purpose of the text is to exert a modifying influence upon [the reader's] disposition, and so, clearly, the text cannot and will not merely reproduce it" (p. 153). Iser reports that in the 17th century "reading [a novel] was regarded as a form of madness, because it meant becoming someone else" (p. 156); when we read, the divide between subject and object disappears and this "causes a different kind of division, between the reader himself" (p. 155). This split leads to tension between a desire to regain coherence and an awareness that this can't be the same old, habitual coherence. Once one reads a good narrative, there is no going back. The reader's desires have been stimulated, and they have become more aware of the inadequacy of their presumptions.

TACTICS AND STRATEGIES OF ONTOLOGICAL WORK

White suggests that one of the basic tasks of a narrative is to invite the reader or listener to imagine a ground or field of action that is then populated by particular kinds of entities, specific objects and people with certain types of relationships:

> That is to say, before a given domain can be interpreted, it must first be construed as a ground inhabited by discernable figures. The figures, in turn, must be conceived to be classifiable as distinctive orders, classes, genera, and species of phenomena. Moreover, they must be conceived to

bear certain kinds of relationships to one another, the transformations of which will constitute the "problems" to be solved by the "explanations" provided on the levels of emplotment and argument in the narrative. (White, 1973a, p. 30)

White provides a detailed account of how this construal of the ground and the figures on it is accomplished. First, the author uses linguistic "tropes" in a *tactical* manner. In addition to these tactics, an author uses three kinds of *strategies* to *explain* "the point of it all." The first strategy is the "argument" of the narrative: the arrangement of objects and events in causal relationships as a "world hypothesis." The second strategy is the "plot": the way these relationships are organized as a story. The third strategy is the author's "evaluation" of what transpires: this provides an explanation through what White calls "ideological implication."

There is no need here to go into details of White's account of the way the tropes of *metaphor, metonymy, synecdoche,* and *irony* are used to establish and populate a ground. A *trope* is a figurative or metaphorical use of a word or expression: a "turn of phrase." Most simply it is the substitution of one word for another: "He was a wolf!" This substitution is a way of suggesting a relationship between different kinds of phenomena. The words may be selected from the same domain or from a different one. As a result, the reader is invited to construe a relationship between things as they appear and things as they really are, and this functions to characterize objects and events and to establish relations among them. Drawing on the work of Giambattista Vico and Kenneth Burke, White proposed that tropes establish the "deep structural forms" of narrative realism. They provide a "prefiguration" that "permit[s] characterization of objects in different kinds of indirect, or figurative, discourse" (White, 1973a, pp. 33–34). A writer's (or speaker's) choice of tropes provides the "metahistorical basis" of their narrative account.

Then the tropes in a text work together to form an author's *strategies* of explanation. These strategies deal with "the point of it all," "what it all adds up to." White draws here on work by the American philosopher Stephen C. Pepper (1942), who identified four fundamental "world hypotheses" that support thinking and talking about reality. The world hypothesis in a narrative is the way the basic elements prefigured by the tropes interrelate and define a specific *kind* of reality. Pepper argued that although many world hypotheses are possible, only four have proved consistently useful. These are *formism, contextualism, mechanism,* and *organicism.* Each uses a basic analogy that Pepper called its "root metaphor."

THE ARTICULATION OF UNDERSTANDING AS EXPLANATION

The purpose of attending to these linguistic tactics and strategies is to enable us to articulate and correct our tacit understanding of an interview. I am not proposing that analysis should be a listing or cataloging of tropes, plot elements, and world hypotheses. Qualitative analysis, even when done badly, is a process of interpretation, and one of the most important lessons we learn from hermeneutics is that interpretation is always grounded in understanding. The purpose of analysis ought to be to improve and develop our understanding. When we read an interview transcript, we have at least some degree of understanding. We recognize the language, grasp the topic, and have a sense of what we are being told. At the same time, if someone were to ask us to explain what we have understood, we would probably find this difficult. Understanding needs to be developed, corrected and improved, articulated, and shared. Analyses that develop coding schemes and then translate the interviewee's words into objective language do little, or nothing at all, to increase our understanding.

This articulation is interpretation. As Heidegger said, interpretation is "the working out of possibilities projected in understanding" (Heidegger, 1927/1962, p. 189). Furthermore, this articulation of understanding has a social character because we interpret in order to enable others to share our understanding. Interpretation is not inherently "subjective," if by this one means personal, idiosyncratic, and one-sided. It is a dialogue with others in which one seeks to clarify one's own position by spelling out the details of one's understanding and laying out the evidence on which it is based.

No kind of analysis that includes a step where something "emerged" or "stood out" from the data can meet this standard. It has no response to the request, "Tell me how that particular meaning or theme emerged." Heidegger put it well: "If, when one is engaged in a particular concrete kind of interpretation ... one likes to appeal to what 'stands there,' then one finds that what 'stands there' in the first instance is nothing other than the obvious undiscussed assumption of the person who does the interpreting" (Heidegger, 1927/1962, p. 192).

There are both positive and negative sides to articulation (Packer, 1989). On the negative side, the interpreter tries to make apparent what is unnoticed because it is so familiar. One way to do this is to focus on breakdowns, misunderstandings, and apparent absurdities. Kuhn (1974/1977, p. xii) – who came to describe his study of science as a hermeneutic activity – has recommended that one "look first for the apparent absurdities in the text and ask yourself how a reasonable person could have written them." In contrast, the positive side to articulation attends to what stands out, what the text points to. This is precisely the *indexicality* that Garfinkel drew our attention to.

The kind of analysis I am recommending involves what Ricoeur called the dialectic of interpretation and explanation. Researchers develop their preliminary understanding in a systematic way – on the basis of attention to linguistic elements, tactics, and strategies – into an articulated interpretation. As our understanding is articulated, it will be elaborated, corrected, and communicated. As we carry out this analysis, we are applying what the text says to our current situation and recontextualizing it.

In Chapter 1, I noted that most researchers, both qualitative and quantitative, believe that qualitative research can provide only descriptions, not explanations. This was the view of Shavelson's committee, and Dilthey also drew a sharp distinction between explanation (*eklaren*) and understanding (*verstehen*) and assigned the former to the natural sciences and the latter to the human sciences. In such a view, because interpretation builds on understanding, it can provide a *description* but never an *explanation*.

But we have seen that an interpretation can in fact provide an explanation. Ricoeur insists that a rigid distinction between explanation and understanding "is quite impossible to identify in the dialogical situation we call conversation. We explain something to someone else in order that he can understand. And what he has understood he can in turn explain to a third party. Thus understanding and explanation tend to overlap and pass into each other" (Ricoeur, 1976, p. 72). The difference between understanding and explanation is that "in explanation we explicate or unfold the range of propositions and meanings, whereas in understanding we comprehend or grasp as a whole the chain of partial meanings in one act of synthesis" (p. 72). Explanation is an articulated reading that is both transparent and warranted – characteristics that Shavelson's committee valued (Feuer, Towne, & Shavelson, 2002). It is transparent when it spells out the details; it is warranted when it lays out the evidence. Where Dilthey saw only a dichotomy, Ricoeur sees a *dialectic* between understanding and explanation, with interpretation playing a mediating role. We can answer a "how" question with a description that provides an explanation.

TRANSCRIBING AS FIXING

Somebody talks. Through skills whose organization we never describe, the talking is transformed into a text: "Hi, how are you?" In the place of ongoing courses of accomplished bodily movements, the talking done, we now have a picture, the talk seen – silent, still sights. And we do something we often do with sights. They are divided up. Names are given the various parts. The parts are subdivided by names into more parts. We have nouns, pronouns, adverbs, prepositions, for instance. Parts of speech? Nonsense. They are parts of sights. (Sudnow, 1979, pp. 34–35)

But there is something else we need to think about. Iser and White studied written narratives, but an interview is an interaction that generates oral language. Certainly what is said is usually recorded and transcribed, but can these kinds of text really be considered in the same way?

Researchers generally think of transcribing the audio recording of an interview as a simple matter of typing what was said verbatim (*word for word*). But producing a written record of spoken language is more complex than this. Verbal features such as hesitations, false starts, pauses, overlaps, self-corrections, and nonlexical expressions such as "um" and "uh," are often left out. Transcripts typically omit paralinguistic (nonverbal) features of discourse such as pitch, volume, and intonation, and of course gestures, body posture and movement, and facial expressions. A transcript usually gives no indication of the physical layout of the setting in which the interview took place, yet this may provide important cues to understanding. Transcribers must make complex decisions about how to use the orthography of written language, including spelling and punctuation. Bogdan and Biklen (1992, p. 131) note that "[c]apturing the punctuation that gets at the meaning of what you heard is especially difficult, so considerable difference can arise when two typists type the same transcript." These "theoretical and cultural underpinnings of the transcription process" have been explored by Ochs (1979, p. 72), who empha-sizes that "transcription is a selective process reflecting theoretical goals and definitions" (p. 44). Even the layout of a transcript on the page "influences the interpretation process carried out by the reader (researcher)" (p. 47).

The idea that "the most accurate rendition of what occurred, of course, is on the tape" (Bogdan & Biklen, 1992) is mistaken. Audio and video recording are selective processes, partial representations in the dual sense of being both incomplete and made from a particular viewpoint (or listening point). Reading a transcript is certainly not the same as listening to a tape recording. But hearing a recording is not the same as conducting an interview.

Certainly the unfolding, fleeting discourse of an interview must be trans-formed into something more permanent that can be examined carefully and repeatedly. The solution to these complexities is not to avoid recording and transcribing but to understand the changes these introduce and recognize that "each is only a partial representation of speech. Furthermore, and most important, each representation is also a transformation" (Mishler, 1986, p. 48).

The most detailed examination of the transformations that are introduced when speech is "fixed" as writing has been carried out by Ricoeur. He explored the implications of the fact that "fixing" is necessary before human activity – talk or action – can be studied systematically and scientifically (Ricoeur, 1971/ 1979). Fixing introduces significant transformations, and it might seem that these must be distortions, but a central point of Ricoeur's argument is that they are necessary and have positive consequences.

Ricoeur began by comparing discourse in general – both written and spoken language – with language considered as an abstract system (as struc-turalism does; see Box 5.1). This comparison enabled him to identify four "traits" of discourse that distinguished it from the system of a language: its temporality and its relations to a speaker, an audience, and the world (see Table 5.2). Armed with these four traits, Ricoeur explored how they differ in spoken and written language.

BOX 5.1. *Structuralism*

Structuralism is a search for underlying structures that organize a text, discourse, or even the sounds of language. Ferdinand de Saussure (1857–1913) was a Swiss linguist whose ideas laid the foundation for many significant developments in modern linguistics, and his work provided the basis for structuralism. It is an approach to the analysis of social phenomena that aims to produce formal rules or structures that underlie the surface features (Pettit, 1975). Saussure's structuralist linguistics (*Course in General Linguistics*, 1916/1959; see Harris, 1987) was very successful in identifying the structures of the phonemes of a language.

Saussure's central notion was that language may be analyzed as a formal system of differential elements, apart from the messy details of real-time production and comprehension. Saussure viewed a language as a system of signs in which each sign is defined not by what it refers to but by its differences from the other signs. These differences can be expressed as a set of binary distinctions. The connection between a sign (signifier) and what it refers to – the signified – is conventional and arbitrary. The signifier for a tree, for example, is "tree" in English and "arbol" in Spanish. The word "'tree" is defined by its relation to other words, such as "bush," "plant," or "animal." Speech (*parole*) is an external manifestation of this sign system: language (*langue*). Saussure envisaged a "semiology" (from the Greek *semeîon*, "sign") that would be the study of these signs and their systems. Linguistics, the study of verbal signs, would be just one branch (see Culler, 1976).

Structuralism has been most successful in phonology, where it has been possible to define the basic sounds or "phonemes" of many languages in terms of a basic set of distinctive features. For instance, in English the sounds /p/ and /b/ differ in their "voicing." In addition, linguist Noam Chomsky (1957) offered a structuralist analysis of language as a formal grammar, a system of recursive rewriting rules.

Claude Levi-Strauss (1908–2009) provided a structuralist analysis of myths (Levi-Strauss, 1958/1963). Beneath their apparent heterogeneity, he maintained that constant universal structures are present, the basic elements of which are "mythemes," analogous to phonemes. For Levi-Strauss, every myth is the product of rules of combination that operate like a formal grammar. These rules (binary opposition, etc.) are universal and native to mind. In a similar manner, Vladimir Propp (1895–1970) analyzed folk tales and identified the basic elements of their narratives, the "narratemes," including character types such as the hero, the helper, and the villain, and functions such as departure, guidance, struggle, trickery, mediation, and complicity (Propp, 1928/1977).

TABLE 5.2. *The Transformations of Fixing*

	Comparison of Language and Discourse		Speech Fixed as Writing		Action Fixed	
	Language as System	Speech, Discourse	Speech	Writing	Action	Fixed Action
Temporality	Virtual, outside time.	Situated in time.	Fleeting event, appearing and then disappearing; hence the need for fixing.	Fixes not the event itself but the speech as *said*.	Fleeting.	Action *is* fixed in everyday life: it is registered; it leaves a trace.
Relation to Subject (i.e., speaker, author, agent)	Has no subject.	Refers to the speaker. Pronouns: I, we.	"Immediacy." Intention (of speaker) and meaning (of words) overlap.	Meaning (i.e., force) of words and conscious intention are disassociated. What text means is more important than what author means. Original context lost; need for interpretation.	Meaning (of act) and intention (of agent) correspond.	Detached; develops unforeseen consequences. "Our deeds escape us and have effects we did not intend." This makes responsibility problematic.
Relation to World	Refers only to other signs. Has no world.	Refers to the world it claims to describe.	Refers to its situation. Ostensive reference. We understand a project.	Still "about" something: a world of nonsituational reference. Understanding lights up the world of the reader; the text projects a world.	Has *relevance* to a local situation.	Has *importance*. Meanings can be actualized, fulfilled in new situations.
Relation to Audience (i.e., recipient, hearer)	Has no audience. Is addressed to no one.	Is addressed to a specific other.	Particular other in a narrow, dialogic situation.	Available to anyone who reads it.	Directed to a recipient who is present.	Meaning is "in suspense." Can be reinterpreted. Action is an "open work."

Temporality is the first and central point of contrast. It is the temporality of discourse that makes fixing necessary. "In living speech, the instance of discourse has the character of a fleeting event. The event appears and disappears. That is why there is a problem of fixation, of inscription. What we want to fix is what disappears." Speech is fleeting but a text lingers, enduring over time, and can be read again and again.

Second, the move from speech to writing transforms the relationship of discourse to the speaker. Spoken discourse has an immediate self-referentiality: we can say that the subjective intention and the meaning of the speaker's words overlap. To ask a speaker "What do you mean?" is essentially to ask of the speaker's words "What does that mean?" But, when one reads a text, the meaning of the text and the intention of its author become dissociated: asking about the author's intentions is not the same as asking what the written text means. Once the text is removed from the circumstances of its original production, questions about the author's intention become hard to answer, as Gadamer pointed out. And Ricoeur agreed with Gadamer that the meaning of the text has become separate and multiple: "What the text says now matters more than what the author meant to say, and every exegesis unfolds its procedures within the circumference of a meaning that has broken its moorings to the psychology of its author."

The third trait is the relationship discourse has to an audience. Spoken discourse is typically directed toward a specific other who is present with the speaker in a shared situation. But once discourse is fixed, "[t]he narrowness of the dialogical relation explodes. Instead of being addressed just to you, the second person, what is written is addressed to the audience that it creates itself." Anyone who speaks the language in which a text is written can read it. Although fixing might be said to impose an "alienation" on discourse, it also makes possible its "escape," a transcendence of the narrow here and now of face-to-face interaction. Once fixed, written discourse has a "relevance" that goes far beyond its original topic and setting.

The final trait, the relationship to a world, is also different in spoken and written discourse. Speech refers to the situation in which it is produced with a variety of ostensive devices. As we saw in Chapter 3, gestures, pointing, and grammatical forms such as verb tense and adverbs (here, now) can be used to make "indexical" reference to the "landmarks" that surround the interlocutors. A written text, in contrast, certainly makes reference to something, but its reference is not simply and directly to the immediate context. Like White and Iser, Ricoeur proposed that when we read a text its references can "open up" a world for us: "the world is the ensemble of references opened up by the texts." Like Gadamer, he suggested that "[t]o understand a text is at the same time to light up our own situation." Both spoken and written discourse "project a world," but we first become aware of this with written texts. This is because "[o]nly writing, in freeing itself, not only from its author, but from the narrowness of the dialogical situation, reveals this destination of discourse as projecting a world" (Ricoeur, 1971/1979, p. 79). It is in this sense that Ricoeur proposes that: "What we understand first in a discourse is not another person, but a project, that is, the outline of a new being-in-the-world" (p. 79).

The importance of Ricoeur's conclusions for our exploration of the analysis of interviews lies in his recognition that these transformations have a *positive* character. Not only can fixed discourse – such as a recorded and transcribed interview – still be understood; it actually has a *greater* relevance, a *wider* range of application, and an *increased* power to project a world. If the interview transcript seems at first an inappropriate object for analysis because it leaves out crucial paralinguistic and situational information, this is because we continue to assume that the goal of analysis is the reconstruction of the speaker's subjectivity. It is indeed the case that from a transcript we will not learn about the subjectivity of the speaker, but one of the main points that has emerged in previous chapters is that we won't learn about them from the "original" interview either. From the transcript, we can learn about the form of life they live in, who they have become in this form of life, and about ourselves. Because fixing transforms the relations between discourse and world, speaker, listener, and temporality, these effects of an interview transcript are *more* powerful than those of the original interview.

Like Gadamer and the others whose work we have considered in this chapter, Ricoeur proposed that "the meaning of a text" should be "conceived in a dynamic way as the direction of thought opened up by the text" (Ricoeur, 1971/1979, p. 92). What a text offers us is not a window into the author's subjectivity but "the disclosure of a possible way of looking at things" (p. 92). There is no single meaning that could be described objectively, using some technical kind of analysis. Understanding is always an application, an event in which reading the text enables us to see another way of being in the world and see our own way of being in a new way.

CONCLUSIONS

In this chapter, I have proposed a fresh approach to the analysis of interview transcripts on the basis of the fact that a reader is always actively interacting with a text, informed by prior experiences and preconceptions, and that the text contains intersubjective structures, rhetorical tactics, and narrative strategies. The outcome of this interaction is the text's meaning, an event that is different for each reader but is always an appropriation relevant to the reader's current situation.

This approach to analysis involves careful attention to the language of an interview account, treated not as a "form" that can be discarded once its "contents" have been extracted. We must avoid, or transcend, the distinction between form and content. Analysis should be holistic, but this does not mean that it looks for themes that emerge (a holistic analysis of content) or only at plot structure (a holistic analysis of form). It looks for what White called "the content of the form": the way an account is narrated and organized so as to prefigure a world for the reader – and a new way of seeing our familiar world. It invites the reader to conceptualize the interviewee's form of life and "ontological complicity" with this form of life. (I am borrowing this term from Maurice Merleau-Ponty and Pierre Bourdieu, whose work we will consider in later chapters.) To understand the interview transcript is to gain access to "a mode of being in the world that the text opens up in front of itself by means of its non-ostensive references" (Ricoeur, 1971/1979, p. 94), for what an interviewee says provides "instructions" that guide us in the construction of such a world. They offer "a structured indicator to guide the imagination of the reader" (Iser, 1980, p. 9). Reading an interview transcript opens up "new modes of being – or if you prefer Wittgenstein to Heidegger – new forms of life" (Ricoeur, 1971/1979, p. 94). "In this way we are as far as possible from the Romanticist ideal of coinciding with a foreign psyche. If we may be said to coincide with anything, it is not the inner life of another ego, but the disclosure of a possible way of looking at things, which is the genuine referential power of the text" (p. 92). An analysis that is sensitive to this "genuine referential power" enables us to see things in a new way, to articulate the "metaphysical grid" in which "an ontological spread is opened up" (Caputo, 1987, p. 114). "The literary text performs its function, not through a ruinous comparison with reality, but by communicating a reality which it has organized itself" (Iser, 1980, p. 181).

The goal of such an analysis is to learn from the interview a fresh way of seeing things, not to reconstruct the subjective experience of the interviewee or what some phenomenon "means to them." It is a process of interpretation that begins with the *tacit* understanding a researcher has reading the text of the interview transcript. Through attention to the elements of narrative that Iser and White draw our attention to, its blanks and negations, its tropes and plots and world hypotheses, this tacit understanding can be articulated, corrected, and communicated. The effect on the researcher, the meaning *for them*, is explicated, along with *how* that effect came about. In short, "Here is my reading of the text. And here's what this reading is responding to."

In this way, we can learn from our interviewees. When we understand what they tell us, and when we interpret and so articulate this understanding in a description of their words, we explain how they are inviting us to view the world we share with them. When a text unfolds a new way of seeing, a critique of how things are becomes possible, along with a critique of the illusions of the interpreter (Ricoeur, 1973/1990). Reading and understanding a text can give to the reader "a new capacity for knowing himself . . . a moment of dispossession of the egoistic and narcissistic ego."

But in writing of culture, critique, and transformation I have taken a step outside the bounds of the qualitative research interview. It has become evident in the course of these first chapters that our understanding of what someone tells us in an interview draws unavoidably from factors that are not personal or individual but *intersubjective*. Language itself is an intersubjective phenomenon, and the researcher's knowledge of language plays a crucial role in the conduct and analysis of an interview.

The coding of interviews, for instance, takes for granted a considerable amount of shared knowledge – membership in a speech community, in a tradition, even in an "interview society." All of these are drawn on as resources when interviews are analyzed (as well as when they are conducted and transcribed). But none of them is acknowledged explicitly. Each ought to be not only a resource for investigation but also a *topic* of investigation. The researcher's location in a specific cultural and historical setting plays a crucial role in the way an interview is conducted and analyzed. But what is a "shared" language? What is a culture or a tradition? These phenomena play an inescapable role, but so far we have taken them for granted.

The practice of interviewing in order to study "subjectivity" turns out to have led us inexorably to something quite different. A seemingly straightforward personal and individual source of data has turned out to involve shared public conventions, practices, and ways of knowing. At the same time, seemingly simple notions such as "subjectivity," "experience," and "meaning" have turned out to be surprisingly slippery. I have suggested that we reconceptualize subjectivity as one of the effects a text can have on its readers – as ontological complicity. Once we accept this, our ontological presuppositions are altered: the conception of researcher and research participant as subject and object (even though the latter is another subject) gives way to a conceptualization of two human beings encountering one another in a doubly intersubjective way, within a cultural context and in a manner mediated by linguistic devices. Jurgen Habermas, for one, has suggested that it is indeed better to begin with intersubjective social structures and social practices rather than individual experiences and interpretations. When we try to start from consciousness and experience and work outward from these to communication and language, we find ourselves going in circles. What is required is a decentering of subjectivity and a reversal of priorities. Communicative and intersubjective relations should be treated as the bases on which subjects, and subjectivities, are formed.

Subjectivity would not be eliminated from investigation with this change in analysis, but it would be displaced. Our interest in subjectivity would actually be deepened: we could begin to study the *constitution* of both the subjects and their subjectivity.

PART II

ETHNOGRAPHIC FIELDWORK – THE FOCUS ON CONSTITUTION

In the first part of this book, I have followed Thomas Kuhn's lead and examined the common practices of qualitative research today in order to see what epistemological and ontological commitments were embedded in them. In the conduct and analysis of semistructured interviews we found the assumption that subjectivity is an inner, mental realm that contrasts and yet coexists with the objectivity of an outer world. We found subjective experience contrasted with objective knowledge. The latter is abstract and general, so that subjective kinds of knowing must be extracted from their context, their indexicality must be repaired, and commonalities must be found across individuals in order to arrive at objective statements. We also found contradictory metaphors for language. Language is a conduit. It is a repository of concepts, names for objects and events. It is a joint production.

In the last two chapters, I started to explore a more adequate ontology and epistemology. I traced the history of hermeneutics from the aim to understand a spoken or written text by reconstructing the author's subjectivity to the view that the meaning of a text is an event, the effect it has on a reader or listener. This event is an *application* of the text by the reader to their current situation.

If we apply this to interviews, we see them as a joint production of discourse that, in its use of linguistic devices, especially tropes such as metaphor, invites a new way of seeing the world – or a new world to see. Analysis becomes a matter of examining this *poesis* to articulate how it works, to explicate how the discourse was skillfully designed to have various effects.

I suggested in the introduction that qualitative researchers are not aiming high enough and are not asking sufficiently interesting questions. When it comes to the qualitative research interview, the most common way to obtain empirical material, the aim has been to describe subjective experience, "to understand the world from the subjects' points of view, to unfold the meaning of people's experiences, to uncover their lived world prior to scientific explanation" (Kvale, 1996, p. 1). It has become clear that we *could* have been asking, *should* have been asking, how we learn about people when we talk to them, when we read what they have said. We *could* have been asking what we can learn about our world from the people we interview, each of whom may participate in a form of life perhaps quite different from our own. What can we learn about ourselves?

The result would not be one final interpretation, but nor would it be multiple interpretations among which no choice would be possible. Such an analysis offers not "the meaning" but an articulation of how what has been said engages us with its intersubjective structures and has an effect on us, moving us, changing us, meeting some expectations and challenging others. The aim would not be to replace the interviewee's words with our own but to explore a way of reading their words that offers an answer to a question about constitution.

The project to use interviews to add a missing subjective dimension is well intentioned but it does not make a genuine break with traditional research. Defining itself entirely in opposition to the mainstream, the search for a "subjective realm" has perpetuated the subject–object dualism of conventional psychological research. In Chapter 15, I will place this approach to the analysis of interviews in the larger context of a program of inquiry for accomplishing the tasks of a reenvisioned qualitative research.

But there is another limitation to research that employs only interviews: that people literally cannot tell us about something that should be of great interest to researchers – their forms of collective activity. This is "a dimension of human activity that cannot be contained in the consciousness of the isolated [i.e., individual] subject" (Prior, 1997, p. 64). There are limits to what can be learned from an interview, and a time comes when we want to consider not just what people *say* but what they *do*.

In the second half of this book, consequently, we turn to the study of practical activity: investigation of the social practices and interactions of participants in a form of life. We will explore various approaches, especially ethnographic fieldwork and ethnomethodology. We begin, however, with efforts to redefine social science and place "constitution" at the center of inquiry, and this will lead us into two very different conceptions of constitution, in both social science and philosophy.

6

Calls for Interpretive Social Science

We cannot measure such sciences against the requirements of a science of verification: we cannot judge them by their predictive capacity. We have to accept that they are founded on intuitions which we all do not share, and what is worse that these intuitions are closely bound up with our fundamental options. These sciences cannot be *wertfrei* [value-free]; they are moral sciences in a more radical sense that the eighteenth century understood. Finally, their successful prosecution requires a high degree of self-knowledge, a freedom from illusion, in the sense of error which is rooted and expressed in one's way of life; for our incapacity to understand is rooted in our own self-definitions, hence in what we are.

Taylor, 1971, p. 57

In the 1970s, a number of calls were made for a new kind of *interpretive* social science that would have ethnography at its center (e.g., Bernstein, 1976; Dallmayr & McCarthy, 1977; Rabinow & Sullivan, 1979). In this chapter I will examine three of these calls: Charles Taylor's proposal for an interpretive approach to political science, Anthony Giddens's hermeneutically informed sociology, and Clifford Geertz's interpretive anthropology. In each case, interpretation – hermeneutics – was regarded as an important, even central, element. In each case, immersion in the social practices of a community – that is to say, ethnographic fieldwork – was considered crucial (though at the same time, as we shall see later, ethnography itself was in crisis). In each case – in sociology, political science, and anthropology – it was claimed that the new approach would resolve core dualisms that had plagued the discipline. And, in each case, it was said that this would be because we would study the key relationship of *constitution* between humans and the world. This chapter begins our exploration of this notion of constitution, a notion that will lead us to a new way of practicing fieldwork. Once again, it will turn out to be important to get the ontology right.

INTERPRETATION AND THE HUMAN
SCIENCES: CHARLES TAYLOR

Charles Taylor (b. 1931) is a Canadian philosopher who has written on the philosophy of science and on political philosophy. He emphasized the importance of "intersubjective" phenomena in the social sciences in an important article, "Interpretation and the Sciences of Man" (1971). He began with the observation that the social sciences have tried to base their investigations on simple acts that seem to be objectively describable; for example, political scientists study voting behavior. But immediately there is the complication that

> these actions also have meaning for the agents which is not exhausted in the brute data descriptions, and which is often crucial to understanding why they are done. Thus, in voting for the motion I am also saving the honor of my party, or defending the value of free speech, or vindicating public morality, or saving civilization from breakdown. It is in such terms that the agents talk about the motivation of much of their political action, and it is difficult to conceive a science of politics which does not come to grips with it. (Taylor, 1971, p. 19)

To try to solve this problem, traditional social science takes "the meanings involved in action as facts about the agent, his beliefs, his affective reactions, his 'values,' as the term is frequently used" (p. 19). Although these kinds of facts are assumed to be in a "subjective realm," they seem to be accessible to scientific inquiry if we use familiar tools such as questionnaires, surveys, and opinion polls. The researcher can then explore correlations between observed behavior and statements of agreement and disagreement, or ratings on a Likert scale.

Taylor argued that this approach fails because it cannot provide access to what he called "inter-subjective meanings" (Taylor, 1971, p. 28). These are "the meanings and norms implicit in ... practices" (p. 27). As Taylor sees it, traditional social science misses an entire and unnoticed kind of phenomenon, that of meaningful *practices.*

Intersubjective meanings are "not just in the minds of the actors but are out there in the practices themselves, practices which cannot be conceived of as a set of individual actions, but which are essentially modes of social relation, of mutual action" (Taylor, 1971, p. 27). These intersubjective meanings "are the background to social action" and are "rooted in [peoples'] social relations." They cannot be equated with, or reduced to, subjective meanings:

> The actors may have all sorts of beliefs and attitudes which may be rightly thought of as their individual beliefs and attitudes, even if others share them; they may subscribe to certain policy goals or certain forms of theory about the polity, or feel resentment at certain things, and so on. They bring these with them into their negotiations, and strive to satisfy

them. But what they do not bring into the negotiations is the set of ideas and norms constitutive of negotiation [itself]. These must be the common property of the society before there can be any question of anyone entering into negotiations or not. Hence they are not subjective meanings, the property of one or some individuals, but rather inter-subjective meanings, which are constitutive of the social matrix in which individuals find themselves and act. (Taylor, 1971, p. 27)

Of course we could just continue taking for granted the particular intersubjective practices and institutions in which we live. But the costs of this are high. As Taylor put it, "the result of ignoring the differences in inter-subjective meanings can be disastrous." The most likely result is "that we interpret all other societies in the categories of our own" (p. 34). We impose the assumptions of our own form of life on people from another. Mainstream social science suffers from an "inability to recognize the historical specificity of this civilization's inter-subjective meanings" (p. 40), and "once we accept a certain set of institutions or practices as our starting point and not as objects of further questioning, [then] ... we give up trying to define further just what these practices and institutions are" (p. 29). Furthermore, these intersubjective meanings should not be confused with consensus. In fact, a shared background of intersubjective practices is often what makes *conflict* possible. In such cases, a common meaning is viewed from different perspectives. Taylor was writing in Canada, in the context of conflict between French and English Canadians.

Taylor pointed out that traditional political science is dualistic. It tries to study two distinct realms, the objective reality of human behavior and social institutions, and the subjective reality of the beliefs and values that make up the culture that individuals carry around inside their head. He argued that the dominant view of culture in sociology and political science is a "subjectivist view" in which culture is "in the mind." It "tries to conceive of [culture] as a set of (cognitive, or affective, or evaluative) 'orientations' of individuals to the (neutral) world of institutions and behavior" (Taylor, 1980, p. 36). "Behavior and institutions on the one hand, culture on the other, represent the two sides of the canonical split: the neutral, absolutely-described reality on one side, and the colorless reaction, pro or con, to this reality on the other" (p. 36). This split is an old one, as Taylor recognized. We find it in explanations of people's behavior in terms of absolute functions such as survival or social equilibrium, and also in the positivist notion that social science should be value-neutral. What is needed, Taylor suggested, is a kind of social science that "avoids the two equal and opposite mistakes: on the one hand, of ignoring self-descriptions altogether, and attempting to operate in some neutral 'scientific' language; on the other hand, of taking those descriptions with ultimate seriousness, so that they become incorrigible" (Taylor, 1983/1985, pp. 123–124).

Intersubjective Practices and Constitution

Taylor proposed that we can overcome the split between objective and subjective if we recognize that intersubjective practices play what he called a "constitutive" role. I have already insisted that this notion of *constitution* is centrally important, so we need to follow the details of Taylor's argument. He proposed that social reality cannot be identified apart from the language used to describe it, and "[t]he language is constitutive of the reality, is essential to its being the kind of reality it is" (Taylor, 1971, p. 24). There is a "mutual dependence" between the two or, better, "the distinction between social reality and the language of description of that social reality" is artificial (p. 24). To explain this point, Taylor drew upon the writing of philosopher John Searle, who wrote of the difference between "regulative" rules and "constitutive" rules. Searle explained:

> As a start, we might say that regulative rules regulate antecedently or independently existing forms of behavior; for example, many rules of etiquette regulate inter-personal relationships which exist independently of the rules. But constitutive rules do not merely regulate, they create or define new forms of behavior. The rules of football or chess, for example, do not merely regulate playing football or chess, but as it were they create the very possibility of playing such games. The activities of playing football or chess are constituted by acting in accordance with (at least a large subset of) the appropriate rules. Regulative rules regulate a pre-existing activity, an activity whose existence is logically independent of the rules. Constitutive rules constitute (and also regulate) an activity the existence of which is logically dependent on the rules. (Searle, 1969, pp. 33–34)

Searle went on to suggest that: "Regulative rules characteristically have the form or can be comfortably paraphrased in the form 'Do X' or 'If X do Y.' Within systems of constitutive rules, some will have this form, but some will have the form 'X counts as Y,' or 'X counts as Y in context C'" (pp. 34–35). However, because "constitutive rules come in systems, it may be the whole system which exemplifies this form ['X counts as Y in context C']" (p. 36). And "constitutive rules, such as those for games, provide the basis for the specifications of behavior which could not be given in the absence of the rule" (p. 35).

Taylor, like Searle, was proposing that intersubjective practices *constitute* the social reality of a form of life, in the same way that the rules of chess define the chess pieces and their legal moves. This includes constitution of the *members* of that form of life for "implicit in these practices is a certain vision of the agent and his relation to others and to society" (p. 35) – what *counts* as voting, for example, and *who* counts as a voter.

But how can we study intersubjective practices? Can we study this process of constitution? Taylor concluded that political science and the other social sciences need to become "hermeneutical sciences" (Taylor, 1971, p. 57). What was needed was "a hermeneutical science of man which has a place for the study of intersubjective meaning" (p. 49). Mainstream social science is unable to do this. The data for such a science would be "readings of meanings . . . [and] the meanings are for a subject in a field or fields" (p. 52). Such a science would inevitably "move in a hermeneutical circle." It could not break out of this circle by appealing to "brute data," free from interpretations, or to universal reasons. A degree of "insight" would be necessary, and this insight "is unformalizable."

In sum, Taylor proposed an approach to inquiry that, through immersion in intersubjective public practices and with a hermeneutic approach, would avoid dualisms like subjectivity–objectivity and agency–structure, and would explore the mutual constitution of a form of life and its members. Such inquiry would break radically with empirical-analytic inquiry because it would build systematically on our tacit intuitions and embodied ethical evaluations as members of a social group. Taylor was well aware that the kind of investigation he had in mind would be considered "radically shocking and unassimilable to the mainstream of modern science" (Taylor, 1971, p. 57).

INTERPRETIVE SOCIOLOGY: ANTHONY GIDDENS

In the early 1970s, Anthony Giddens, a British sociologist who was the first Professor of Sociology at Cambridge University and Director of the London School of Economics, also argued that the idea that social science could be based on positivist philosophies of natural science "must surely be reckoned a failure" (Giddens, 1976, p. 13). No general laws had been identified, and the public generally felt that sociology and the other empirical-analytic social sciences merely told them what they already knew. It was time we recognized, Giddens insisted, that "the logical empiricist view of science represents only one possible philosophy of science" (Giddens, 1979, p. 238). In particular, it was time to abandon the positivist view "that questions of philosophy can be clearly distinguished from the main body of social theory" (p. 239). Mainstream sociology had adopted "a mistaken self-interpretation of its origins vis-a-vis the natural sciences" (p. 240, emphasis removed). There was a need to rethink the basis of social scientific inquiry.

Giddens argued that empirical-analytic sociology had failed to recognize that society is *made* by humans and that this "production" of society "is a skilled performance" (Giddens, 1976, p. 15). When we act, members of a society draw on tacit knowledge of how it works. One of the problems with traditional social science, Giddens argued, is that it, too, uses this tacit knowledge to understand what people are doing but never stops to examine it. As a result, traditional sociology has tended to portray the everyday member of society as merely following prescribed social roles and passively "internalizing" values.

A New Sociology

What sociology needed, Giddens proposed, was a new way of conceptualizing the relationship between society and its members and a new way of studying this relationship. This would be what he called a "hermeneutically informed social theory" (Giddens, 1982, p. 5), an exploration of how society is produced and reproduced through human agency, through explication of the multiple forms of life that make up society.

To articulate this new sociology, Giddens drew from a variety of philosophers and social theorists, including some we have already considered, such as Wittgenstein, Garfinkel, and Gadamer, and others we have yet to meet, such as Schutz and Habermas. He argued that one of the factors in "the toppling of the orthodox consensus" (Giddens, 1982, p. 4) had been a growing interest in hermeneutics, facilitated in part by Ludwig Wittgenstein's influence on the English-speaking social sciences. Peter Winch, for example, in his book *The Idea of a Social Science and Its Relation to Philosophy* (Winch, 1958), proposed an approach to social science, based on Wittgenstein's work, that was very different from the positivist model. Like Taylor, Giddens insisted on the importance of rejecting functionalist and positivist forms of social scientific investigation. He, too, was trying to escape dualisms such as those of "agency" and "structure."

Giddens pointed out that "an important dualism" had been present both in sociology and in philosophy. "'Action' and 'structure' normally appear in both the sociological and philosophical literature as antimonies" (Giddens, 1979, p. 49). Sociology has always struggled with this fundamental division between "agency" and "structure." Should the sociologist's focus be on the individual as an active agent, or on the societal institutions and structures in which individuals find themselves? How does one reconcile the notion that members of a society are active agents with the notion that social institutions constrain and even determine individual action? The distinction between "microsociology" and "macrosociology" reflects this division, and the repeated efforts to bridge or combine the two reflect its persistence. Giddens pointed out that some sociologists and philosophers have focused on human action, whereas others have focused on the circumstances in which people act. Both have presumed a dualism of subject and object, the distinction between a "purposeful, capable social actor" and a "social structure or social system" (Giddens, 1982, p. 29). Whether it was the first or the second that was emphasized, the existence of both was taken for granted, together with the "gap" between them.

The new approach to sociology that Giddens recommended would require rethinking the character of society. He argued that the term "social structure" has been used in two distinct ways. One is "structure" in the sense of the fixed relations of the parts to a whole, analogous to the "anatomy of a body, or the girders of a building" (Giddens, 1979, p. 62). The other is the way interactions among members of a society are patterned. Giddens proposed that we clearly distinguish these two and think of "structure" as the "rules and resources instantiated in social systems, but having only a 'virtual existence'" (Giddens, 1982, p. 9). A "social structure" is like a language, which in Giddens's view has a set of rules that we never see but we assume exists and makes it possible for us to speak. The social structure is what enables members of a society to organize their interactions. "Structure thus is not to be conceptualized as a barrier to action, but as essentially involved in its production" (Giddens, 1979, p. 70, emphasis removed). Members of a society do not find their activity *confronted* by its organization; rather, this organization is a resource that they can draw on. Society is *not* like a body, or a completed building; social structure exists only virtually, as something we imagine.

Giddens proposed, then, that we view society and social institutions as systems rather than structures, as "structured social practices that have a broad spatial and temporal extension" (Giddens, 1982, p. 9). A "system" is made up of "relations of interdependence . . . best analyzed as *recurrent social practices*" (Giddens, 1979, p. 66, emphasis original). These relations are "patterns of social activity reproduced across time and space" (Giddens, 1987, p. 11). Systems *have* structures, or at least structural properties, but they are not structures.

Human Agents and Social Institutions Are Constituted

In this way, the new sociology would reconceptualize the relationship between subject and object. Giddens proposed a new way of describing the relationship between people and society, in terms of "structuration":

> The structured properties of society . . . "exist" only in their instantiation in the structuration of social systems, and in the memory-traces (reinforced or altered in the continuity of daily social life) that constitute the knowledgeability of social actors. But institutionalized practices "happen," and are "made to happen" through the application of resources in the continuity of daily life. Resources are structured properties of social systems, but "exist" only in the capability of actors, in their capacity to "act otherwise." This brings me to an essential element of the theory of structuration, the thesis that the organization of social practices is fundamentally *recursive*. Structure is both the medium and the outcome of the practices it recursively organizes. (Giddens, 1982, pp. 9–10, emphasis original)

Giddens, like Taylor, appealed to a notion of constitution, and proposed that a *duality*, rather than a dualism, is present in all interactions: the "duality of structure" (p. 36). Society must be viewed as *both* the basis for *and* the product of human action. This is the "recursive" character of social life: "the structural properties of social systems are both medium and outcome of the practices that *constitute* these systems" (pp. 36–37, emphasis added). To say that something is recursive is to say that a part requires the operation of the whole: a recursive computer program, for example, will call itself so that it executes multiple times. Giddens's example was that when I speak, language is the medium that enables me to do so. At the same time, by speaking I contribute to the reproduction of the language. In the same way, every member of society knows a lot about how that society works, and this knowledge is central to the reproduction of society in and through its members' actions. As Giddens put it, "structure is both enabling and constraining" (p. 37). He insisted that people are active and intelligent agents, even though we inevitably act without complete knowledge of the possible consequences of our actions, or even of our own motivation. Because people are always acting and interacting within existing social institutions, their actions play a role in the continued existence of these institutions – and in their transformation:

> One consequence of the preceding arguments is that the personal, transient encounters of daily life cannot be conceptually separated from the long-term development of institutions. The most casual exchange of words involves the speakers in the long-term history of the language via which their words are formed, and simultaneously in the continuing reproduction of that language. (Giddens, 1982, pp. 10–11)

The failure of sociology to recognize this relation of duality has led, in Giddens's view, to the division between "micro" and "macro" sociologies. Central to Giddens's theory of structuration was his insistence that

> neither subject (human agent) nor object ("society," or social institutions) should be regarded as having primacy. *Each is constituted in and through recurrent practices.* The notion of human "action" presupposes that of "institution," and vice versa. Explication of this relation thus comprises the core of an account of how it is that the structuration (production and reproduction across time and space) of social practices takes place. (Giddens, 1982, p. 8, emphasis original)

Giddens used a powerful image to illustrate this point: "[S]ocial systems are like buildings that are at every moment constantly being reconstructed by the very bricks that compose them" (Giddens, 1987, p. 12, emphasis removed).

This theory of structuration (Giddens, 1979) was intended to provide a view of the human agent that avoided both objectivism and subjectivism. It was a conception of the human subject that neither reduced it to the abstract structures of semiotic systems (as structuralism does) nor elevated it to the center, "a starting-point for analysis" (Giddens, 1982, p. 8) in the form of a constituting consciousness (as some kinds of phenomenology do). Consciousness, Giddens proposed, must be "de-centered" without being "dissolved." Consciousness cannot be taken as a given, but we must also not "dissolve subjectivity into the abstract structures of language" (p. 8). Giddens considered it crucial to explore the *origins* of consciousness and subjectivity. To avoid the errors of both objectivist social theory, which has treated human behavior as determined by anonymous social forces, and also subjectivist approaches, which have merely tried to describe universal structures of consciousness, we must "*recover* [the] subject, as a reasoning, acting being" (Giddens, 1982, p. 8). People act not because society *determines* their behavior in a causal manner but because they *understand* their social circumstances. Society (the object) and its members (the subjects) form a duality of mutual dependence. "Structure forms 'personality' and 'society' simultaneously" (Giddens, 1979, p. 70).

If we are to explore how this mutual constitution occurs, we need to pay attention to two "components of human conduct," which Giddens called our "capability" and our "knowledgeability." By "capability" Giddens meant the fact that a person "could have acted otherwise," a notion that he considered closely linked to the significance of power in social theory. "Knowledgeability" refers to both conscious knowledge and the tacit modes of knowledge in "practical consciousness" and the unconscious. The latter, in Giddens's view, were missing from traditional (functionalist and structuralist) sociological analyses.

A Hermeneutically Informed Methodology

This new sociology would need to employ a different kind of methodology (see Box 6.1). Giddens proposed that it must be "hermeneutically informed": "We *cannot* approach society, or 'social facts,' as we do objects and events in the natural world, because societies only exist in so far as they are created and recreated in our own actions as human beings.... [W]e cannot treat human activities as though they were determined by causes in the same way as natural events are" (Giddens, 1987, p. 11). A *completely* hermeneutic approach would be going too far, he argued, but "contemporary hermeneutics is very much in the forefront of developments in the theory of the text, and yet at the same time has relevance to current issues in the philosophy of science" (Giddens, 1982, p. 60). The task was to study meaningful action and explore its linkage to social institutions, to study the social world that is produced and reproduced by human subjects. The social scientist "studies a world, the social world, which is constituted as meaningful by those who produce and reproduce it in their activities – human subjects. To describe human behaviour in a valid way is in principle to be able to participate in the forms of life which constitute, and are constituted by, that behaviour. This is already a hermeneutic task" (Giddens, 1982, p. 7).

To accomplish this task, Giddens recommended fieldwork or ethnography – "immersion in a form of life" – as the proper basis for sociology. He recognized that this could not mean becoming a "full member" of society; what was

BOX 6.1. *New Rules of Sociological Method*

Émile Durkheim (1858–1917), often considered the founder of sociology, published a book titled *The Rules of Sociological Method* (Durkheim, 1895/1982). Giddens echoed this title with his 1976 book *New Rules of Sociological Method*. He included a set of themes for the new sociology, based on the recognition that "human beings transform nature socially, and by 'humanizing' it they transform themselves" (Giddens, 1976, p. 160). Like Taylor, Giddens emphasized the need to avoid dualism and recognize constitution. The themes were:

1. Sociology is not concerned with a "pre-given" universe of objects, but with one which is constituted or produced by the active doings of humans....
2. The production and reproduction of society thus has to be treated as a skilled performance on the part of its members....
3. The realm of human agency is bounded. Men produce society, but they do so as historically located actors, and not under conditions of their own choosing....
4. Structures must not be conceptualized as placing constraints upon human agency, but as enabling. This is what I call the duality of structure....
5. Processes of structuration involve an interplay of meanings, norms and power....
6. The sociological observer cannot make social life available as a "phenomenon" for observation independently of drawing upon his knowledge of it as a resource whereby he constitutes it as a "topic for investigation." ...
7. Immersion in a form of life is the necessary and only means whereby an observer is able to generate such characterizations....
8. Sociological concepts thus obey what I call a double hermeneutic....
9. In sum, the primary tasks of sociological analysis are the following: (1) The hermeneutic explication and mediation of divergent forms of life within descriptive metalanguages of social science; (2) Explication of the production and reproduction of society as the accomplished outcome of human agency (Giddens, 1976, pp. 160–162, emphasis removed).

required was "to know how to find one's way about" in the "ensemble of practices" (Giddens, 1976, p. 161). He emphasized that "the condition of generating descriptions of social activity is being able in principle to participate in it. It involves 'mutual knowledge,' shared by observer and participants whose action constitutes and reconstitutes the social world" (Giddens, 1982, p. 15).

More than this, Giddens argued that the logic of inquiry of the social sciences involves a *"double hermeneutic."* The first hermeneutic task is to study a social world that is produced and reproduced by human subjects. "But social science is itself a 'form of life', with its own technical concepts" (Giddens, 1982, p. 7). Postpositivist philosophy of science has, Giddens noted, made it clear that "science is as much about 'interpreting' the world as about 'explaining' it; and that these two forms of endeavour are not readily separable from one another" (p. 12). Even natural science turns out to involve a hermeneutic, though one that "has to do only with the theories and discourse of scientists, analysing an object world which does not answer back, and which does not construct and interpret the meaning of its activities" (p. 12). Social science is different: there is a "logical tie" between ordinary language and technical language. It has become clear that "social theory cannot be insulated from its 'object-world', which is a subject-world" (p. 13). The logical positivists saw this but considered it a problem, "a nuisance, something which gets in the way of testing the predictions whereby generalizations are validated" (p. 14), a threat to validity that should be minimized if objectivity is to be achieved. In contrast, Giddens argued that:

> The fact that the "findings" of the social sciences can be taken up by those to whose behavior they refer is not a phenomenon that can, or should, be marginalized, but is integral to their very nature. It is the hinge connecting two possible modes in which the social sciences connect to their involvement in society itself: as contributing to forms of exploitative domination, or as promoting emancipation. (Giddens, 1982, p. 14)

This "hinge" is common to both social scientific inquiry that contributes to domination and to inquiry that contributes to emancipation. Its existence means that:

> Social theory is inevitably critical theory.... [T]hose working in the social sciences cannot remain aloof from or indifferent to the implications of their theories and research for their fellow members of society. To regard social agents as "knowledgeable" and "capable" is not just a matter of the analysis of action; it is also an implicitly political stance.... Human beings ... are not merely inert objects of knowledge, but agents able to – and prone to – incorporate social theory and research within their own action. (Giddens, 1982, pp. 15–16)

A full discussion of "critical theory" must wait until Part III. The important point here is that Giddens insisted that knowledge in the social sciences differs from knowledge in the natural sciences: "Laws in the social sciences are intrinsically 'historical' in character: they hold only given specific conditions of 'boundedness' of knowledgeably reproduced systems of social interaction" (Giddens, 1982, p. 15). This is not a weakness of social science: on the contrary it shows the potential that investigation has to change the "conjunctions of intended and unintended consequences" that reproduce causal relations.

Giddens went so far as to propose that sociology "necessarily has a subversive character" (Giddens, 1987, p. 2). It "unavoidably demonstrates how fundamental are the social questions that have to be faced in today's world" (p. 2). Sociology should not remain "academic" if this means being disinterested and scholarly. Like the other social sciences, sociology is "inherently controversial" (p. 3). Giddens acknowledged that problems remained. What counts as a "valid" description? How are "alien cultures" to be studied, and how are they to be critiqued? What is participation "in principle"? Giddens himself has not carried out empirical research, so we cannot look to his example for how to conduct inquiry or solve these problems. But it is clear that, for Giddens, sociology, like the social sciences as a whole, needs to be focused on the duality of agent and society – the relationship that I have been calling constitution – and can and should have an influence on that duality.

INTERPRETIVE ANTHROPOLOGY: CLIFFORD GEERTZ

The third proposal for a new kind of social science was made in anthropology. Clifford Geertz, who helped found the School of Social Science at the Institute for Advanced Study at Princeton University, described the basic aim of his approach to cultural anthropology as simplifying the concept of culture. Ironically, this concept has always been a problem for anthropology. It has been called a "sponge" (Bennett, 1998), and a famous book listed 164 definitions of culture (Kroeber & Kluckhohn, 1952). The discipline can be said to have been founded on the problematic distinction between nature and culture (Bennett, 1998), and a series of related dichotomies continues to be a topic of debate and the focus of efforts to redefine anthropology. Is culture objective or subjective? Is it to be found in people's beliefs and values or in their material artifacts? In Geertz's view:

> The interminable, because unterminable, debate within anthropology as to whether culture is "subjective" or "objective," together with the mutual exchange of intellectualist insults ("idealist!" – "materialist!"; "mentalist!" – "behaviorist!"; "impressionist!" – "positivist!") which accompanies it, is wholly misconceived. Once human behavior is seen as . . . symbolic action . . . the question as to whether culture is patterned conduct or a frame of mind or even the two somehow mixed together, loses sense. (Geertz, 1973, p. 10)

It doesn't make sense, Geertz insisted, to debate whether culture is objective or subjective. The elements of culture can be as solid as rocks or as insubstantial as dreams, but they are always "things of this world." The important question is what they *say*: "what it is, ridicule or challenge, irony or anger, snobbery or pride, that, in their occurrence and through their agency, is getting said" (p. 10). Anthropologists only get confused, and confuse others, Geertz insisted, when they think that culture is in the mind or that it is merely brute behavioral events. Culture is not in people's heads but is "the public world of common life," a collection of public practices, events, and institutions. "Though ideational, it does not exist in someone's head; though unphysical, it is not an occult entity" (p. 10). In the same way, Geertz suggested, a Beethoven quartet is neither the skills nor beliefs that one needs to play it, nor the written score, nor a particular performance, nor the way people listen to it.

Geertz proposed a new "interpretive anthropology" that would explore "what . . . is getting said" based on the presumption that "culture is public." His undergraduate degree was in English literature, and he brought many of the concerns and insights of literary criticism to anthropology. His approach to the study of culture was to treat it as "an assemblage of texts" that could be read and interpreted. His new vision of anthropology, based on an "interpretive theory of culture" (Geertz, 1973), was that it is something to be *understood*. Geertz argued that studying culture is not analogous to dissecting an organism (as functionalists had supposed), or like deciphering a code (as semioticians had suggested), or like ordering a system (as structuralists had proposed). It is much more like penetrating a literary text, an approach that in the 1970s Geertz was able to say, with characteristic understatement, "has yet to be systematically explored." A central assumption of this new anthropology was that "societies, like lives, contain their own interpretations. One has only to learn how to gain access to them" (1972, p. 29). The access involved reading and interpreting cultural texts that are (and here Geertz cited Aristotle's *On Interpretation* [1962]) "saying something of something."

Deep Play

In 1972, Geertz published an account of something he argued was a central element of the culture of the Indonesian country of Bali: the cockfight. It provides a helpful illustration of interpretive anthropology and Geertz's practice of fieldwork. Although illegal, the cockfight was a popular public event in which two fighting roosters, with metal spurs strapped to their legs, were pitted against one another in a fight to the death. Geertz proposed that the Balinese cockfight is a "paradigmatic human event." Perhaps it was not "the master key" to the culture of Bali, but he considered it to be a highly important "commentary" on central aspects of Balinese life. He began his interpretation with the metaphorical identification that the Balinese – especially the men – have with the fighting birds. The Balinese, he wrote, are "cock crazy" – and the cock symbolizes "the narcissistic male ego." Heaven, for the Balinese, is like "a man whose cock just won," while hell is like "a man whose cock just lost." Geertz noted that the name of this bird had the same symbolic connection with a body part for the Balinese as it does for Westerners.

Geertz's reading began with the logistics of the cockfight. There were elaborate and detailed rules, written and handed down from one generation to the next and applied to each game by an "umpire." The fight had a complex and precise form, though at the same time it was "rage untrammeled" (Geertz, 1972, p. 10). This led Geertz to what he called the "central pivot" of the game: the system of gambling and betting that gave the cockfight its thrill and disclosed its connection to the "excitements of collective life." The center bet, between the owners of the two cocks, "makes the game" and defines its "depth." Around this was a whirling collection of bets made among the spectators. An evenly matched game, in which the two birds had roughly equal odds of winning, was unpredictable, and this was the kind of game that Geertz called "deep." Such a game was played not for money – because it was so risky that to bet on it was irrational – but for "esteem, honor, dignity, respect . . . status." The phrase that for Geertz captured the essence of the Balinese cockfight, when well organized, was "deep play." The 18th-century English philosopher Jeremy Bentham first used this term. The betting turned the cockfight into "deep play" when the odds seemed irrational but to bet was extraordinarily appealing because it offered the possibility of huge gains in status, honor, and social standing – and of course at the same time put these at risk.

Geertz summarized the consequences of the depth of the match:

The deeper the match . . .

1. the closer the identification of cock and man (or, more properly, the deeper the match the more the man will advance his best, most closely-identified-with cock).
2. the finer the cocks involved and the more exactly they will be matched.
3. the greater the emotion that will be involved and the more general absorption in the match.
4. the higher the individual bets center and outside, the shorter the outside bet odds will tend to be, and the more betting there will be overall.
5. the less an "economic" and the more a "status" view of gaming will be involved, and the "solider" the citizens who will be gaming.

(Geertz, 1972, p. 22)

To play such a game was "to lay one's self, allusively and metaphorically, through the medium of one's cock, on the line." As deep play, the cockfight was, Geertz argued, a "simulation of the social matrix" and in particular of the struggle for prestige that was a central part of everyday life in Balinese culture. It offered an avenue to the "inner nature" of Balinese culture. Geertz wrote that it "renders ordinary, everyday experience more comprehensible by presenting it in terms of acts and objects that have had their practical consequences removed and been reduced (or, if you prefer, raised) to the level of sheer appearance, where their meaning can be more powerfully articulated and more exactly perceived" (p. 23). The fight was "'really real' only to the cocks" (p. 23). The staging of this bloody, literally animalistic (but metaphorically human, masculine, and sexual) fight for status was disquieting but fascinating because it expressed and revealed something typically hidden in Balinese culture. Geertz emphasized that the Balinese find animalistic behavior absolutely revolting. The cockfight offered the anthropologist an access to Balinese culture because it offered the Balinese access also. It was a great work of art, Geertz suggested, equivalent to *"King Lear* and *Crime and Punishment"* (p. 23). It enabled the Balinese to understand the jealousy, envy, and brutality that were parts of their form of life. It "provides a metasocial commentary"; it is "a Balinese reading of Balinese experience, a story that they tell themselves about themselves" (p. 26).

Generating Subjectivity

Geertz proposed that an event such as the cockfight offers its participants a way of understanding themselves and their place in the world. The job of anthropology is not merely to describe such cultural events but to disclose the "commentary" they offer. The cockfight was a way of "saying something of something," and the anthropologist's task – an interpretive one – was to figure out what was being said: a task of "social semantics" (Geertz, 1973, p. 26). In Chapter 9, we will explore in more detail how Geertz thought interpretive investigation should be carried out. For now I want to focus on the important point that, for Geertz, in such an investigation the anthropologist is able to see the way that subjectivity is *created* by culture:

> Enacted and reenacted, so far without end, the cockfight enables the Balinese ... to see a dimension of his own subjectivity.... Yet, because that subjectivity does not properly exist until it is thus organized, art forms generate and regenerate the very subjectivity they pretend only to display. Quartets, still lifes, and cockfights are not merely reflections of a preexisting sensibility analogically represented; they are positive agents in the creation and maintenance of such a sensibility. (Geertz, 1972, p. 28)

Events like the cockfight both illustrate and create subjectivity among those who participate. The cockfight is "a story they [the Balinese] tell themselves about themselves" (p. 26). For a Balinese, participation "opens his subjectivity to himself" (p. 28). It offers "a kind of sentimental education" (p. 27).

Like Taylor and Giddens, and at around the same time, Geertz was proposing a new conception of the relationship between people and their social worlds. In his view, and the view of those who followed his lead, cultural anthropology needs to be interpretive if it is to offer insight into something fundamental and important, the way that "human nature is continuously transformed by the never-ending attempt of particular groups of human beings – Balinese, Moroccans, Northern European Calvinists, Satmar Hasidim – to understand themselves and create a social world that manifests their self-understandings" (Shweder, 2007, p. 202).

CONCLUSIONS

A popular view in the social sciences has been, and continues to be, that society is made up of objective structures that influence or even determine people's behavior, attitudes, and beliefs. In this view, human agency is pitted against social structure. These calls in the 1970s for a fresh approach to inquiry in three of the central disciplines in the social sciences – political science, sociology, and anthropology – had several things in common. First, each rejected the efforts to base a model of inquiry on the principles of logical positivism. Second, each aimed to overcome the dualism that it traced to the positivist model of science. Third, each recommended that the focus of study in the new approach should be social practices and that interpretation was a necessary component of this study. Fourth, each presumed that "immersion" in a form of life is crucial if one is to understand it. And fifth, each focused on the importance of the *constitutive* relationship between humans and our forms of life.

The proposal that social practices have a "constitutive" role is of central importance to our understanding of qualitative research, but what is constitution? A dictionary will define it as "the composition of something . . . the forming or establishing of something." Taylor, Giddens, and Geertz each offered a different account of what is constitutive and what is constituted. Taylor suggested that intersubjective meanings are "constitutive of the social matrix in which individuals find themselves" (Taylor, 1971, p. 36). The phrase "find themselves" is nicely ambiguous; it can mean merely that individuals discover that they happen to be in a particular social matrix. Or it could mean that in such a matrix an individual has specific resources to work out, or discover, the kind of person that they are. Taylor's subsequent book *Sources of the Self* (Taylor, 1989) explored the latter interpretation. Geertz also wrote of the "social matrix" of a culture and how a game like the Balinese cockfight is a "simulation" of the larger matrix that offers the Balinese an interpretation of themselves, enabling an individual "to see a dimension of his own subjectivity" (Geertz, 1972, p. 28). For Geertz, the texts of a culture "generate and regenerate the very subjectivity they pretend only to display" (p. 28). For Giddens, individuals and social institutions are each "constituted in and through recurrent practices" (Giddens, 1982, p. 8, emphasis removed);

human subjects and the objects they find around them in the social world have equivalent status; neither should be viewed as primary.

The *substantive claim* made in these calls for a new kind of inquiry then is that there is a constitutive relationship between cultural or social practices and human subjectivity. The proposal is that the traditional empirical-analytic approach to inquiry in the social sciences is unable to grasp this relationship, so a new form of inquiry is needed that will study "the ways subject and object, self and other, psyche and culture, person and context, figure and ground, practitioner and practice live together, require each other, and dynamically, dialectically, and jointly make each other up" (Shweder, 1991, p. 1). Each of these pairs has been a dualism that has impeded social science.

But how exactly can a new social science overcome these troublesome dualisms? What role does the notion of "constitution" play in avoiding dualism? And how does fieldwork enable us to explore constitution? Confusingly, Taylor, Giddens, and Geertz each offered a different solution to dualism. Taylor suggested that the intersubjective is a *third* realm that provides the basis for both subjectivity and objectivity. Giddens saw agent and structure in a relationship of duality distinct from dualism. Geertz proposed that to view culture as a public text would put an end to worries about whether it is in the mind or in material artifacts. If we are to resolve these differences, we will need to retrace the history of the concept of constitution. In the next two chapters I will explore two different ways that the concept of *constitution* has been used, their connections with various kinds of dualism, and links to the popular notion of "social construction." But the emphasis on fieldwork, the calls for immersion and participation, and the attention to intersubjectivity were all important recommendations, and in other chapters in Part II we will return to ethnography and other ways of studying practical activity.

7

Dualism and Constitution: The Social Construction of Reality

Edit and interpret the conclusions of modern science as tenderly as we like, it is still quite impossible for us to regard man as the child of God for whom the earth was created as a temporary habitation. . . . Man is but a foundling in the cosmos, abandoned by the forces that created him. Unparented, unassisted and undirected by omniscient or benevolent authority, he must fend for himself, and with the aid of his own limited intelligence find his way about in the indifferent universe. Such is the world pattern that determines the character and direction of modern thinking.

Becker, 1932/1961, pp. 14–15

It seems common sense that each of us has a mind in which we construct conceptions of the world around us. Our "subjectivity" is not merely ideas in our heads; it is the way the whole world appears to us. The "mental" is taken to be something inner, personal, and subjective. In addition, an information-processing model is accepted throughout the social sciences. In this model, the brain is seen as a computer, actively processing data received through the sensory organs, forming complex internal models or theories about the external world, and deciding how to act on the basis of these models. This, too, has come to seem obvious and natural.

The problems with such views are not so obvious, but humans have not always thought about themselves in this way. Of course, no single person could be responsible for such a model, but one person in particular was a highly influential spokesperson: the 18th-century German philosopher Immanuel Kant. Although he wrote over 200 years ago, many still consider Kant the most important philosopher of all time. The accusation of dualism is usually directed toward Descartes (e.g., Burwood, Gilbert, & Lennon, 1999), but the model of human being that the social sciences assume, and the dualisms in which they have become caught, are due much more to Kant. This chapter explores Kant and his influence in order to see how this model

BOX 7.1. *John Locke*

John Locke (1632–1704) was a political theorist who in 1669 participated in drafting one of the first state constitutions in the Americas. In *An Essay Concerning Human Understanding* (1690/1975), he argued for a system of checks and balances in government, and the right for people to resist unjust authority and as a last resort even engage in revolution. He believed that every adult has a duty to judge how their society is best preserved, and a legitimate government is one based on a *social contract* – the explicit consent of its citizens. Empiricism is the philosophy that our knowledge is built entirely on what we perceive with our senses, and Locke proposed that each of us is a *tabula rasa*, a blank slate on which experience writes. Reflection is the capacity to examine these ideas, and if we want to be free from authority and superstition we need to use our reason to judge and decide for ourselves.

Locke began his philosophical work with the conviction, typical of the early Enlightenment, that there is a relationship between the epistemological question of how knowledge is possible and the ethical question of how we should act and try to live. He aimed to show that a rational understanding of man's place in nature required men to live like Christians. He believed that men's appetites could be tamed through the demonstration that there is a valid standard for human conduct independent of what is found attractive. Such a standard could become apparent to each of us through the diligent and careful use of our reason. But Locke was unable to find this link between knowledge and ethics, and he came to view his own theory of practical reason as a failure. He concluded that rational understanding doesn't require any specific kind of life. Faith, he wrote toward the end of his life, is a form of trust, not counter to reason but instead beyond reason's scope.

arose and why, and how it both requires a process of "constitution" and trivializes this as something individual, primarily cognitive and intellectual.

KANT AND THE PROBLEM OF GROUNDING KNOWLEDGE AND ETHICS

To understand why Kant's model continues to have such a strong impact, we need to begin with a little historical context. In the 18th century, the period known as the Enlightenment, the writings of René Descartes (1596–1650) had a powerful impact on thinkers such as John Locke (see Box 7.1) and David Hume (see Box 7.2). This was a time of political foment, including the American Revolutionary War (1775–1783) and Declaration of Independence (1776), the French Revolution (1789–1799), and the Napoleonic Wars

BOX 7.2. *David Hume*

Like Locke, Scottish philosopher David Hume (1711–1776) emphasized in *A Treatise on Human Nature* (1739–1740/1978) and *An Enquiry Concerning Human Understanding* (1748/2000) that experience is the origin of our ideas (see Ayer, 1980; Stroud, 1977). Hume advocated experimental investigation of human thoughts, feelings, and behavior rather than armchair theorizing. From his investigations, he concluded that human intellect is dominated by feelings – "reason is, and only ought to be the slave of the passions" (Hume, 1739–1740/1978, p. 415). Beliefs are products of interactions between the material world and "principles of human nature." Our beliefs about cause and effect, for example, are based merely on our perception of events that tend to occur one after another. There is no necessary connection between these events, merely a "constant conjunction." Our belief that causality is a necessary relationship in the world is an illusion. (Today many would still say that knowledge of causality, central to science, is merely probabilistic.) A "science of man" would study our ideas as "atoms of the mind" and, by disclosing their regularities, expose such "fictions." For example, we tend to believe that there is an external world, but reason cannot justify such a belief (pp. 187ff). And because the whole point and purpose of moral judgments is to guide our actions, if this guidance cannot be provided by reason but only by the passions, it followed that "[m]orality, therefore, is more properly felt than judg'd of" (p. 470). Hume considered the passions to be natural and fixed; they could not be educated or corrected. "A passion is an original existence" (p. 415) for which truth and reason are irrelevant. Consequently, Hume's justification for morality was a purely utilitarian one. It is human nature to have certain desires and needs, and these are satisfied by obeying moral rules.

(1804–1815) (Gay, 1969, 1977). It was also the time of a revolutionary recon-ceptualization of mankind's place in the natural and social world, one that promised liberty from tyranny and mythology and encouraged people to think and decide for themselves what was true and false, just and unjust. The physics of Isaac Newton (1643–1727) was providing a fresh, exciting example of rational inquiry. The medieval view that humans live in a mean-ingful world created by God was being replaced by the Newtonian vision of the universe as material, mechanical, and lawlike. Now humans became seen as one of the animals – albeit one with a capacity for reason – living in a clockwork universe, and this meant that each individual had to find meaning and value for him- or herself.

How was an individual able to do this? One answer to this question was Descartes's, that reason is the source of knowledge. Descartes, a mathe-matician and a monk, had argued that through reason each individual can decide on the validity of their own knowledge, knowledge about self (*Cogito, ergo sum – I think, therefore I am*), about the world, and about God. During the Middle Ages, people had believed they were formed in the image of God, and even for Descartes, poised with one foot still in the medieval world, God was the final guarantor that he was not deceived.

The other answer, offered by Locke and Hume, was that sensory experience is the basis for knowledge. Sensation seemed an important basis for the new scientific study of humans, for "moral sciences . . . relate to man himself; the most complex and most difficult subject of study on which the human mind can be engaged" (Mill, 1843/1987, pp. 19–20). The new view was – and has largely continued to be – that a human is a creature with an objective and universal nature, the same in all places and times, who can be explained in causal, even mechanistic, terms. A "study of the human mind" that will "go back to the origin of our ideas . . . and thus establish the extent and the limits of our knowledge" (Locke, 1690/1975) has been the aim of human science ever since.

The new emphasis on science, the discovery of new lands with different cultures and different species of plants and animals, and the dawning recognition that humans, too, are animals, provided an exciting new vision of the world and the place of humans within it. But there was a problem: the new view of "man" clashed with the new view of "knowledge." We humans have ideas in our minds – but how do we know whether these ideas truly "conform" to objects in the world?

This was something troubling in the celebration of human liberation from superstition and servitude. The "philosophical anthropology" of the times, "which promoted man from servitude, ironically enough demoted him at the same time – from his position little lower than the angels to a position among the intelligent animals" (Gay, 1977, p. 159). Could a creature as finite as a human being really recognize the true and the good?

Everyone agreed that *matter* and *mind* are two fundamentally different kinds of substance. How then could they relate? Neither the rationalist Descartes nor the empiricists Locke and Hume had a satisfactory answer.

It was Kant who offered a solution. He recognized the difficulties and tried to solve them and in a sense elevate humans again. He added a key element: the proposal that humans draw on *both* perception and reason in order to construct ideas that are *representations* of the world. In doing so, Kant unintentionally gave rise to the dualistic image of human beings that today has become common sense and continues to dominate both traditional empirical-analytic research and much qualitative research.

TRANSCENDENTAL IDEALISM: IMMANUEL KANT

Immanuel Kant (1724–1804) was born into a strict Lutheran family in Königsberg in East Prussia (now Kaliningrad, Russia) and traveled no more than 100 miles from there during his lifetime. He is considered one of the most influential thinkers of modern Europe, and the last major philosopher of the Enlightenment. In 1770, at the age of 46, Kant, already an established scholar, was awakened from what he called his "dogmatic slumber" (Kant, 1783/1977, p. 5) by Hume's skeptical empiricism. Hume argued that causality is merely our perception of events that tend to occur one after another. This proposal horrified Kant, for in his view it undercut the whole basis of science, which he considered a search for *certainty*, for *necessary* truth. If Hume's skepticism were correct, the empirical sciences could only observe and describe regularities and offer no way to be certain of anything. At the same time, Kant was deeply troubled by Hume's proposal that human action can be explained in mechanistic terms. In a world that was increasingly breaking with religious custom and tradition, it seemed that Hume had destroyed the foundation for moral values, too. The new scientific account viewed humans as finite creatures, each born to a specific time and culture. Our values and beliefs are limited by the language we speak, the society we live in, our short life span, and our personal interests and desires. How we act depends on our desires and the norms and values of our particular culture. In such circumstances, how can we be certain that our knowledge is valid or our conduct ethical? Yet surely what is meant by knowledge and by morality is precisely this certainty.

Kant has been described as "both a typical and supreme representative of the Enlightenment; typical because of his belief in the power of courageous reasoning and in the effectiveness of the reform of institutions ... ; supreme because in what he thought he either solved the recurrent problems of the Enlightenment or reformulated them in a much more fruitful way" (MacIntyre, 1966, p. 190). Kant treated these problems of knowledge and ethics as philosophical rather than religious, and his response was an analysis in the form of a "critique." ("Critique" comes from the Greek *kritike*, or "art of discerning, or critical analysis.") It had three parts, dealing in turn with our knowledge of the natural world (*The Critique of Pure Reason*, 1781/1965), our actions in society (*The Critique of Practical Reason*, 1788/1956), and our appreciation of beauty (*The Critique of Judgement*, 1790/1952). Kant's conclusion was that there was still a basis for secure knowledge and ethical action, as well as for aesthetic judgment: it lay in the *relationship* between the human mind, with its capacity for rational thought, and the world we experience. Kant proposed that the individual human mind has a natural capacity for reason. By "reason" he had in mind Newton's mathematical physics: reason lays down principles that are consistent, categorical, and universal. This capacity enables the mind to "constitute" various forms of knowledge.

Kant argued that our experiences are not merely ideas but *representations*, related outwardly to objects and inwardly to a subject. He proposed that these representations are governed by "faculties" of the mind that define the *conditions* for human knowledge. Our knowledge has a sensory basis, but our sensations are organized by the concepts (or "forms" or "categories") that our mind brings to them: "Thoughts without content are empty, intuitions without concepts are blind" (Kant, 1781/1965, p. 93). These concepts are innate and universal, the same for all people and all times. They include space, time, causality, and object – each of which seems to be a property of the world but in fact, Kant argued, is a concept the human mind *brings* to experience. They are "modes of representation" (p. 382), logically necessary conditions for any experience to be possible. When we observe an object, our experience is actively constituted in the very act of perception. Our knowledge of the world is the result of this "*constitution* of our reason" (p. 10, emphasis added).

Kant called this his "Copernican Revolution." Just as Copernicus had showed that although the sun seems to rotate around the earth, in fact the earth rotates around the sun, Kant argued that although the world seems prior to our experience of it, in fact the mind is primary, and the world we experience circles around it. Whereas Descartes and Hume had appealed to a natural *harmony* between objects and our knowledge of them, dictated by God or by nature, Kant proposed that nature *submits* to human knowing, to innate and universal mental concepts. In effect, subjectivity expands to fill the whole world that we experience around us. "The understanding is itself the lawgiver of nature" (Kant, 1787/1965, p. 126); "The 'laws of nature' are nothing other than the *rules* according to which we *constitute* or *synthesize* our world out of our raw experience" (Solomon, 1983, p. 75, emphasis original).

Kant's position combined "transcendental idealism" with "empirical realism" (Kant, 1787/1965, p. 346; see Allison, 1983; Collins, 1999). In his view, a real world does exist, though we can never know anything definite about "things in themselves." All we can know are our experiences of these things, how they appear to us; we can know only the "phenomena," not the "noumena" (from the Greek, *noien*, to conceive or apprehend) that underlie them. Kant was certain that we must infer that "things in themselves" actually exist, but to go any further and try to say anything positive about them would be speculative metaphysics. Equally, the "transcendental" activity of mind is also something that Kant believed we could never be aware of; it will always operate behind the scenes.

Kant's position on ethics was similar to his position on knowledge. Kant was both a Newtonian and a Protestant, and just as he tried to reconcile empiricism and rationalism, he intended to reconcile science and religion. Newton's clockwork universe, in which everything has a mechanical cause, seemed to leave no place for God. But to call an action *moral* was to say it was not caused but freely chosen. Here, too, Kant's critique led him to the conclusion that the basic principles of moral conduct are not based on experience but are supplied by the human mind. Ethical principles are universal and necessary because they are rational. Here, too, reason has a constitutive capacity. Each individual can, and should, question the norms and values of their cultural tradition and reason about what is truly moral. And every rational being will reach the same conclusions because everyone has the same innate rational capacity.

Just as transcendental rationality allows humans to have indubitable knowledge of the laws of the physical universe, in the realm of human action all rational creatures can recognize a universal morality (Solomon, 1983, pp. 77ff). Each person can identify those moral rules that can be formally "universalized." A well-known example is what Kant called the "categorical imperative," which includes rules such as "treat others never simply as means, but as ends." Rational duties such as this are the basis for action that is truly disinterested. Kant believed that an act that is moral must be done for its own sake, not to satisfy the desires of either oneself or another person. The capacity for reason means each individual can figure things out for themselves rather than simply accepting what they have been told, and they can decide for themselves what it is right to do.

Just as we must conclude that there is a world in itself, in Kant's view humans must logically conclude that a divine, all-powerful God exists. God has a place in the in-itself, standing outside space and time, and so can legitimately play a role in religious faith and human morality. The universal human capacity for reason does not make us moral, but it gives us the *potential* to be moral. We can and must work on ourselves to *become* "universal subjects." People tend to base what they know and value on habit, convention, faith, emotion, and authority figures, but Kant insisted that this is not the whole story. He emphasized "rational autonomy" – "the central, exhilarating notion of Kant's ethics" (Taylor, 1975, p. 32). In Kant's view, each person has a capacity for radical self-determination.

In the third critique, Kant offered his analysis of judgment: how we judge what is agreeable, beautiful (such as a work of art), sublime, and noble, and how we judge goals and purposes. He described the "genius" that makes possible the creation of a work of beauty. Judgment, he proposed, provides the link between theoretical and practical knowledge. The gulf between what *is* and what *ought to be* is bridged by the faith that nature has a teleology with which our moral projects can coincide.

In a nutshell, Kant offered a new account of the *relationship* between subjectivity and objectivity that he believed preserved the possibility of objective knowledge, ethical action, and aesthetic evaluation. The Enlightenment's new notion that each individual forms ideas in their mind left important questions unanswered. How can individuals have *valid* internal, subjective knowledge of the external, objective physical universe, and how can we *know* whether our knowledge is valid? How can an individual *act* in a way that they can be sure is ethical? Empiricists like Locke and Hume had emphasized perception; rationalists like Descartes had emphasized reasoning. Kant's answer was that the two are linked. Our ideas are neither *caused* by external objects nor are they *copies* of these objects. They are *representations* of physical reality, products of the mind's capacity for rational synthesis, and as such are *constitutive* of that reality. An individual can have valid knowledge because there is a universal human capacity for reason that provides universal mental concepts. An individual can act morally because, once again, the universal human capacity for reason allows them to identify those duties that are logically necessary. Reason – and for Kant the best example of reason was mathematics – *constitutes* the world, both natural and moral, in which an individual lives.

Kant's Legacy

> Before Kant, epistemology struggled with a separation between thought and reality occasioned by essential differences between the two: thought, consisting of concepts that are general and continuous, and reality, consisting of flux. Since Kant, epistemology has had to deal with a separation between thought and reality created by human understanding: natural reality is always perceived in terms of human categories of thought and never in itself. . . . Human ways of perceiving and thinking add something to reality that was not there in the original. As a consequence, human knowledge seems not to stand in an empirically valid relationship with reality. (Rawls, 1996, p. 431)

Kant's proposal, that the two fundamental human faculties of perception and reason are intimately linked and that reason actively synthesizes data from the senses to form mental representations of the world, has become second nature to us. This is the view of individual cognition that, as Kuhn pointed out, has "guided Western philosophy for three centuries" (Kuhn, 1962, p. 125).

For the Enlightenment philosophers, including Kant, humans were fundamentally paradoxical: natural creatures driven by desire but with the capacity to be thoroughly rational and ethical. Humans had the "strange stature of a being whose nature (that which determines it, contains it, and has traversed it from the beginning of time) is to know nature, and itself, in consequence, as a natural being" (Foucault, 1966/1973, p. 310). Kant's attempt to resolve this paradox "was both brilliant and perplexing" (Sullivan, 1989, p. 8). It was brilliant in the way it combined the rational and the natural. It was perplexing because Kant's solution to epistemological and ethical skepticism only works if every individual mind has an innate capacity for the *same* reason. Kant believed that there is only one set of categories that every mind uses to

BOX 7.3. *Piaget's Genetic Epistemology*

Jean Piaget, the famous Swiss scholar of children's development, took up the Kantian project. Piaget (1896–1980) called himself a "genetic epistemologist" rather than a psychologist. Piaget's great book on infancy, *The Construction of Reality in the Child* (1937/1955), explored precisely the four concepts that Kant had focused on: space, time, causality, and object. Piaget did not believe that these concepts are innate; he considered that Kant had gone too far in claiming this (Piaget, 1970/1988). To Piaget, the great cognitive project of human infancy is the "construction" of an understanding of these four concepts, albeit embodied and "sensorimotor" rather than cognitive. The well-known Piagetian notion of "object permanence" is one of the end products of this project of "organization of reality" (Piaget, 1937/1955, p. xiii); there are comparable constructions in each of the three other areas. By the end of infancy, the stage has been set for the "semiotic function," the ability to use and understand representations, both mental and material (Piaget, 1945/1962). At each subsequent stage of development – the preoperational stage and the stages of concrete and formal operations – knowledge is actively "constructed" by the child in the form of increasingly complex mental representations or "schemes," through processes of assimilation and accommodation. Piaget proposed that mind does not appear until the end of infancy, but he argued that its formation is logically necessary. Mental action is built on the basis of practical sensorimotor schemas, which the child progressively replaces with increasingly formal and abstract kinds of representations.

Piaget is generally considered a central figure in "constructivism" (Phillips, 1995). It is certainly very clear from Piaget's writings that he considered the child to be constructing *knowledge* of the world.

represent the world, and each of us is born with it. This notion of a *transcendental* reason was unacceptable to many people even at the time, but even the critics accepted Kant's basic definition of a human being: they accepted that individuals know the world by forming mental representations. The search for a convincing basis for valid knowledge and ethical action did not end with Kant, but his model of man (Kant's "anthropology") defined the terms of the problem from that day forward.

The Search for Constitution

If it is not satisfactory to claim that the validity and universality of mental representations are guaranteed by an innate capacity for universal reason, as Kant proposed, two alternatives seem to remain. One is that the guarantee can be found elsewhere, perhaps in sensorimotor knowledge (see Box 7.3) or in cultural categories. The other possibility is that no guarantee can be found and that epistemological and moral skepticisms are unavoidable. The stakes here are high. For those who accepted the basic terms of Kant's model, what seemed to be needed was further exploration, both empirical and conceptual, of the central notion of *constitution*.

In the rest of this chapter, I will trace one line central to these explorations. In the chapter that follows, I will trace a second line. Both explore what is often now called the "social construction of reality," but in radically different ways. The first line considers constitution to be an *epistemological* process in which each individual constructs *knowledge* of the world. It leads from Kant to Edmund Husserl, Alfred Schutz, and Peter Berger and Thomas Luckmann. The second treats constitution as an *ontological* process in which the very constituents of reality – objects and subjects – are constituted. This path leads from Georg Hegel to Martin Heidegger, Maurice Merleau-Ponty, and Harold Garfinkel (and also Karl Marx and others we will consider in Part III). The first line accepts Kant's anthropology and insists on remaining "ontologically mute" (Gergen, 2001). I will argue that it fails to either escape from or resolve the important epistemological and ethical problems that Kant recognized. The second line rejects Kant's model and reconsiders the question, "What is a human being?" Far from trying to avoid making ontological claims, it insists that it is crucially important to get the ontology right.

TRANSCENDENTAL PHENOMENOLOGY: EDMUND HUSSERL

We would be in a nasty position indeed if empirical science were the only kind of science possible. (Husserl, 1917/1981, p. 16)

The Austrian philosopher Edmund Husserl (1859–1938), like Kant, was interested in "the relationship, in particular, between the subjectivity of knowing and the objectivity of the content known" (Husserl, 1900/1913, p. 42). But although Husserl also described himself as a transcendental idealist, he was critical of Kant for what he considered his mysticism. Kant provided no room for a scientific study of constitution. Kant presumed the existence of things-in-themselves that could never be directly experienced and a transcendental process of constitution that could never be brought into consciousness because it *is* consciousness. Husserl took a different view; he insisted that we *can* become aware of the ways in which mind structures experience, if we conduct the right kind of investigation. Husserl called this investigation "transcendental phenomenology." Kant had used the term "phenomenon" to refer to things as they appear to us (the word comes from the Greek verb *phainein*, to appear or show), and Husserl called his philosophy "phenomenology" to emphasize that it was the study of appearances, not real entities (see Box 7.4).

BOX 7.4. *What Is Phenomenology?*

Phenomenology has had an important influence on qualitative research, but not many people realize that it has two very different forms. Phenomenology is usually defined as a descriptive approach to human experience, one that starts from the recognition that "things" (objects, events, processes, other people) are always given to us in human experience and understanding. But the definition in *Webster's Dictionary* – "the description of the formal structure of the objects of awareness and of awareness itself in abstraction from any claims concerning existence" – applies to only one version of phenomenology.

These two kinds of phenomenologies will be described in detail in this chapter and the one that follows, but a brief overview may be helpful.

"Transcendental phenomenology" was formulated by Edmund Husserl (1859–1938). It required a reflective attitude of introspection, an examination of the contents of one's own consciousness. It called for a "bracketing" of all ontological presuppositions: all claims about the reality of the objects of consciousness. Husserl's famous slogan "To the things themselves!" meant an objective and detached examination of our consciousness of things, free from all biases and preconceptions.

Husserl's phenomenology was transformed radically by his student Martin Heidegger, who, influenced by Georg Hegel, insisted that human beings are in-the-world, involved and caring. Heidegger's "hermeneutic phenomenology" was an investigation that began with our practical engagement with tools and equipment and articulated this involved understanding into an explicit interpretation. It valued involved participation over detached reflection and replaced introspection with the detailed study of concrete situations (what Heidegger called "circumspection").

Husserl proposed that we spend most of our time simply accepting one spatiotemporal reality, even if some parts of it may turn out from time to time to be surprising, doubtful, illusory, or a hallucination. He called this the "natural attitude": an attitude in which "corporeal physical things with some spatial distribution or other are *simply there for me*" (Husserl, 1913/ 1983, p. 51, emphasis original). In this attitude, we unthinkingly accept that the world is simply *present,* and we experience its existence without thematizing it or thinking or theorizing about it. This is where any investigation must start: "We begin our considerations as human beings who are living naturally, objectivating, judging, feeling, willing, *'in the natural attitude'*" (p. 51, emphasis original).

But, like Kant, Husserl believed that this experience is in fact "constituted" by human consciousness. All phenomena are shaped by the experiencing subject – by what Husserl called "transcendental subjectivity." Transcendental phenomenology is the kind of investigation that brings to light how subjectivity "continues to shape the world through its concealed internal 'method.'"

Bracketing Ontological Claims

To study this constituting activity requires breaking with the natural attitude. Husserl declared that "*[i]nstead of remaining in this attitude, we propose to alter it radically*" (p. 57, emphasis original). He believed that we can grasp how we are constituting reality when we reflect, and so Husserl's phenomenology starts with reflection. It requires a simple but radical shift to the "phenomenological attitude."

The shift is accomplished by what Husserl variously called "bracketing," "parenthesizing," the "reduction," or the "epoché" (Greek: εποχη: the suspension of judgment or the withholding of assent). We need to suspend our naive belief that objects are unaltered by our consciousness of them. We need to "put it out of action," "exclude it," or "parenthesize it." The phenomenological attitude is a "definite, *specifically peculiar mode of consciousness*" (p. 59, emphasis original) in which the everyday world does not vanish: "It is still there, like the parenthesized in the parentheses, like the excluded outside the context of inclusion" (Husserl, 1913/1983, p. 59). But in this new attitude we resist making any ontological assumptions or claims that the objects or events we experience are real, existing outside our consciousness, and independent of us: "I am *not negating* this 'world' as if I were a sophist; I am *not doubting its factual being* as though I were a skeptic; rather I am exercising the 'phenomenological' epoche which also *completely shuts me off from any judgment about spatio-temporal factual being*" (Husserl, 1913/1983, p. 61, emphasis original). The result of this phenomenological reflection is an articulation of "eidetic structures" (from the Greek *eidos* or "form"). These are the "essence" of what appears in our consciousness of the world. Objects now appear to us not as independent entities in an outer world but as "unities" of "sense" or "meaning" in the "inner world" of the conscious individual.

At first glance, Husserl's bracketing seems to resemble Descartes's "method of doubt," which questioned all sensory experience and tried to reconstruct knowledge on the basis of the "*cogito.*" But Husserl didn't reject sensory experience; he just rejected the assumption that the senses tell us about objects as they really exist: "Husserl's doubt is sharply focussed: it is aimed at eliminating all ideas related to the *existence* of objects our consciousness tells us about; to be exact – the existence of objects apart from, and independently of, their presence in our consciousness" (Bauman, 1981, p. 118). Like Descartes, however, Husserl intended to achieve an ultimate, final, objective knowledge – that of pure consciousness.

The Sciences Are Not Disinterested

Husserl was critical of empirical science because it accepts the natural attitude and merely studies the details of some part of the factual world, then makes a "surreptitious substitution" of mathematical ideals and imperceptible entities for "the only real world, the one that is actually given through perception, that is ever experienced and experienceable – our everyday life-world" (Husserl, 1936/1970, p. 48). Scientific research *presupposes* the everyday life-world but then treats it as a *derivative* of the world of mathematics, which is assumed to be more real. Husserl believed that science, despite its claims of objectivity and neutrality, was driven by human interests and concerns, and he was convinced that these needed to be eliminated if truly objective knowledge were to be obtained. Husserl's transcendental phenomenology had no place for theory, either philosophical or scientific, and avoided theoretical preconceptions just as much as ontological assumptions: "In like manner all theories and sciences which relate to this world, no matter how well they may be grounded positivistically or otherwise, shall meet the same fate" (Husserl, 1999, p. 65).

For Husserl, the investigation of constitution must be free from the distortions of human concerns, worries, and interests if it is to put empirical science on a firm foundation. As he saw it, phenomenological investigation disclosed the fundamental structures of consciousness and so was more objective than science itself. In Husserl's view, "phenomenology is, in fact, a *purely descriptive* discipline, exploring the field of transcendentally pure consciousness by *pure intuition*" (Husserl, 1913/1983, p. 136, emphasis original). It is "the reflective study of the essence of consciousness as experienced from the first-person point of view." It excluded, Husserl insisted, all the social, practical, cultural factors that he felt interfered with the ability to be objective about the formal structure of our human experience:

> [W]ith the exclusion of the natural world, the physical and psychophysical world, all individual objectivities which become constituted by axiological and practical functionings of consciousness are excluded, all the sorts of cultural formations, all works of the technical and fine arts, of sciences (in so far as they come into question as cultural facts rather than as accepted unities), aesthetic and practical values of every form. Likewise, naturally, such actualities as state, custom, law, religion. (Husserl, 1913/1983, p. 131)

An Endless Road

Husserl's transcendental phenomenology, then, was to be a new kind of science, more far-reaching than typical empirical science, one that explored the key constituting activity of consciousness. His conception of the way consciousness functions is strikingly similar to the dominant model today in cognitive science (Dreyfus & Hall, 1982), and his work provided an influential example for many people who were looking for a new kind of social science or a new kind of philosophy. But his project to ground knowledge in an indubitable foundation ran into serious problems. The task of turning consciousness on itself, to identify eidetic structures that were objective and certain because they were completely abstract and detached, was more difficult than Husserl anticipated. His effort to cut free from all historical and social entanglements, to find a core to experience that was independent of society, history, or culture, turned out to be endless. Husserl was continually dissatisfied with his progress and repeatedly announced fresh attempts to start all over again on the path to "pure consciousness."

The basic problem with the project of transcendental phenomenology was that to the extent that Husserl was able to escape from the mundane world of everyday life he was leaving behind the natural context of communication, of practical concerns, of social interaction with others. If he found an eidetic structure, how could he communicate it without using a natural language that belonged to one culture or another? Who would he share it with, if he had bracketed the social world? Why would people care, if he had excluded all human concerns from his investigation?

> What is epoche, what is the whole series of phenomenological reductions, if not an effort to peel away successive layers of content, to arrive at the end at the tough nucleus which is explicable only from itself, and not reducible any more to either tradition, or culture, or society? But how do we know that such a nucleus exists? What kind of evidence can we ever get that it does? (Bauman, 1981, p. 121)

To many, Husserl's work has demonstrated that the search for Kant's transcendental activity of constitution is futile. He was dedicated to following the path Kant had pointed out, but his dedication showed that the path led nowhere: "We can now be sure that there is nothing at the end of the road which – as Husserl hoped and we, tentatively, hoped with him – led to the station called certainty" (Bauman, 1981, p. 129). At least, there was nothing at the end of *this* road, which tried to follow constitution deep into the mind in the belief that it is a *transcendental* activity.

PHENOMENOLOGY OF THE SOCIAL WORLD: ALFRED SCHUTZ

> We have to distinguish between the scientist *qua* human being who acts and lives among his fellow-men his everyday life and the theoretical thinker who is, we repeat it, not interested in the mastery of the world but in obtaining knowledge by observing it. (Schutz, 1970, p. 259)

A different direction was taken by Alfred Schutz (1899–1959), a philosopher, sociologist, and professional financier who drew on Husserl's phenomenology to develop a "sociology of understanding." Schutz modified Husserl's project in significant ways and avoided some of the difficulties that Husserl ran into. Like Husserl, Schutz believed that the everyday world is constituted by human subjects. Unlike Husserl, he believed that this activity of constitution is carried out not by transcendental subjectivity but by "mundane subjectivity." "Mundane" here has the sense not of something boring or tedious but something worldly and everyday: mundane is the opposite of transcendental. Schutz set out to study precisely those "existential" aspects of human life that Husserl believed needed to be put in brackets. Studying phenomena involved "reducing them to the human activity which has created them" (Schutz, 1954, p. 10).

Schutz emphasized the way the complex structures of the everyday "life-world" (*Lebenswelt*) are constituted in and by the consciousness of the individual ego. His form of phenomenology was the investigation of these structures:

> By the term "social reality" I wish to be understood the sum total of objects and occurrences within the social cultural world as experienced by the common-sense thinking of men living their daily lives among their fellow-men, connected with them in manifold relations of inter-action. It is the world of cultural objects and social institutions into which we are all born, within which we have to find our bearings, and with which we have to come to terms. From the outset, we, the actors on the social scene, experience the world we live in as a world both of nature and of culture, not as a private but an intersubjective one, that is as a world common to all of us, either actually given or potentially accessible to everyone; and this involves intercommunication and language. (Schutz, 1963a, p. 236)

Central to this, in Schutz's view, were the ways in which we experience the life world as intersubjective. Social reality "is a world common to all of us" (Schutz, 1970, p. 163); it is from the start, Schutz insisted, an "intersubjective world," "a preconstituted and preorganized world whose particular structure is the result of an historical process, and is therefore different for each culture and society" (p. 79). This was the world that "the wide-awake, grown-up man who acts in it and upon it amidst his fellow-men experiences with the natural attitude as a reality" (p. 72). People cope somehow with the everyday problem of understanding other people; the phenomenological social scientist can do no better than study how they do this.

The Social Sciences Take the Life-World for Granted

Like Husserl, Schutz accused the empirical-analytic social sciences of taking for granted the reality of this everyday life-world but then trying to replace it with formal models. Researchers take for granted that they *understand* what someone is doing and saying and then busy themselves looking for explanations. But their understanding, and the life-world that makes it possible, ought to be a *topic* of inquiry. A social science that ignores the way the social world is understood and interpreted by the actors within it can only end up imposing the scientists' abstract constructs because "this type of social science does not deal directly and immediately with the social life-world common to us all, but with skillfully and expediently chosen idealizations and formalizations of the social world" (Schutz, 1954, p. 6).

In contrast, a phenomenological social science studies how people – both actors and social scientists – *make sense* of the social world. For Schutz, phenomenology takes up this neglected topic by exploring how social reality is constituted and maintained by human common sense. The study of everyday social reality must be based on the way people understand and conceptualize it. Social science ought to deal in what Schutz called "second level constructs," interpretations of the "pre-interpretations" or common-sense "constructs" people have of the social world:

> The observational field of the social scientist – social reality – has a specific meaning and relevance structure for the human beings living, acting, and thinking within. By a series of common-sense constructs they have pre-selected and pre-interpreted this world which they experience as the reality of their daily lives. It is these thought objects of theirs which determine their behavior by motivating it. The thought objects of the social scientist, in order to grasp this social reality, have to be founded upon the thought-objects constructed by the common-sense thinking of men, living their daily life within their social world. Thus, the constructs of the social sciences are, so to speak, constructs of the second degree, that is, constructs of the constructs made by the actors on the social scene, whose behavior the social scientist has to observe and to explain in accordance with the procedural rules of his science. Thus, the exploration of the general principles according to which man in daily life organizes his experiences, and especially those of the social world, is the first task of the methodology of the social sciences. (Schutz, 1963a, p. 242)

Suspending Belief in the Life-World

Schutz followed Husserl in emphasizing the importance of bracketing ontological claims "not by transforming our naive belief in the outer world into a disbelief... but by suspending belief" (Schutz, 1970, p. 58) in order to focus on these "common-sense constructs": "The method of phenomenological reduction, therefore, makes accessible the stream of consciousness in itself as a realm of its own in its absolute uniqueness of nature. We can experience it and describe its inner structure" (p. 59).

Phenomenological sociology differed from traditional sociology in neither taking the social world at face value nor accepting scientific idealizations and generalizations about this world but instead studying the meaning of social phenomena for actors, their processes of idealizing and generalizing, the activities of consciousness by which people make sense of everyday reality. "The safeguarding of the subjective point of view is the only but sufficient guarantee that the world of social reality will not be replaced by a fictional nonexisting world constructed by the scientific observer" (Schutz, 1970, p. 271).

Knowledge Is Practical

Schutz recognized that an individual's involvement in the everyday life-world is first of all practical. An individual moves from one "project" to another, and the life-world is primarily a place of practical "routine." An "interest at hand" motivates all our thinking, and we experience other people's actions in terms of their motives and goals. All interpretation of the everyday life-world is based, in Schutz's view, on the "stock of knowledge" that each individual has acquired. There is a "social distribution" of this "common-sense knowledge" (Schutz, 1970, p. 239); an individual's stock is never complete, and it depends on their position in society, job, interests, and so on. Each individual's stock of knowledge is, for Schutz, a "system of constructs." "Any knowledge of the world, in common-sense thinking as well as in science, involves mental constructs, syntheses, generalizations, formalizations, idealizations" (p. 272). But it is for the most part *practical* knowledge and as such, Schutz proposed, it will be incoherent, inconsistent, and only partially clear. Frequently meaning becomes apparent only retrospectively, and this is what makes phenomenology necessary:

> [W]e no longer naively accept the social world and its current idealizations and formalizations as ready-made and meaningful beyond all question, but we undertake to study the process of idealizing and formalizing as such, the genesis of the meaning which social phenomena have for us as well as for the actors, the mechanism of the activity by which human beings understand one another and themselves. (Schutz, 1970, p. 269)

Schutz insisted that any description of action needs to refer to the "subjective meaning" it has for the actor. He described how we understand action as spontaneous activity oriented toward the future, so the span and unity of an action is determined by the "project" of which it is part. What is projected in action is the completed act, the goal of the action. One isn't just "walking towards the window" (let alone just "putting one foot in front of the other"); one is "going to open the window." Schutz called this the action's "in-order-to motive." The goal of an action defines its subjective meaning and is a necessary part of any description.

Typification and Language

Our stock of knowledge consists in large part of being able to identify what types of things we are dealing with. Schutz viewed the process of *"typification"* as an essential part of all social knowledge:

> The world, the physical as well as the sociocultural one, is experienced from the outset in terms of types: there are mountains, trees, birds, fishes, dogs, and among them Irish setters; there are cultural objects, such as houses, tables, chairs, books, tools, and among them hammers; and there are typical social roles and relationships, such as parents, siblings, kinsmen, strangers, soldiers, hunters, priests, etc. Thus, typifications on the commonsense level . . . emerge in the everyday experience of the world as taken for granted without any formulation of judgments or of neat propositions with logical subjects and predicates. (Schutz, 1970, p. 120)

Each of us has a generalized knowledge of types of things and their typical styles. Language is the "typifying medium *par excellence*" (p. 96). "Language as used in everyday life . . . is primarily a language of named things and events" (p. 117).

Schutz pointed out that the life-world is composed of *multiple* realities, each of which is a distinct "finite province of meaning." Primary among these is the intersubjective world of everyday life, the "world of the natural attitude with its dominant pragmatic motives," but there are many others: the worlds of dreams, of fantasy, of science, of religion. Each region requires its own kind of epoché: "Individuals suspend doubt, not belief, in the Lebenswelt." Each province has its distinct "style of lived experience"or cognitive style, a distinct accent to reality, distinct structures and spatial and temporal relations, and its own systems of relevance and schemes of interpretation. We "leap" among these worlds. For Schutz, the "world of scientific theory" is merely one of these multiple realities.

Reality: Subjective or Intersubjective?

We have seen that for Schutz the goal of phenomenological sociology was "explaining the thought-objects constructed by common sense" in terms of "the mental constructs or thought-objects of science" (Schutz, 1970, p. 272). The social phenomenologist, in his view, proceeded by a process of "subjective interpretation" that aimed to grasp the "subjective point of view" of the individual. Schutz acknowledged that different people will have different constructs, generalizations, and typifications, but his interest was in the general process of forming them rather than in individual differences. Phenomenology was not a matter of understanding another person in their uniqueness or their specific situation. The goal was "the subjective point of view," but for subjectivity in general.

Here Schutz ran directly into the contradiction of Kant's anthropology. Can objective knowledge truly be based on subjective constructs? To his credit, he recognized the difficulty: "Indeed, the most serious question which the methodology of the social sciences has to answer is: How is it possible to form objective concepts and an objectively verifiable theory of subjective meaning-structures?" (Schutz, 1963a, p. 246).

His answer was that it is the "procedural rules" of science that enable the researcher to develop objective constructs. The scientist is "not involved in the observed situation" because to him or her this is "merely of cognitive interest" (Schutz, 1963a, p. 246). The scientist has "replaced his personal biographical situation by what I shall call ... a scientific situation" and achieves "detachment from value patterns" that operate for the people studied. The scientist imagines "ideal actors" to whom he or she ascribes "typical notions, purposes, goals" in order to construct "a scientific model of human action" (p. 247) that can have "objective validity." "The attitude of the social scientist is that of a mere disinterested observer of the social world" (Schutz, 1963b, p. 335).

But Schutz had become trapped by the implications of his own criticism of scientific sociology. He was surely correct to argue that the traditional sociologist takes for granted their form of life when they study the actions of people around them, and they should make it a topic of inquiry. But a truly "mere disinterested observer" would not be able to use the life-world as a resource. He or she would not be able to understand the actions of other people or communicate their findings.

Kant's contradiction is evident in Schutz's phenomenology in a second way. Schutz set out to describe social reality and show how it is constituted. He insisted that social reality is intersubjective, but he viewed it as constituted by individual and subjective processes such as typification and generalization. At the same time, he insisted that the "interpretive schemes" with which we understand our ongoing experiences are social and intersubjective, not personal and subjective. He had problems, however, in describing exactly how this was so. Schutz, like Husserl, was searching for universal structures, though they were structures *of* the life-world rather than "transcendental" structures somehow *underlying* the world of everyday life. But the status and character of these structures were unclear. How did individuals come to share the same schemes? Did they not change historically? And, if so, how could they be universal? These problems stemmed from the fact that the task Schutz set for himself was to describe how social reality is *experienced*, how an individual *knows* it. He was limited to exploring "the social cultural world as *experienced* by the common-sense *thinking* of men living their daily lives" (Schutz, 1963a, p. 236, emphasis added). What social reality actually *is*, as an *objective* reality, lay out of reach, just like Kant's things-in-themselves. Ironically, Schutz's study of the constitution of social reality was not able to grasp reality at all.

THE SOCIAL CONSTRUCTION OF REALITY: PETER BERGER AND THOMAS LUCKMANN

> Only a few are concerned with the theoretical interpretation of the world, but everybody lives in a world of some sort. (Berger & Luckmann, 1966, p. 15)

There are now hundreds of books with the term "social construction" in their title, but the first was *The Social Construction of Reality: A Treatise in the Sociology of Knowledge*, by Peter Berger and Thomas Luckmann (1966). It soon became highly influential. Like Schutz (with whom they studied), Berger and Luckmann set out to study the reality of everyday life as experienced in "the commonsense of the ordinary members of society," not the objective reality described by natural science or by the social sciences as they are usually practiced. But, unlike Schutz, Berger and Luckmann insisted that society exists as *both* objective and subjective reality, and an adequate sociology must grasp both aspects. By including both the subjective and the objective reality of the social world, they aimed to recast constitution as a truly *social* process, intersubjective rather than merely subjective. To do this, sociology needed to move in and out of the phenomenological attitude, and this was reflected in the organization of their book:

> Thus some problems are viewed within phenomenological brackets in Section I [*The Foundations of Knowledge in Everyday Life*], taken up again in Section II [*Society as Objective Reality*] with these brackets removed and with an interest in their empirical genesis, and then taken up once more in Section III [*Society as Subjective Reality*] on the level of subjective consciousness. (Berger & Luckmann, 1966, p. vi)

The overall task was "a sociological analysis of the reality of everyday life, more precisely, of knowledge that guides conduct in everyday life" (p. 19) that would solve or avoid the problems that Schutz had encountered. Berger and Luckmann insisted that sociology is a science (and can be "value-free"), but it is one that must deal with "man *as* man"; it is a humanistic discipline:

> [S]ociology must be carried on in a continuous conversation with both history and philosophy or lose its proper object of inquiry. This object is society as part of a human world, made by men, inhabited by men, and, in turn, making men, in an ongoing historical process. It is not the least fruit of a humanistic sociology that it reawakens our wonder at this astonishing phenomenon. (Berger & Luckmann, 1966, p. 189)

Sociology Neglects Members' Knowledge

Like Schutz, Berger and Luckmann were critical of the idea that sociology can be a science of social institutions and processes that pays no attention to how these are understood by the people who participate in them. They insisted that sociology must pay attention to the "knowledge" that members of society have of their own circumstances. And they directly confronted the problem of the constitution of the social world, remarking on the fact that "the constitution of reality has traditionally been a central problem of philosophy" but "there has been a strong tendency for this problem, with all the questions it involves, to become trivialized in contemporary philosophy" with the result that the problem has moved from philosophy to the social sciences and "the sociologist may find himself, to his surprise perhaps, the inheritor of philosophical questions that the professional philosophers are no longer interested in considering" (Berger & Luckmann, 1966, p. 189).

Berger and Luckmann insisted that understanding constitution, the social construction of reality, is a necessary part of every kind of sociology. They proposed that "the analysis of the role of knowledge in the dialectic of individual and society, of personal identity and social structure, provides a crucial complementary perspective for all areas of sociology" (Berger & Luckmann, 1966, p. 168). Their aim was to provide "a systematic accounting of the dialectical relation between the structural realities and the human enterprise of constructing reality – in history" (p. 186). They were clear that "[t]he basic contentions of the argument of this book are implicit in its title and subtitle, namely, that reality is socially constructed and that the sociology of knowledge must analyze the processes in which this occurs" (p. 1).

Berger and Luckmann proposed that "the sociological understanding of 'reality' and 'knowledge' falls somewhere in the middle between that of the man in the street and that of the philosopher" (Berger & Luckmann, 1966, p. 2). The man in the street takes his specific reality for granted. The philosopher, in Berger and Luckmann's view, aims to identify a genuine underlying reality. The sociologist, in contrast to both, cannot take either kind of reality for granted in part because she or he knows that different people inhabit different realities. This is why studying the construction of reality requires moving in and out of a phenomenological attitude.

Putting on and Removing Brackets

Like Husserl and Schutz before them, Berger and Luckmann believed that a phenomenological sociology required bracketing the ontological assumptions of everyday life and science:

> The method we consider best suited to clarify the foundations of knowledge in everyday life is that of phenomenological analysis, a purely

descriptive method and, as such, "empirical" but not "scientific" – as we understand the nature of the empirical sciences. The phenomenological analysis of everyday life, or rather of the subjective experience of everyday life, refrains from any causal or genetic hypotheses, as well as from assertions about the ontological status of the phenomenon analyzed. (Berger & Luckmann, 1966, p. 20)

The products of their phenomenological analysis resembled Schutz's in several respects. Like Schutz, Berger and Luckmann emphasized the existence of different "spheres of reality." Among these multiple realities, the reality of everyday life is "reality par excellence," experienced in a wide-awake state with the highest tension of consciousness. Here the world appears already "objectified," full of objects defined *as* objects "before I arrive on the scene." It is taken for granted, simply the world of "here" and "now," an intersubjective world that I accept is shared with others. Generally this reality is routine and unproblematic, and hitches are quickly resolved.

PRAGMATIC, RECIPE KNOWLEDGE

Everyday life is dominated by the pragmatic motive, and a prominent ingredient in the social stock of knowledge is recipe knowledge – "that is, knowledge limited to pragmatic competence in routine performances." People's knowledge about everyday life is structured in terms of *relevances*: "It is irrelevant to me how my wife goes about cooking my favorite goulash as long as it turns out the way I like it" (Berger & Luckmann, 1966, p. 45). And "my relevance structures intersect with the relevance structures of others at many points, as a result of which we have 'interesting' things to say to each other" (p. 45). Other people are experienced in several different modes: in the prototypical case of face-to-face encounters, and in a continuum of progressively anonymous contacts apprehended by means of "typificatory schemes" (as "an ingratiating fellow," "a salesman," "an American"). Knowledge is socially distributed – different people have different kinds of expertise – and knowledge of *how* it is distributed is an important part of that stock of knowledge.

Also like Schutz, Berger and Luckmann considered face-to-face conversation to be the "the most important vehicle of reality maintenance" (Berger & Luckmann, 1966, p. 152), and they argued that an individual's subjective reality is constantly maintained, modified, and reconstructed by "the working away of a conversational apparatus." Much of this work is implicit: "[M]ost conversation does not in so many words define the nature of the world. Rather, it takes place against the background of a world that is silently taken for granted" (p. 152). It is precisely because casual conversation *is* casual that a taken-for-granted world and its routines are maintained. The reality of something never talked about becomes "shaky." Things talked about, in contrast, are allocated their place in the real world. Conversation in face-to-face interaction is the principal way that language objectifies and realizes the world "in the double sense of apprehending and producing it" (p. 153).

Moments in Social Construction

Berger and Luckmann also agreed with Schutz that a science of the social world should not take its reality for granted, but they went further than he did in emphasizing the *historical* dimension of the construction of reality and the "active dialectical process" whereby people maintain, modify, and reshape the social structure as they are, at the same time, formed and shaped in their identity in social relationships:

> Man is biologically predestined to construct and to inhabit a world with others. This world becomes for him the dominant and definitive reality. Its limits are set by nature, but once constructed, this world acts back on nature. In the dialectic between nature and the socially constructed world the human organism itself is transformed. In this same dialectic man produces reality and thereby produces himself. (Berger & Luckmann, 1966, p. 183)

In their analysis, "society is understood in terms of an ongoing dialectical process composed of the three moments of *externalization, objectivation,* and *internalization*" (p. 129). Each of these moments "corresponds to an essential characterization of the social world. *Society is a human product. Society is an objective reality. Man is a social product*" (p. 61, emphasis original). Externalization is how "social order is a human product, or, more precisely, an ongoing human production.... It is important to stress that externalization ... is an anthropological necessity.... Human being must ongoingly externalize itself in activity" (p. 52). Objectification is "the process whereby the externalized products of human activity attain the character of objectivity" (p. 60). Internalization is "the process whereby the objectivated social world is retrojected into consciousness in the course of socialization" (p. 61).

To illustrate these three moments in the dialectical process of the social construction of reality, Berger and Luckmann invited the reader to imagine two people who come from "entirely different social worlds" but are marooned together on a desert island. As they interact, they produce "typifications" of each other's behavior ("Aha, there he goes again") and also assume the reciprocity of this typification process. Typifications become the basis for role playing, and these roles over time become habitualized. This is the beginning of institutionalization, the process of *externalization* in which a microsociety is created as a product of human activity.

If the pair have children, there is a qualitative change in their situation, as their "institutional world" (Berger & Luckmann, 1966, p. 58) is passed along to the new generation and "perfects itself" in the form of historical institutions that, now crystallized, have a reality "that confronts the individual as an external and coercive fact" (p. 58). Now we can speak of an objective social world "in the sense of a comprehensive and given reality confronting the individual in a manner analogous to the reality of the natural world" (p. 59). The microsociety has become *objectified* because it already existed prior to the children coming to act within it. The children, growing up in this microsociety and taking it for granted, are socialized into its habitual ways. This is the *internalization* in which humans become social.

Objective and Subjective Reality?

Whereas Schutz had tried to explain the constitution of the social world in terms of individual consciousness, Berger and Luckmann aimed to bridge the gap between subjective experience of the social world and its objective reality. They introduced new and important considerations: the social relations in which social reality is constructed, their historical dimension, and the mutual constitution of person and world. It is not hard to see why their book has had a powerful and lasting impact. But, at the same time, their analysis moved uneasily between the "subjective" and "objective" aspects of society, aspects they tried to connect by appealing to processes of "externalization" and "internalization." They alternated between phenomenological investigation and objective analysis without explaining how the two methods can be reconciled given their criticism of traditional sociology. And as the phrase "sociology of knowledge" indicates, Berger and Luckmann continued to view "reality" primarily in terms of what people *know*, although they included "everything that passes for 'knowledge' in society," including practical and commonsense knowledge:

> It will be enough, for our purposes, to define "reality" as a quality appertaining to phenomena that we recognize as having a being independent of our own volition (we cannot "wish them away"), and to define "knowledge" as the certainty that phenomena are real and that they possess specific characteristics. (Berger & Luckmann, 1966, p. 1)

The Social Construction of Reality is filled with rich observations of a variety of phenomena such as schooling, religious conversion, and everyday interaction, and it explores their implications for our sense of what is real and who we are. But the conceptual framework of the book still approached the problem of the constitution of reality in Kantian terms, with an emphasis on what we *experience* as real. Unifying the "objective" and "subjective" aspects of social reality turned out to be a more difficult task than Berger and Luckmann had anticipated. Their account of constitution as a dialectical process of social construction did not end the search that Kant had begun for the basis for valid knowledge and ethical action.

TABLE 7.1. *Conceptions of Constitution, Part 1*

	Husserl	Schutz	Berger and Luckmann
Constitution	Transcendental phenomenology Transcendental subjectivity shapes the world though its concealed internal "method."	Phenomenology of the social world Constitution of social world by mundane subjectivity.	Sociology of knowledge "Social construction of reality"; the mutual, dialectical constitution of society by people, and of people by society.
Subjectivity	Transcendental ego: a level of consciousness common to all people. Knowing the world, but disinterested and detached.	Mundane subject: knowing the world, but with everyday practical interests and concerns.	Mundane subject: knowing the world, with everyday practical interests and concerns, also in dialectical interaction with the world and so changed by it. Historically developing.
Methodology	Pure intuition: the transcendental attitude. Bracketing assumptions about being. Pure description.	"The subjective point of view."	Phenomenological analysis, without causal hypotheses or ontological claims. Plus objective study.
Products of Investigation	Abstract and universal categories: eidetic structures.	The structures of the life-world. "Typifications"; "Systems of relevance."	Grasp of externalization, objectification, and internalization.

CONCLUSIONS

Husserl, Schutz, and Berger and Luckmann all accepted Kant's proposal that reality as we know it is constituted (see Table 7.1). They shared the assumption that *knowing* the world gives it sense and order, though they differed on where this process of knowing was located. For Kant it had been the activity of transcendental reason, with universal categories of space, time, causality, and object. For Husserl, transcendental subjectivity brings eidetic structures to the "hyletic" (sensory) data of perception, so that *every* object of our experience – trees, cats, tables and chairs – is a mental construct. For Schutz, our individual mundane "commonsense constructs," our "typificatory schemes," define the meaning of social phenomena and enable us to make sense of the world and get along in it. Berger and Luckmann placed two individuals face to face, in a reciprocity of *mutual* typification, but they, too, emphasized ways of *knowing* – "recipe knowledge" and so on – to explain how the world is *experienced* as real.

None of these approaches was able to make constitution do the work that Kant wanted it to do. None was able to build a bridge between individual subjective experience and objective reality. Each of them was critical of traditional inquiry for taking the objective reality of the world for granted, but none of them was able to demonstrate how this world is actually constituted by subjective experience.

Hand in hand with their focus on knowledge and mental representation was the effort by all these people to avoid making any ontological claims. This began with Kant's insistence that although we must infer that things-in-themselves do exist, to go any further and say anything more definite about them would be speculative metaphysics. Husserl found even this limited claim unnecessarily metaphysical (or mystical). Even more stringently than Kant, he avoided making any claims about the actual existence of the objects of experience. His interest was limited to bringing to light the mental machinery, the cognitive apparatus, that makes these objects *appear* real.

For Berger and Luckmann, too, "[t]he phenomenological analysis of everyday life, or rather of the *subjective experience* of everyday life, *refrains* from any causal or genetic hypotheses, as well as from assertions about the *ontological status* of the phenomenon analyzed" (Berger & Luckmann, 1966, p. 20, emphasis added). But, at the same time, Berger and Luckmann asserted that society exists as *both* subjective reality and objective reality. These confused claims have the consequence that it is unclear whether the "construction of reality" they described is an epistemological or ontological process.

The root problem is that, far from avoiding all ontological assumptions, each of these analyses presumed a basic ontological distinction between subjectivity and objectivity, between the world as the individual experiences it and the world as it really is, between appearance and reality. This dualism of "the two realities" is inscribed in the structure of Berger and Luckmann's book, divided into sections on *Society as Objective Reality* and *Society as Subjective Reality*.

Once one accepts the Kantian dualism of things-in-themselves and things-as-they-appear, it seems that one can study only an individual's *sense* of reality, their *experience* of reality. Berger and Luckmann recommended, in fact, that the words "reality" and "knowledge" always be placed within quotation marks.

But we were promised an explanation (or at least a description) of how *reality* is constituted (or constructed), not how a *sense* of reality comes about. To be told we are dealing not with reality but with "reality" is disappointing. Such epistemological scepticism may seem apt when we are talking about *social* reality because it may seem reasonable to say, as philosopher John Searle does in his book *The Construction of Social Reality* (1997), that a piece of paper is not *really* money; we just come to *believe* that it is money. The problem is that if we are speaking only about individuals' beliefs, there is no more basis to say that it is *really* a piece of paper.

The insistence that one is not making ontological claims is diagnostic of a hidden ontological dualism. The key symptom is the appearance of doubles: "subjective reality" and "objective reality"; "noumenon" and "phenomenon"; "appearance" and "reality." With such a dualist ontology, we are still in the terrain of Kant's representational model of human being. With this model, we can only explore how the world can *appear* objective to an individual subjectivity. We can never solve the problem of how to test the validity of such an appearance. This kind of constitution – a construction of *knowledge* of the world – can never successfully draw a distinction between what is valid knowledge and what is mere opinion.

8

Constitution as Ontological

Consider that immortal ordinary society evidently, just in any actual case, is easily done and easily recognized with uniquely adequate competence, vulgar competence, by one and all – and, for all that, by one and all it is intractably hard to describe procedurally. Procedurally described, just in any actual case, it is *elusive*.
Garfinkel, 1996, p. 8

In this chapter, I want to change your ontology! We saw in the last chapter how Husserl, Schutz, and Berger and Luckmann tried to study the kind of constitution that Kant had identified, in which individual perception and reason together form representations of an external reality. We discovered how their ontological dualism prevented them from doing more than study the *experience* of "reality" while, paradoxically, trying to bracket all claims about what actually is real. This chapter follows a different path, one that explores the *conditions* for the capacity to form subjective representations. I begin with Georg Hegel's response to Kant, then continue with Martin Heidegger, Maurice Merleau-Ponty, and finally Harold Garfinkel. Their work amounts to a different kind of phenomenology, one that explores a *nondualist ontology*, a "radical realism." Here constitution is viewed not as a matter of forming concepts or representations but as the forming of objects and subjects, an *ontological* rather than epistemological process. The focus shifts from conceptual knowledge, studied with a detached, theoretical attitude, to practical, embodied know-how, studied in an involved way. Knowhow provides a way to *see* the world. By the end of the chapter, I hope to have convinced *you* to see people and objects as inextricably one with their forms of life, and to see reason and thinking as cultural, historical, and grounded in practical know-how.

The analyses in the last chapter started from the assumption that we are naturally creatures with minds, inner spaces in which representations are formed, and asked how these representations are structured and under what circumstances they are valid.

Yet these human and social sciences – in both their experimental and qualitative forms – have been unable to escape from a persistent anxiety that makes evident the problems in Kant's anthropology. If Kant were correct that individual subjectivity is active – that each person creates their own subjective model of the world – how could this be reconciled with the view that science deals with the "objectivity" of things in the world: physical things (the natural sciences), organic things (biological sciences), or human things (the human sciences)? Since the Enlightenment, the new human sciences – sociology, anthropology, and psychology – have busied themselves studying people's representations. They have mirrored the work of the biological and physical sciences – whereas those studied objective reality, the new sciences studied, in large part, subjective reality. They didn't stop to ask whether representation was the whole story or where the capacity for representation came from.

However, "mind" and "world" have been located in two separate realms, once we assume that humans are naturally and fundamentally individuals, each with a mind that forms representations, then skepticism about the world, about other minds, and about the validity of knowledge and the basis for ethics becomes unavoidable.

What More?

Kant's analysis provided plenty of work for those who followed, and wanted to improve, his work. It should now be obvious that we *cannot* solve the epistemological and ethical problems that troubled Kant within the representational model of man. We need a different model: a different ontology. The more fundamental question that must be asked is: How is mind possible? How is it that we *become* people who can represent the world in an inner space? We must explore "what *more*" there is to human beings above and beyond the capacity to form mental representations.

The work of the people who have raised this question has been the basis for a completely different exploration of constitution. They have turned Kant's analysis upside down and explored the possibility that some more basic way that humans are involved in the world constitutes both that world *and* the human capacity for representation. Like the people in the previous chapter, they have explored the relationship between mind and world, between representation and represented, but with very different conclusions (see Box 8.1).

THE PHENOMENOLOGY OF *GEIST*: GEORG HEGEL

In pressing forward to its true existence, consciousness will arrive at a point at which it gets rid of its semblance of being burdened with something alien, with what is only for it, and some sort of "other," at

BOX 8.1. *Durkheim's Sociology*

Emile Durkheim (1858–1917) is considered by many to be the founder of sociology. The contradictions in Kant's analysis prompted Durkheim to reject Kant's conclusion that the categories of knowledge are products of an innate and universal individual capacity for reason. Dissatisfied with Kant's solution to the epistemological problem of how valid knowledge is possible, he called it a "lazy man's solution" (Durkheim, 1912/1995, p. 172). He set out to find the basis for valid knowledge and the fundamental categories of thought in the concrete empirical details of social practices (Rawls, 1996). Durkheim believed that key practices can be found in all societies, so these categories are universal without being innate. In a similar way, society provides the basis for moral action. Durkheim's project has frequently been misinterpreted as a form of idealism and cultural relativism, but Durkheim did not intend to treat society merely as a source of collective ideas. He viewed society as a set of enacted practices that can give rise to objective knowledge.

Durkheim wanted to avoid Kant's division between nature and society. In his view, each society defines "social facts" that are as objective as physical nature. Durkheim also rethought Kant's distinction between individual and society: he contrasted traditional societies, which had "mechanical solidarity" in their social relations, with modern societies, which have "organic solidarity" because their division of labor makes people thoroughly interdependent. He proposed that traditional societies involve "collective consciousness," whereas in modern societies an "individual consciousness" develops. In his view, the modern individual is a product of modern society. Later in this chapter, we will see how Harold Garfinkel's work has been described as a continuation of Durkheim's project (Box 8.2).

BOX 8.2. *Durkheim's Project Continued*

Garfinkel can be seen as continuing the project that Emile Durkheim began (see Box 8.1). To Durkheim, Kant's "emphasis on the individual and individual perception of natural forces" made the epistemological problem appear unsolvable. Durkheim sought to replace the individualist approach of traditional philosophy with an approach solidly embedded in enacted social practice" (Rawls, 1996, p. 431). This approach "treats concrete social processes as natural processes whose function is to make general categories of thought available to their human participants" (p. 433). Approached this way, sociology would be able to solve the problem that philosophy had failed to solve. "In rejecting the individual as a starting point, the way is opened for

(continued)

BOX 8.2 *(continued)*

Durkheim to explain the origin of the necessary basic concepts in terms of concrete social processes, something that had never been tried before" (Rawls, 1996, p. 433; see Rawls, 1998).

Garfinkel has continued this project, seeing social practices as the origin not just of a *sense* of social order but of order itself. Unlike Durkheim, Garfinkel has focused on ordinary everyday practices rather than institutionalized ritual practices, which are much less common today than in Durkheim's time. But Garfinkel's interest is the same: understanding social practices as the place where order is achieved. Order includes "achieved phenomena of logic, meaning, method, reason, rational action, truth, evidence, science, Kant's basic categories, or Hume's, or the primordials of anyone else" (Garfinkel, 1996, p. 11). All these, he insists, have their "origins, sources, destinations, locus, and settings" (p. 11) in ordinary interaction.

Many sociologists believe that social interaction cannot possibly produce categories that have the *necessity* that Kant considered crucial. It is usually assumed that if reason and knowledge have a social basis, this means they will inevitably be relative. For most social scientists, "social consensus, structure, or shared practices, they argue, lead persons to believe certain things or think in certain ways. Because persons share the same beliefs, they act in ways that reinforce those beliefs. The resulting consensus creates the appearance of a valid relation between thought and reality where there can in fact be none" (Rawls, 1996, p. 474).

But Durkheim believed that sociology can explain how genuinely *valid* knowledge arises from participation in social practices. He emphasized "the essential role played by social processes in creating the human faculty of reason" (Rawls, 1998, p. 900). He proposed that it is only by participating in ritual social practices that individuals develop the capacity to form valid representations of the world, including the fundamental categories Kant had emphasized: time, space, object, and causality. The origin of these is neither raw experience nor individual cognition; their validity and logic are *social* in origin. Durkheim believed that the categories arise from practices that have a *moral* force. Social practices give rise to real social forces that participants experience together, and the consequence is that individuals in a society come to share what Durkheim called central "categories of the understanding."

a point where appearance becomes identical with essence, so that its exposition will coincide at just this point with the authentic Science of Spirit. And finally, when consciousness itself grasps this its own essence, it will signify the nature of absolute knowledge itself. (Hegel, 1807/1977, p. 57)

The story begins with the German philosopher Georg Wilhelm Friedrich Hegel (1770–1831). At first, Hegel intended merely to develop Kant's philosophy, but he came to see that it had profound difficulties. Kant's view that the mind constitutes an individual's experience of an objective world by providing the transcendental concepts of space, time, causality, and object seemed to Hegel to effectively *double* both object and subject. The object was doubled into *noumenon* (thing-in-itself) and *phenomenon* (appearance), while the subject was divided into an empirical subjectivity and a transcendental ego. Kant himself was satisfied that he had shown how subject and object are linked at the level of experience, though they appear to be distinct. But at the level of reflection, subjectivity (the transcendental ego) and objectivity (the thing-in-itself) were still completely separate in Kant's account. Kant's analysis seemed to imply that we are truly and fully human only when we accept this separation from natural and social reality and that this is how we best exercise our capacity for reason.

In Hegel's view, Kant also failed to bridge the gap between knowledge (the realm of science) and action (the realm of politics, morality, and religion) (Solomon, 1983, pp. 77ff). For example, the notion of the world-in-itself permitted Kant to conceive of God as standing outside space and time but still as a necessary figure in human faith and morality. To Hegel, this ended up separating components whose relationship Kant had been trying to explain.

Hegel's genius was not to try to eliminate these tensions and contradictions but to interpret them as aspects of an evolving unity. They became opposing sides in his famous dialectic of *thesis, antithesis, synthesis* (though Hegel himself never used these terms). The resulting philosophy has had a profound impact on many schools of thought, including existentialism, Marx's historical materialism, and psychoanalysis. Hegel's writing is notoriously difficult, and there are many different interpretations of his ideas (e.g., Rockmore, 1997; Solomon, 1983; Taylor, 1975). Here I will give only a brief summary of two of his central proposals: that human reason is a cultural and historical phenomenon and that consciousness follows a path toward more complex and adequate ways of knowing both self and world. These proposals open up a fresh way of thinking of humans in which the mind is reconceptualized as the way we are involved in the world.

Reason Has a History

Hegel proposed that Kant had not been sufficiently critical of his own critique. Kant's error, said Hegel, lay in his appeal to a rationality that lay outside human practice on a transcendental plane. Kant had failed to explain how he could adopt his own critical position. He had claimed that reason provides the conditions for the possibility of experience but had failed to explore the conditions for the possibility of reason. Hegel's response was to put reason – and the reasoner (the philosopher, the thinker) – back in their proper place in the tide of human affairs; that is to say, in history. Reason, Hegel argued, also has a history. Any investigation of the conditions for knowledge must start from a position *within* this historical process of coming to know.

One simple way to put this is that what we call reason – whether it is logic, mathematics, or the differential calculus – has been figured out over time. It only seems timeless and eternal once it is complete. Mind itself, for Hegel, is not part of a universal, timeless human nature but has developed over history and will continue to develop. The human mind is worldly and secular, not transcendental or spiritual.

Similarly, Hegel proposed that Kant had failed to grasp the concrete character of moral problems and dilemmas. He shared Kant's view that to be moral we must be rational and make free choices, but he believed that individual ethical choice cannot be separated from social contexts. Self-conscious moral action, Hegel proposed, is based on social practices and institutions. He developed a *concrete* ethics in which he described the ethical ideals of his particular society. The morality that Kant had argued was universal was in reality a middle-class, Western morality. If values become universal, Hegel argued, it will only be because communities expand and become international.

Hegel insisted that any attempt to base knowledge or morality on the individual will inevitably fail. Such ethical and epistemological theories are possible only because we are members of a community, but because they start from the individual they will be formal and empty. Analyses such as Kant's presuppose a background of social practices that they fail to examine or question. They assume that the individual is merely an isolated atom, outside society and culture, and only reinforce the "alienation" (*Entfremdung*) of the individual in modern society. Hegel argued that knowledge is always the product of participation in an organized, ethical community, and the basis of morality is to be found in the "reason" of this community. This organized community life, what he called *Sittlichkeit*, is the practices and customs each of us is born into. "*Sittlichkeit* is morality as established custom, not a set of principles. [It] is shared activity, shared interests, shared pleasures" (Solomon, 1983, p. 534). In modern society, Hegel suggested, these practices are the basis for individualism and a modern bourgeois morality that divides public life from private life and personal values from community values, and pits each individual's interests against those of other people (p. 491).

Consciousness Follows a Path

Hegel sought a way of both recognizing and resolving the opposition and conflict between subjectivity and objectivity and (what amounted to the same thing) the opposition between idealist and empiricist theories of knowledge. He suggested that there is "subjectivity at the level of objectivity" (Hyppolite, 1946/1974, p. 83). What does this mean? Hegel, just like Husserl, saw consciousness as intentional:

> When we experience, say, a table within consciousness, we understand our perception to refer to a table beyond consciousness, in the same way phenomenologists such as Brentano and Husserl use the concept of intentionality as the property of consciousness to be directed towards something. In the process of knowing, the distinction between what appears and what is, is overcome. At the limit, when we fully know, knowing becomes truth. (Rockmore, 1997, p. 30)

Husserl appreciated that when we experience a table, we understand that we perceive a real table that is partly beyond our present experience. It has, for example, a hidden side. But Hegel saw also that experience grows and changes, and he proposed that, in the process of knowing, the distinction between the table as we experience it and the table as it is can be overcome. Our experience can become increasingly *adequate* to the object. Achieving this adequacy requires being able to distinguish between the object experienced and how we experience it, and this in turn requires self-knowledge and self-consciousness.

So Hegel acknowledged things-in-themselves, but unlike Kant he argued that we can come to know them. Such knowledge is "scientific" knowledge (though science for Hegel was part of philosophy). Whereas Kant had offered an analysis only of how things *appear*, Hegel argued that we can know how things *are*. The distinction between "our view of the object within consciousness" and "the object of that view within consciousness" is a distinction that we can become consciously aware of. Whereas Kant had discounted any claim about things-in-themselves as "speculative metaphysics," Hegel maintained that such claims can be rational and grounded:

> Kant illustrates the effort, widespread in modern philosophy, to know an independent external object through an analysis of the relation between the knowing subject and its object. Yet there is no way to grasp the relation of whatever appears within consciousness to an independent external reality. Hegel's solution is to replace this relation through a very different relation between a subject and an object that falls entirely within consciousness. Knowledge is not a process of bringing our view of the object into correspondence with an independent external object, but rather a process of bringing our view of the object

within consciousness into correspondence with the object of that view within consciousness. (Rockmore, 1997, pp. 28–29)

This difference between Hegel and Kant with respect to our ability is know things-in-themselves is important because, as we have seen, many contemporary constructivists believe they must avoid saying anything specific about reality. Hegel offered a constructivism in which ontology plays a central role. He offered an ontology in which knowledge is constituted but in which the knowing subject and the known object are constituted, too. Whereas Kant had taken for granted the existence of the individual subject who represents the world, Hegel studied both the conditions of experience *and* the conditions for the possibility of the subject who experiences.

As Husserl would do, Hegel called his approach "phenomenology." The term reflected his view that philosophy should examine knowing as it actually occurs and study consciousness as it actually exists. His *Phänomenologie des Geistes* (1807) was the study of how consciousness or mind appears to itself. The title has been translated both as *Phenomenology of Mind* and *Phenomenology of Spirit*; the German word *Geist* can mean mind, spirit, or even ghost. Hegel's working title was *Science of the Experience of Consciousness*. Whatever the translation, *Geist* should be understood as both subject and object, a unified subject/object. For at least one modern commentator, Hegel's "concept of spirit is roughly a view of people in the sociocultural context as the real subject of knowledge" (Rockmore, 1997, p. 4).

Hegel proposed that there is a reflexive capacity to consciousness: an immediate, noncognitive relation of the self to itself. Consciousness always relates to an object and at the same time *distinguishes* itself from that object: this apple is an object for *me*; it is a being for *my* awareness. Knowing is not a relationship to something outside consciousness but a relationship *within* consciousness.

If the distinction between subject and object emerges *within* consciousness, it follows that consciousness cannot be something within the *subject* (in the head, or made up of mental states). For Hegel, consciousness is a relationship *between* a subject (knowing and acting) and an object (known and acted on), a relationship that is always social and can only develop fully in specific kinds of social practices and institutions. Hegel insisted, moreover, that to recognize this one cannot find a position *outside* the natural attitude, such as Husserl's transcendental attitude. We can describe consciousness only from within our natural, everyday experience. And because this experience *develops*, there is no single fixed and unchanging natural attitude; each of us progresses through a series of attitudes. Hegel believed that he was standing at the end of the process of the development of consciousness, able to look back and describe it.

The *Phenomenology of Mind*, then, offered "an exposition of how knowledge makes its appearance" (Hegel, 1807/1997, p. 49). It was a description (phenomenological) of the way human beings come to know, of "the path of the natural consciousness which presses forward to true knowledge." It described how "the series of configurations which consciousness goes through along this road is, in reality, the detailed history of the *education* [*Bildung*] of consciousness itself to the standpoint of Science" (p. 50, emphasis original), even though at times it seems "a highway of despair" (p. 135): "Hegel's phenomenological self-reflection surmounts dogmatism by reflectively reconstructing the self-formative process (*Bildungsprozess*) of mind (*Geist*)" (McCarthy, 1978, p. 79).

The historical unfolding of human consciousness is expressed in reason. Hegel described this unfolding as a dialectical process in which understanding moves from certainty to uncertainty and contradiction and then on to certainty again. Limited kinds of understanding are progressively incorporated into a whole. The first kind is "sense-certainty": immediate sensuous experience of the here and now. This becomes what Hegel calls perception, then understanding. This is followed by self-consciousness and then consciousness of others. Next comes consciousness of society as an objective reality, and finally consciousnesss of how society is produced through human activity. Natural consciousness passes through this series of stages or phases, of natural skepticism, doubt, and despair, and finally becomes self-critical consciousness. First, we take things to be just the way they appear to be. Then, we come to experience a distinction between things as they appear and things as they are. We eventually become conscious of the way our own consciousness has been shaped by our biography and by our own society – we come to see society as an objective reality. Then we become conscious of the way society itself is a product of human activity. And finally we become aware of ourselves as a manifestation of something grander and know that individual consciousness is not self-sufficient or complete. "For Hegel, the highest form of knowledge turns out to be self-knowledge, or knowing oneself in otherness and otherness as oneself" (Rockmore, 1997, p. 188): "Beginning with the natural consciousness of the everyday life world in which we already find ourselves, phenomenological reflection traces its own genesis through the successive stages of the manifestation of consciousness" (McCarthy, 1978, p. 79).

Knowing is first "in-itself," then "for-itself," and finally "in-and-for-itself." An object is first (for sense-certainty) mere being, then (for perception) a concrete thing, then (for understanding) a force – always seemingly in-itself. Then, with self-consciousness, this in-itself turns out to be a mode in which the object is for me: the "I" is a connecting of the object's in-itself and for-me. That is to say, the appearance/reality distinction presumes an "I" to and for whom reality appears. Self-consciousness has a double object.

In Hegel's view, there is both a direction, a teleology, to this process and an end to it. Knowing is a "dialectical movement which consciousness exercises on itself, both on its knowledge and on its object" (Hegel, 1807/ 1967, p. 55). Knowing is not a single event but a process extended over time. Hegel was "an epistemological optimist" (Rockmore, 1997). He saw consciousness developing from a state of immediacy toward a knowing that is aware of itself and finally to a knowledge that is "absolute." Hegel maintained that the dialectic would proceed to a point where "the partiality of perspectives can be progressively overcome" (Held, 1980, p. 177). In Hegel's account, this "absolute knowledge" is the final working out, the final development, of *Geist*.

As Hegel viewed it, this dialectic is both the way history unfolds *and* the process of individual thinking. It is both because these two – history and thought – are not distinct. Remember that both the human mind and Geist itself are found *in* nature and *in* history. Hegel called the "governing principle" of thought "determinate negation." It is a "continuous criticism and reconstruction of the knowledge of subject and object as their relation to one another" (Held, 1980, p. 176). It "consists precisely in surmounting old forms of consciousness and in incorporating these moments into a new reflective attitude" (p. 176). Understood this way, Hegel's phenomenology itself is an exercise in thinking: it is a critical reflection that explores the conditions of *its own* possibility – the historical and cultural process by which it has come about. It is reasoning that doesn't take itself for granted, reflection that asks how reflection can be possible. This is the approach, the method, necessary to trace the development of *Geist*.

A New Model of Human Being

At the heart of Hegel's *Phenomenology* is a powerful historical narrative that weaves together cultural history and individual development. Darwin would not publish *On the Origin of Species* for another 50 years, but today we can add evolution to a picture in which humans have evolved from simpler lifeforms that developed from insensate matter. We are substance that became first self-reproducing, then sentient, then conscious, then self-conscious, then conscious of the concrete conditions of its own consciousness. Hegel imagined this evolutionary journey as ultimately culminating in a consciousness that can know this process of its own formation and self-formation and overcome the apparent distinction between itself as subject and the world as object by *transforming* the world to *make* it rational.

ONTOLOGICAL HERMENEUTICS: MARTIN HEIDEGGER

World is not something subsequent which we calculate as a result from the sum of all beings. The world comes not afterward but beforehand, in the strict sense of the word. Beforehand: that which is unveiled and understood already in advance in every existent Dasein before any apprehending of this or that being. . . . We are able to come up against intraworldly beings solely because, as existing beings, we are always already in a world. (Heidegger, 1975/1982, p. 165)

Hegel's grand system was not the final word. The philosopher Martin Heidegger (1889–1976) objected to what he called Hegel's "onto-theo-ego-logy" (Heidegger, 1980/1988): his treatment of time as basically spatial. This might seem a strange thing to say, given Hegel's emphasis on history. But Heidegger's point was that the historical movement of Hegel's phenomenology comes to an end in timelessness, in the totality of a final system in which no change will be necessary so none will be possible. "Hegelian time lacks what is truly proper to time: contingency, freedom, exposure to the future" (Caputo, 1987, p. 18). In Hegel's account: "[t]he eternal logical structure of Geist is always the same. Appreciating the ceaseless activity of Geist is essential for understanding history, the rise and fall of political and social institutions, the development of the stages of consciousness. However, from the perspective of logic, of Geist as *Nous* or Reason, Geist displays an eternal, necessary, rational structure" (Bernstein, 1971, p. 22).

Heidegger set out to "appropriate" and "radicalize" Hegel (Heidegger, 1975/1982, p. 178). He argued that both philosophy and science have forgotten the *world* in which we live. This sounds like Husserl and Schutz, but Heidegger considered this world to be where human beings *are* rather than something around us. For Heidegger, the world is the "ground" for all the entities – whether people or objects – encountered within it. Heidegger set out to clarify what it is to be human on the basis of this insight. Human being is not a mind or a self but "being-in-the-world," a unitary structure of our complete involvement in the totality of a form of life.

Being Is an Issue for Human Beings

Heidegger began with the observation that it is only for humans that "being is an issue." Only people ask the questions, "What is that?" and "Who am I?" It is somehow fundamental to human being – to the human way of being – that we try to understand (*verstehen*) and interpret the kinds of entities that we deal with every day. It is often said that with Heidegger hermeneutics became ontological. That is to say, he proposed that interpretation is not simply a special way of dealing with texts but something intrinsically human. To be human *is* to understand and interpret, so interpretation is not a special method but a fundamental aspect of human being. Understanding is a matter of grasping an entity as a certain kind of being and at the same time to have a grasp of what it is to be human. (We saw in Chapter 4 how Gadamer, a student of Heidegger, drew on this idea that interpretation is grounded in understanding.)

Heidegger seems to have been a thoroughly unpleasant person. He betrayed his mentor, Edmund Husserl, breaking off contact when Husserl was excluded from the university by the Nazi Party, and he betrayed his wife by having an affair with his student Hannah Arendt. He not only sympathized with the Nazi regime but also refused to repudiate either the regime or his own actions after the Second World War. This raises the question of whether a person's work, whether it be philosophy or any other activity, should be judged in terms of how they live. In Heidegger's case, the answer is surely yes. Heidegger's philosophy was a philosophy of existence – it was precisely a philosophical exploration of how to live. When its author failed so conspicuously, we must consider his philosophy with critical care.

Yet Heidegger was attempting something interesting and difficult, rethinking one of the central questions of philosophy. That he failed should perhaps not cause surprise, though certainly regret. He proposed that philosophy had consistently misunderstood what it is for something to *be*. It had focused on *beings* – individual entities – instead of *being*, just assuming that being has only two possibilities, "matter" and "mind." Heidegger proposed instead that actually there are many different *ways* for both people and things to be, ways that are made possible by history and culture.

A Phenomenology Focused on Ontology

Heidegger's conceptions of phenomenology and of the constitutive relationship in human being were very different from those of his teacher Husserl. In *Being and Time* (1927), Heidegger raised what he called "the question of the meaning of being." This sounds like some kind of existentialist question, but for Heidegger it meant: What makes being possible? What makes it possible for a thing – or a person – to be? Heidegger's answer was that things and people become what they are only against a ground, a taken-for-granted background, of cultural and historical practices. For Heidegger, a phenomenological analysis means the investigation of what underlies all particular entities and allows them to show up *as* entities.

Heidegger drew a distinction between existence and being. He was a realist: he didn't believe that if all humans died the universe would stop existing. But when he insisted that being is an issue only to humans, he meant that if there were no humans around, entities would have no being. The being of an entity is made *possible* by the human practices in which it circulates. A dollar, for instance, is constituted by specific economic practices that occur only in certain societies and developed at a particular historical juncture. Outside such contexts, no piece of paper with printing on it would *be* a dollar. This is an ontological claim, not an epistemological claim. We may in addition *know* things about this dollar and say things about it. But these beliefs and assertions are not what make it a dollar: an individual may know nothing about it, yet it is still what it is.

There is no way to grasp what something *is* outside of a human context. If all humans were to die, the cup in front of me would still exist, but it wouldn't *be* anything. It wouldn't be a cup because being a cup is a matter of involvement in practices like drinking, and with no humans there would be no such practices. And it wouldn't even be a piece of "matter" because being matter is also based on involvement in the practices of a culture of scientists. We have learned from Kuhn that the understanding of matter changed dramatically when the paradigm of Newtonian physics was replaced by the paradigm of Einsteinian physics. As Kuhn pointed out, different scientific paradigms understand differently the being of the entities they deal with. It is tempting to think that there is a neutral description of things outside particular cultural practices, perhaps in terms of atoms, quarks, or some fundamental particles. But this doesn't make sense; there is no "view from nowhere" because "being" is what is an issue for humans. Humans care about what something is. No humans, no concern. No concern, no way to be.

Clearly this is not idealism, either transcendental or naïve. It is not the view that the world that we take to be real is "actually" just ideas in our minds. Heidegger's view was that what is real is what our public cultural practices define as real. Each culture defines specific ways to be; for example, in U.S. culture there is "a market" and "commodities," and "consumers" and "voters." So it is clear that, far from avoiding or bracketing all ontological claims, Heidegger's phenomenology *focused* on ontological matters and undertook an ontological analysis of them.

For Heidegger, the grasp humans have of the entities around us (and of ourselves) comes not from contemplation and intellectual conceptualizations, as Kant and Husserl thought, but from practical activity. For Husserl, the slogan "to the things themselves!" meant adopting the disinterested attitude of transcendental subjectivity. For Heidegger, it meant "pick up the cup!" It is in our everyday practical activity that we have the most direct access to things, and understand what they are. We experience the world not by thinking about it but in practical engagement, in concrete activities such as hammering. Human beings are *in* the world in the sense not of spatial inclusion but of practical involvement. We are involved; we care. For this reason, Heidegger said that human being – the human way of being – is *Dasein* (German, literally "being there"). Dasein is "being-in-the-world," fundamentally one with a world defined by public practices. Heidegger offered "an understanding of the agent as engaged, as embedded in a culture, a form of life, a 'world' of involvements, ultimately to understand the agent as embodied" (Taylor, 1993, p. 318).

Heidegger insisted that "adequate treatment of the ontology of Dasein is the presupposition for posing the problem whose solution Kant takes as his task" (Heidegger, 1975/1982, p. 56). To understand how humans can know the world, we need first to examine our "basic constitution" (p. 59). Understanding begins with practical activity in the world. When we stand back and contemplate with detachment and objectivity, the result is a distorted view. Know-how, practical coping, is a concrete grasping in which things are what they are:

> In *Being and Time*, *Verstehen* [understanding] is precisely that knowledge which informs Dasein's most concrete involvement with the world. Dasein knows what it is about without having explicit conceptual knowledge to fall back upon. *Verstehen* is the capacity to understand what is demanded by the situation in which Dasein finds itself, a concrete knowledge which gets worked out in the process of existence itself. It is the grasp which Dasein has of its own affairs but which cannot be reduced to formalized knowledge and rendered explicit in terms of rules. (Caputo, 1987, p. 109)

Modes of Engagement

Heidegger offered an important analysis of understanding and interpretation (Table 8.1). He proposed that understanding is always situated in place and time: it has the quality that Heidegger called "thrown-projection," with three aspects. First, humans understand the entities they deal with, and themselves, in terms of a "project," a tacit practical task or undertaking. Second, understanding always involves projection on a context: the background cultural practices that provide what Heidegger called "the meaning of being." And, third, each of us is thrown into a world we did not create or choose. This existential structure of "thrown-projection" shows that *time* is central to being human.

Interpretation develops from this situated understanding. Heidegger distinguished three "modes of engagement." The first is the understanding we obtain in a practical activity, such as hammering. When this activity is going smoothly, when it is routine, we are absorbed in what we are doing, not at all reflective about our activity – we "lose ourselves" in it. If we are using the hammer to build a fence, for example, the tool will be transparent and we will be aware only of our effort to drive in a nail, to get a board in place, or even, if all this is going smoothly, simply to get the fence finished. If we are involved in a routine everyday conversation (buying a cup of coffee, perhaps), then the words, the turns and moves of the dialogue, are transparent and we will be aware only of the aim of the conversation: getting our coffee. In this first mode, Heidegger says that entities are "ready-to-hand" for us. In smooth activity, the world is an invisible background to what we are doing, taken for granted and unnoticed. Our understanding is tacit and unreflective, as much a matter of emotion (which Heidegger viewed as an aspect of being-in-the-world) as of thinking. In this mode, we encounter not objects but tools and equipment that have practical relevance for our projects.

But humans do, of course, have reflective and explicit ways of knowing the world and knowing themselves. Understanding can be "developed" as

TABLE 8.1. *The Relationship between Understanding and Interpretation*

Mode of Engagement	Kind of Knowledge	What Shows Up	Our Attitude
Ready-to-Hand Participation	Understanding	Tools are transparent.	Concerned with a project
	Practical, tacit, and unreflective	The setting is an invisible background.	Situated in space and time
	Thrown-projection	We lose ourselves in our activity.	Absorbed in routine activity
Unready-to-Hand Breakdown	Interpretation	An aspect of the tool stands out. (*The hammer is too heavy.*)	Deliberation: working out what to do
	The working out of possibilities projected in understanding	The setting becomes lit up.	Circumspection: looking around
	A *true* interpretation points out an aspect of the setting relevant to the project.	The project is evident.	Reflection: noticing one's project
Present-at-Hand Detachment	Assertions about properties of objects	Objects seem to have isolable properties. (*The hammer weighs 10 grams.*)	Disinterested contemplation
		The setting goes unnoticed.	Detached observation
		People and objects seem to have distinct kinds of being.	Unconcerned reflection

interpretation. Interpretation, according to Heidegger, is "the working-out of possibilities projected in understanding" (Heidegger, 1927/1962, p. 189). Interpretation is an explication, a making thematic, of what has been understood in practice. Activity never goes completely smoothly; there are always repairs to be made, in human conversations just as much as with tools. When there is a *breakdown* (or when something is missing, when there is a hitch of some kind, or when we make a mistake), various aspects of the world–person–tool relationship become apparent. The broken tool now is noticed, and an aspect of it now stands out. The marker for the whiteboard is

TABLE 8.2. *The Fore-Structure of Interpretation*

Fore-having	Fore-sight	Fore-grasp
A grasp of the whole situation, the totality of involvements	A point of view or perspective from which one has an initial grasp, a sense of a particular thing one is dealing with	An interpretive framework; an articulated system of concepts

"dried out"; the hammer is "too heavy"; the book we wanted to buy is "not cheap enough"; the lecture we are listening to is "too long." In each case, the aspect that stands out depends on the context; it is defined by the activity or project we are engaged in. The hammer is too heavy for *this* particular nailing task; the lecture is too long for *this* sunny day in wintry Ann Arbor. In this second mode of engagement, entities become "unready-to-hand." *What they are* becomes apparent. That is to say, their *being* is evident.

When there is a breakdown, we look around, surveying our circumstances, noticing the project or course of action we are engaged in, in order to start to work out alternatives and begin repair. Heidegger called this looking around "circumspection." He called noticing one's project "reflection"; working out alternatives is "deliberation." The way the tool was grasped in practice now becomes evident as one possibility among many. The "equipmental totality" in which we are operating, and that has provided an invisible background for our activity, now becomes apparent. And the setting is now lit up, as we become aware of other tools that may be helpful.

Occasions of breakdown involve a shift from the first to the second mode of engagement with things and people, a shift from "participation" to "circumspection." The relationship between those two modes can be seen as a hermeneutic circle: the way tools were grasped and understood in practice is now articulated and interpreted.

This means that interpretation is never free from presuppositions. It is never a detached, objective, or neutral observation of an object, event, or text. Interpretation is always based on what Heidegger called a "fore-structure" of interests and tacit assumptions, a fore-having, fore-sight, and fore-grasp (Table 8.2). I have already cited Heidegger's criticism of those interpreters who claim to have no preconceptions and to report only what "emerges" from the text: "If, when one is engaged in a particular concrete kind of interpretation, in the sense of exact textual Interpretation, one likes to appeal to what 'stands there,' then one finds that what 'stands there' in the first instance is nothing other than the obvious undiscussed assumption of the person who does the interpreting" (Heidegger, 1927/1962, p. 192).

A *true* interpretation is one that uncovers and points out some aspect of the current situation that has relevance to the practical task at hand. The claim that "this hammer is too heavy" can be perfectly true, though of course it will be a *local* truth, relevant only to a specific situation. Heidegger argues that *all* truth claims are of this kind. Truth cannot be viewed as a correspondence between a mental representation and a material object. Heidegger proposes instead that truth be conceived as "uncovering."

A third mode of engagement is possible, one of detached contemplation. In this attitude, entities *seem* to be self-sufficient objects with specific, independent properties. We seem to be completely separate from objects like the hammer and to be a completely different *kind* of being. It seems that we know objects only by forming mental representations of them. But it is only in this mode that the apparently distinct realms of "the mental" and "the material" appear. This estranged kind of nonrelationship between subject and object can arise only on the basis of the more fundamental understanding characteristic of practical involvement.

Heidegger argued that the principal error made by philosophers since the ancient Greeks – including Descartes, Kant, and Husserl – had been to give priority to this third mode of engagement when in fact it is "privative" and derived from the two others. Descartes's efforts to "rid myself of all opinions which I had formerly accepted, and commence to build anew from the foundation" (Descartes, 1637, 1641/2003, p. 66) and Husserl's efforts to adopt a transcendental attitude avoiding everyday involvement and thus gaining access to "the things themselves" both illustrate this mistake. This way of knowing has been taken to be "objective," but Heidegger argued that this is an illusion. Contemplation always takes for granted the cultural and historical practices that define both objects and the person contemplating them.

A Basic Relationality

Heidegger proposed that the cognitive processes that Kant described are constituted in and by this more fundamental level of human being, our engagement in and relatedness to the world. Kant's reconstruction of knowledge and ethics took this practical involvement for granted:

> When Kant talks about a relation of the thing to the cognitive faculty it now turns out that this way of speaking and the kind of inquiry that arises from it are full of confusion. The thing does not relate to a cognitive faculty interior to the subject; instead, the cognitive faculty itself and with it this subject are structured intentionally in their ontological constitution. (Heidegger, 1975/1982, p. 66)

Kant had ignored the fundamental involvement of humans in the world, involvement that is practical, emotional, and concerned. Intentionality – the way perception is always a relationship to something in the world – is fundamental to being human. Kant cheated: he "has to make use" of this basic relationality in his analysis of perception and knowledge "without expressly recognizing it as such" (p. 67).

Heidegger reminds us that humans are involved and caring. We are not detached observers of the world but are always embedded in a specific cultural and historical setting, and our understanding of ourselves and the entities we encounter is grounded in our practical activity in this setting. We have here an ontology that emphasizes a "contextualized" relationship between subject and object: both people and the various kinds of objects they deal with are always situated in a world that provides a background against which they can stand out. In *Being and Time*, Heidegger generally treated language as a tool, something ready-to-hand. But, in addition to this instrumental treatment of language, he also at times used a "constitutive" view of language as "not so much a tool on hand for our use as a medium in which man dwells" (Guignon, 1983, p. 118). "On the constitutive view, language generates and first makes possible our full-blown sense of the world" (p. 118). Heidegger would develop this notion of the constitutive power of language in his later writing. Ultimately, in Heidegger's analysis, we will come to understand that we have no fixed nature, that we will die, and that in this sense we are "homeless" on this earth. Facing up to this existential challenge and finding the resoluteness to go on is what Heidegger viewed as coming to have an authentic relation to oneself. It is here, in his analysis of what he considered authentic existence, that we find the most troubling aspects of his philosophy. Here, he claimed, each human being must face the future as a matter of fate or destiny, live each moment with resolve, and seek to retrieve and hand down its heritage. But even if we do not accept Heidegger's conclusions about how we ought to live, his analysis moved along the path toward a new nondualistic way of thinking about *constitution*.

A PHENOMENOLOGY OF EMBODIMENT: MAURICE MERLEAU-PONTY

> Our own body is in the world as the heart is in the organism: it keeps the visible spectacle constantly alive, it breathes life into it and sustains it inwardly, and with it forms a system. (Merleau-Ponty, 1945/1962, p. 203)

The next step on this path was taken in the philosophical writing of Maurice Merleau-Ponty (1908–1961), who also focused attention on what Kant forgot. Merleau-Ponty noted how "Kant's conclusion . . . was that I am a consciousness which embraces and constitutes the world, and this reflective action caused him to overlook the phenomenon of the body and that of the thing" (Merleau-Ponty, 1945/1962, p. 303). Merleau-Ponty focused on the embodied character of human action, perception, and knowledge. For most philosophers and social scientists, the body has been irrelevant (Fraser & Greco, 2005,

p. 1). After all, the body is stuff, matter, and surely what is important to explore is mind? But Merleau-Ponty emphasized that our material embodiment makes us *one with* the world. He proposed that conceptual representation and thought are ways of *perceiving*, and perception is a way of *being*. Like Husserl, Merleau-Ponty conceived of phenomenology as an effort to study a level of experience of the world that is *prior* to that of explicit knowledge. And, like Heidegger, he viewed this level as that of practical activity, of an embodied subjectivity, the "body-subject." By showing the dialectical relationship between the body-subject and the world, his phenomenology avoided the dualism of subjectivity and objectivity. And he offered a new conception of rationality; he argued that reason and meaning exist not in the head but in the world.

Forms of Behavior

Merleau-Ponty started to explore "the relations of consciousness and nature" in *The Structure of Behaviour* (Merleau-Ponty, 1942/1963, p. 3). Kant and his followers, as we have seen, considered the objective world as a mental construction: "an objective unity constituted vis-á-vis consciousness" (p. 4). Scientists tend to view consciousness as a natural phenomenon and look for its causes and effects. Merleau-Ponty's phenomenology provided an approach to the problem that was "underneath" both positions (which he referred to as "intellectualism" and "objectivism"), allowing him to inspect their foundations. Like Hegel, he started "from below" (p. 4) by looking at contemporary research in psychology and physiology and showing that its findings contradict its implicit ontology. He began with the notion of behavior, "neither thing nor consciousness" (p. 127), which takes place within a natural world yet in some sense emerges from an organism. Merleau-Ponty distinguished three fundamental organizations or "forms" of behavior – the "syncretic," the "amovable," and the "symbolic" (p. 93). These are increasingly sophisticated in their capacity to generalize and transform the concrete situation into a *typical* situation (p. 125). Syncretic behavior is "imprisoned in the framework of its natural conditions" (p. 104). A toad will persist in its efforts to grab at a worm placed behind glass. At the amovable level, we see the emergence of *signals*: a chicken can learn simple distinctions, such as between dark and light corn. But the symbolic structures of behavior show flexibility and a "multiplicity of perspectives" (p. 122) that are absent from animal behavior. A chimpanzee "manifests a sort of adherence to the here and now, a short and heavy manner of existing" (p. 126), but symbolic behavior is able to incorporate and restructure the simpler structures of behavior. This is a "third dialectic" (p. 184), in which, again following Hegel, Merleau-Ponty proposed that the freedom to change perspectives gives a new dimension to the structure of behavior and makes possible a new "existential order." Culture emerges in the temporal gap between stimulus and response, and language transcends concrete facts. The human subject, *conscious* of nature, is the product of a dialectic that is *part* of nature.

Merleau-Ponty's radical conclusion was that consciousness is not something intellectual but is practical and perceptual. He proposed that "The mental, we have said, is reducible to the structure of behavior" (Merleau-Ponty, 1942/1963, p. 221). Human action contains an intentionality prior to representation and a kind of understanding prior to cognition. We need to "define transcendental philosophy anew" because it turns out that the meaning that "springs forth" in things "is not yet a Kantian object; the intentional life which constitutes them is not yet a representation; and the 'comprehension' which gives access to them is not yet an intellection" (p. 224). Kant had reduced all our connection with the world to an intellectual, conceptual contact. He had appealed to a kind of reflection in which the thinking subject discovers that they are free. Merleau-Ponty insisted that this consciousness of self "is not given by right" but requires "elucidation" of one's "concrete being" (p. 223). Kant had failed to penetrate to the profound truths of our embodied existence. His philosophy had claimed to "lay bare only what was implicit" but could it not better be said that it had merely entered "as into a lucid dream, not because it has clarified the existence of things and its own existence, but because it lives at the surface of itself and on the envelope of things?" (p. 223).

Slackening the Threads

In *Phenomenology of Perception* (1945/1962), Merleau-Ponty explored this uniquely human kind of organization of behavior in more detail, in a dialogue with rationalism and empiricism and especially with Husserl. He argued – on the basis of detailed descriptions of everyday experience – that rationalists are wrong to maintain that we construct the world in thought. But the empiricist is equally wrong to believe that our knowledge of the world is simply a product of the data of our senses. Both approaches detach the conscious subject from the world. We have seen how Husserl, bracketing the natural attitude, retained the world only as *thought*. Merleau-Ponty tried instead to practice a phenomenology that "slackens the intentional threads which attach us to the world" rather than undoing them entirely and that "reveals the world as strange and paradoxical" (p. xii). The metaphor doesn't seem entirely apt – we are not "attached" to the world; we are in it and of it. But certainly for Merleau-Ponty, phenomenology is a matter of learning to look closely at one's own existence within the world, and phenomenological analysis shows that we both create and are created by the world. He insisted that we "need to reawaken our experience of the world as it appears to us in so far as we are in the world through our body, and in so far as we perceive the world with our body"

(p. 206). By disrupting our everyday absorption in the world, we find that the world is not something that one *thinks* but something one "lives through." Perception is not so much an act of consciousness as an act of the whole body, the living body. For the human body, the world is a system of possibilities, a ground on which are *constituted* all forms of human knowing. Each of us *is* an "opening into the world" in which our perception is both general and anonymous, grasped by the "habitual body." Things are what we can get a grip on, but our grip stems from the fact that our body, too, is a thing of the world. We can only grasp the world from within it. At the same time, the world always precedes, outlives, and in the end transcends every attempt on the part of human analysis to grasp and understand it fully.

For Merleau-Ponty, perception is the "primordial matrix" for the every-day world and also for science and philosophy. He argued that perception is a modality that is neither empirical nor rational. What we perceive is neither simply "present" nor "inferred" but the result of our body's "polarization" of the world, the "correlate" of our body and its sensory systems. For example, the characteristics of time and space, which we normally assume are in the world itself and Kant argued are in the mind, emerge, Merleau-Ponty pro-posed, from our ways of existing in the world. We are certain that an object has a side that is hidden from us, and this hidden side is given in its own way, without being either directly present to the senses or inferred logically. An object "is given as the infinite sum of an indefinite series of perspectival views in each of which the object is given but in none of which is it given exhaustively" (Merleau-Ponty, 1964, p. 15). Perception is perspectival, open, and indeterminate: as we move, fresh perspectives open up and objects disclose themselves in new ways. We are an opening to the world, but each object, too, is in its own way both an opening and a way of hiding.

Reflection and cognition are possibilities for this human way of being-in-the-world. Thought is a taking up of what has been seen. Cognition never replaces perception; the two always work together. Ideas flow from a *subli-mation* of perception, and all cognitive operations presuppose the body's motion and its capacities for gesture and language. Cognition depends on the body (Merleau-Ponty, 1945/1962, p. 127). Thought is grounded in pre-reflective activity and dependent on symbolic behavior. It both preserves and transforms perception, "distilling" its sense while reconstituting its "substance." Cognition seeks to articulate the world thematically in linguistic structures but leaves much behind, especially our opaque and indeterminate bond with the world. The thinker "fixes" and "objectifies" life, but a part of existence always escapes. Whereas traditional philosophy insisted that per-ception is fallible and thinking indubitable, Merleau-Ponty argued that the truths of thought are always dependent on the ways in which the real is evident in perception. Propositional truths are always based on situational truths (Mallin, 1979, p. 199). Our knowledge is always contingent, but this stems from the uncertainty and finitude of life and of the world itself and doesn't mean that we experience merely "appearances."

Visible and Invisible Intertwined

The title of *The Visible and the Invisible* (Merleau-Ponty, 1964/1968) refers to the way an object of perception is given both in the senses, as a partially grasped *particular*, and in the invisible realm of concepts, as an abstract *universal*. In this unfinished book, Merleau-Ponty explored how humans are "inherent" in the world in a way that cannot be reduced to essences or categories. He wanted to find a dimension that "offers us, all at once, pell-mell, both subject and object – both existence and essence – and, hence, gives philosophy resources to redefine them" (p. 130). He struggled to find a way to write about human being as a part of the world, as the "flesh of the world," in a language that would completely break free from the subject–object dichotomy. The "flesh" is the element in which both my body and things themselves are given – an "element of being" like earth, air, fire, and water. To perceive is to be drawn into the tissue of being. When I touch something, my hand itself is touched – sensible things do not exist *within* space and time but *organize* space and time in a "dimensional sensuality." An object is a field of forces, unified by a particular style. The recognition of its style, along with its variations, is the recognition of the universal in the particular.

Both rationalism and empiricism treat the world as completely opaque and consciousness as completely transparent, but Merleau-Ponty insisted that neither is the case. Perception, like existence, is a dialectical process in which a single existential fabric underlies both subject and object so that they are mutually complicit. In perceiving an object, we orient our bodies in the world, assuming a position before the tasks of the world. Body and world are "intertwined"; my body is "folded into" the sensible object. The world and people share a fundamental corporeality: "[T]he thickness of flesh between the seer and the thing is constitutive for the thing of its visibility as for the seer of his corporeity; it is not an obstacle between them, it is their means of communication" (Merleau-Ponty, 1964/1968, p. 178).

This same intertwining holds between the visible and the invisible, the seen and the thought, the sensible and the ideal, the concrete and the abstract. The body is touching and touched, but not at the same instant. There is a difference between the body and itself that offers an "infrastructure" for thought. The invisible – the thought, the conceptual – is not in some separate realm; it is the invisible *of* and *in* the visible. Thought is a transformation of perception, "an ideality that is not alien to the flesh, that gives it its axis, its depth, its dimensions" (p. 152). Perception is primary, but the degree and manner of our openness to perceptual contact can be altered. Thinking and seeing are mutually transforming. Merleau-Ponty proposed that the thinking and seeing body-subject is where being becomes visible to itself.

Language plays a special role in this transformation of perception that is thinking. Merleau-Ponty rejected the conduit metaphor, the idea that language is an "envelope or clothing of thought" (Merleau-Ponty, 1945/1962, p. 211). Language is not thought's clothing but its body; language accomplishes or completes thought. "The spoken word is a gesture, and its meaning, a world" (p. 184). This is why we sometimes struggle to find the right words. Like Saussure, Merleau-Ponty viewed language as an abstraction from the primacy of speech, but he did not accept Saussure's notion that language is a system of arbitrary conventions. "The spoken word is a genuine gesture, and it contains its meaning in the same way as a gesture contains its [meaning]. What I communicate with primarily is not 'representations' of thought, but a speaking subject, with a certain style of being and with the 'world' at which he directs his aim" (p. 213). Language is a public cultural system that can level individuality to the impersonal "one." And language reverses sublimation to provide perception with new structures that organize our dealings with the world: "Silent vision falls into speech, and in return, speech opens a field of the nameable and sayable . . . , it metamorphizes the structures of the visible world and makes itself a gaze of the mind" (Merleau-Ponty, 1964/1968, p. 178).

The Flesh of the World

Like Heidegger, Merleau-Ponty rejected the assumption that the knowing subject is the center of knowledge or existence, and tried to create a language to communicate a fresh understanding of the mutual constitution of subject and object within what he called "the flesh." His focus on embodied activity drew attention to the materiality of the conscious subject and the corporeality of objects and the world. We are *of* the world, not in some separate ontological realm. For Merleau-Ponty, as was the case with Heidegger, both subject and object *emerge* from a more primordial way of being in which the distinction between them does not yet exist. There is an "intentional life . . . which is not yet a representation" and a form of comprehension "which . . . is not yet an intellection" (Merleau-Ponty, 1942/1963, p. 224). My body has an intelligence and intentionality that does not require deliberate thought and decision: "In so far as I have hands, feet, a body, I sustain around me intentions which are not dependent upon my decisions and which affect my surroundings in a way which I do not choose" (Merleau-Ponty, 1945/1962, p. 440).

This "constitution," this bodily know-how, is both used and ignored by science. Kant took it for granted and then ignored it: "[T]he numerical specifications of science retrace the outline of a constitution of the world which is already realized before shape and size come into being. Kant takes the results of this pre-scientific experience for granted, and is enabled to ignore them only because he makes use of them" (Merleau-Ponty, 1945/1962, pp. 301–302). It follows that the task for investigation – and for Merleau-Ponty this would be a phenomenology – is the study of this neglected constitution.

ETHNOMETHODOLOGY: HAROLD GARFINKEL

> EM is concerned with "What More," in the world of familiar, ordinary activities, does immortal, ordinary society consist of as the locus and the setting of every topic of order, every topic of logic, of meaning, of method respecified and respecifiable as the most ordinary Durkheimian things in the world. (Garfinkel, 1996, p. 6)

Harold Garfinkel (b. 1917), Professor Emeritus of sociology at the University of California, Los Angeles, is responsible for another exploration of constitution, a form of sociology he named "ethnomethodology" (Garfinkel, 1967, 1996, 2002). The focus of this "eccentric, original phenomenology" (Manning, 2004, p. 279) is the ongoing work of social interaction in which people create and re-create social order. Ethnomethodology is not a method of inquiry; rather, the "ethnomethods" are the *topic* of inquiry. It is the study (*logos*) of the methods used by folks (*ethnos*) in their commonsense everyday activity. Garfinkel was dissatisfied with the tendency in sociology to view people as merely acting out predetermined social roles. Traditional sociology takes the member of society "to be a judgmental dope of a cultural and/or psychological sort" or a "'cultural dope'" (Garfinkel, 1964, p. 244) whose behavior is determined by preexisting norms or motivations, by the "stable structures" of "culture," "society," or "personality." Such approaches fail to ask of the people "What is *their* game?" in the sense of Wittgenstein's language games.

In contrast, ethnomethodology sees human activity as skilled, intelligent, and improvisatory. Like good jazz, social action is artfully made up on the spot from available resources rather than following prescribed rules. Garfinkel proposed that "persons discover, create, and sustain" the orderly character of society. Society is not an objective structure standing behind this activity but a *product* of "members'" skilled activity. Garfinkel said he wanted to solve the problem of the "moral order" of society, which "For Kant ... was an awesome mystery" (Garfinkel, 1964, p. 225). "A society's members encounter and know the moral order as perceivedly normal courses of action – familiar scenes of everyday affairs, the world of daily life known in common with others and with others taken for granted" (p. 225). This commonsense world is the topic of sociology, yet sociologists rarely ask "how any such common sense world is possible." Its existence is either taken for granted or settled by theoretical mandate (see Box 8.2 on pp. 169–170).

Garfinkel called for the "rediscovery" of this moral order. His central argument was that "a concern for the nature, production, and recognition of reasonable, realistic, and analyzable actions is not the monopoly of philosophers and professional sociologists" (Garfinkel, 1964, p. 250); members of a society are equally concerned with making recognizable social order. The task for the researcher is to treat as problematic "the actual methods whereby members of a society, doing sociology, lay or professional, make the social structures of everyday activities observable" (p. 250). These methods have been a *resource* for sociology; now they must become a *topic*. Garfinkel proposed that these methods are found not in the individual mind but in social practice. We see order whenever we look at traffic on the freeway, a jazz quartet, the science laboratory, or ordinary conversation. Rather than searching for its underlying causes (or motivations) or overlying concepts (or functions), we can and should study just what people do to create this order. The aim of ethnomethodology is to examine, discover, and describe this work and the methods used. Hidden causes and abstract functions are hypothetical and unobservable; more important, they are irrelevant to the practitioners themselves. They are part of the game of worldwide science, not the game(s) of everyday life. Ethnomethodology avoids appealing to hidden factors and instead conducts careful and detailed study of the methods and practices that provide "the routine grounds of everyday life."

Garfinkel's work has been called "as revolutionary as the work of Darwin, Einstein or Crick and Watson. It has fundamentally changed the way that sociologists think about their discipline and about the way that they do their research" (Dingwall, 1988). But ethnomethodology has often been misunderstood. It has been accused of being "sociology without society" (Mayrl, 1973), a "microsociology" that fails to pay attention to the larger structures that make up a society, a method without substance, and as lacking all methodology. It has been accused of being conservative in its lack of attention to power structures, liberal in its focus on individual agency, and positivist in its attention to empirical detail. It has been characterized as inherently subjective and as lacking attention to experience. Even its supporters have misunderstood it, describing it, for example, as aiming "to elucidate the arena of commonsense experience and to 'understand' life-world situations as perceived by concrete social actors or participants" (Dallmayr & McCarthy, 1977, p. 222). It is true that Garfinkel's 1964 paper was couched in terms of the beliefs, expectations, and attitudes of an individual actor – the sense that an actor makes. But since then he has made it clear that the emphasis in ethnomethodology is not at all on how things are *perceived* but how they are *produced and accomplished*. Garfinkel "inverted the phenomenological primacy accorded to subjective experience in favor of studying public activities and common practices through which members achieve the apparent reality of those objects" (Maynard, 1986, p. 348). Ethnomethodology seeks "to treat practical activities, practical circumstances, and practical sociological reasoning as topics of empirical study, and by paying to the most commonplace activities of daily life the attention usually accorded extraordinary events, seeks to learn about them as phenomena in their own right" (Garfinkel, 1967, p. 240).

The basic premise is that this practical reasoning cannot "remain the unexamined medium of one's discourse" (Sharrock, 2004) but must be studied. Like Schutz, Garfinkel has been interested in the mundane reality that Husserl believed should be bracketed. But like Heidegger and Merleau-Ponty, Garfinkel emphasizes that this mundane reality is created in public practices, not in mental activity.

Society as a Product of Members' Activity

The fundamental phenomenon that ethnomethodology aims to study, Garfinkel insists, is exactly what sociology has always set out to study, namely "the objective reality of social facts." But this "fundamental phenomenon" of sociology must be seen not as given or natural but as a *"practical achievement,"* the result of "members' work":

> For ethnomethodology the objective reality of social facts, in that and just how it is every society's locally, endogenously produced, naturally organized, reflexively accountable, ongoing, practical achievement, being everywhere, always, only, exactly and entirely, members' work, with no time out, and with no possibility of evasion, hiding out, passing, postponement, or buy-outs, is thereby sociology's fundamental phenomenon. (Garfinkel, 1988, p. 103)

The objective reality of everyday life is a matter not of shared knowledge but of a "background texture of expectancies," the "expectancies of everyday life as a morality" that is first of all the result of *practical* enterprise: "[E]veryday social life, he tells us, and social life on extraordinary days as well, is a practical enterprise and every man is a practitioner" (Swanson, 1968, p. 122). To understand social reality, then, what is needed is not formal analysis but a focus on the details of everyday practices, for "[t]he witnessably recurrent details of ordinary everyday practices constitute their own reality" (Garfinkel, 1996, p. 8).

Social facts have an objective reality that is achieved, in every society. This achievement is local, ongoing, and practical. It is the work of the members of a society – "with no time out!" Garfinkel's central insight is that "[t]he expectancies that make up the attitude of everyday life are constitutive of the institutionalized common understandings of the practical everyday organization and workings of society as it is seen 'from within'" (Garfinkel, 1964, p. 249). Modification of these expectations will "transform one perceived environment of real objects into another environment of real objects" (p. 249). Play, religious conversation, and scientific inquiry are such modifications, as is psychosis, brain injury, and neonate learning. In an interview, Garfinkel explained, "We have to talk about practices which, as vulgar competence, are necessary for the constitutive production of the everyday phenomena of social order" (Jules-Rosette, 1985).

Actual Events, Not Underlying Patterns

Garfinkel distinguishes ethnomethodology from "the worldwide social science movement," with its "ubiquitous commitments to the policies and methods of formal analysis and general representational theorizing" (Garfinkel, 1996, p. 5). Demographics, definition of variables, quantification, statistical analysis, causal explanation, and so on are "available to all administered societies, contemporary and historical" (p. 5). Without disputing the achievements of "formal analysis," ethnomethodology "asks 'What More?'" What more does this formal analysis depend on (p. 6)? Garfinkel's answer is that "what more?" "has centrally (and perhaps entirely) to do with procedures" (p. 6). *Procedures* in the sense not of processes but of work, of labor, such as improvising jazz at the piano, typing thoughtful words, collaborating in the workplace: "procedural means labor of a certain incarnate methodological sort" (p. 10). Ethnomethodology is about the work of producing a phenomenon and "coming upon" the phenomenon in and through this work; it is a matter of describing how people produce and display, how they demonstrate, the local phenomena of order – "the unremarkable embodiedly ordered details of their ordinary lives together" (p. 11), the "commonplace, local, endogenous haecceities of daily life" (p. 7), where *haecceities* means "thisness."

Garfinkel has no place for the techniques of formal analysis because it aims to reconstruct a *hidden* order that precedes or underlies society in the form of causal mechanisms or rational functions. Like Kant, it takes for granted the work of producing order, using this work itself as a resource but never stopping to consider it. It assumes that order can be accounted for only by adopting a transcendental perspective and using the objectifying techniques of statistical analysis. Garfinkel insists instead that an order is *visible* in the mundane details of everyday interaction, if only we will look. Ordinary society is easy to do, yet it is "strange," "elusive," and "intractably hard to describe." How on earth is society "put together"? The answer to this question cannot be imagined but must be "actually found out" in concrete, first-hand investigations of every specific occasion. The statistical and formal models built by formal analysis "lose the very phenomenon that they profess" (p. 7). Even though they are "exercising the privileges of the transcendental analyst and the universal observer" (p. 8), they still don't show how society is made. Formal social science produces its *own* order, not the order of everyday practice. Ironically, their formal work itself becomes part – an "enacted detail" – of the way ordinary society is put together. These analysts, with their "generic representational theorizing," plan and administer, and make signs that they then have to "interpret" because "the phenomena they so carefully describe are lost" (Garfinkel, 1996, p. 8).

Garfinkel is especially critical of what he calls "the documentary method of interpretation" (Garfinkel, 1967). Karl Mannheim and Alfred Schutz both used this phrase; for Garfinkel it is the common practice in formal analysis of seeing some everyday event of action or talk as evidence for an underlying, hidden organization: "treating an actual appearance as 'the document of,' as 'pointing to,' as 'standing on behalf of' a presupposed underlying pattern" (p. 78). What "appears" is treated as only a sign of the "real" phenomenon, which is accessible only through interpretation. This is clearly what Kant did; both sociologists and ordinary folks do it, too, and the process goes both ways: the underlying pattern gains credibility from the document, while the document is read in terms of the underlying pattern. Whether the underlying pattern is claimed to be culture, social structure, a value system, occupational categories, interactional functions, or roles and rules, it is assumed to be more real, more stable and enduring, than the actual events that are observed! The lay or professional sociologist appeals to "a correspondence of meaning" (p. 79) to "epitomize" the underlying, hidden pattern. Clearly this correspondence is "a product of the work of the investigator and reader as members of a community of cobelievers" (p. 96), but it is treated as "what everybody knows."

This doesn't mean that ethnomethodology is indifferent to social structures. It has "a concern with structure," but "as an achieved phenomenon of order" (Garfinkel, 1996, p. 6). Nor is it "changing the subject" for sociology. Our "immortal, ordinary society" (here Garfinkel cites Durkheim) is the "locus" and "setting" of all our activities. It is here (and now) that any order, reason, logic, typicality, classification, and standardization are achieved. Whereas formal analysis finds no order in the circumstantial concrete details of everyday life, only in the products of its own "analyzing devices" and practices of objectification and analysis, ethnomethodology sees the basis of *all* order, both commonsense and scientific, in concrete everydayness. Garfinkel insists that "there *is* order in the most ordinary activities of everyday life in their full concreteness" (p. 7). In place of the "generic" descriptions that formal analysis provides, ethnomethodology explores the "unexplicated specifics of details in structures, in recurrences, in typicality." Consequently: "Ethnomethodology's fundamental phenomenon and its standing technical preoccupation in its studies is to find, collect, specify, and make instructably observable the local endogenous production and natural accountability of immortal familiar society's most ordinary organizational things in the world, *and provide for them both and simultaneously as objects and procedurally, as alternative methodologies*" (p. 6, emphasis original).

Garfinkel insists that ethnomethodology is not critical of formal analysis but "indifferent to (independent of)" it. But, as Manning (2004, p. 281) says, "This is an artful ploy, for if this version of social life is accurate and valid, FA cannot be." The two are "incommensurably different *and* unavoidably related." The question of their relationship, as two different technologies, is of central interest to ethnomethodology. It offers "alternates" to formal analyses, "not alternatives." Wherever a formal analysis has been conducted, an ethnomethodological alternate will be "findable."

If ethnomethodology is not formal analysis, Garfinkel also insists that "[i]t is not an interpretive enterprise" (Garfinkel, 1996, p. 8). His point here, too, is that what people do and say are not "representations" of something else. "Enacted local practices are not 'texts.'" They have no inner or hidden "meaning" that the analyst must reconstruct. What an element of such a practice *is* is a matter to members, a matter that they will often negotiate. The analyst's task is not to decide what an action means, or even what it is, but to describe what it is *taken to be* in members' work. Attempts to explain social phenomena in terms of consciousness, theory, and representation will *always* lose the phenomena they are interested in: "The lessons are clear: In order to lose the phenomena that the devices describe, give them over to the intentionalities of consciousness. And in order to assure their loss in any actual case, do so with the methods of generic representational theorizing" (p. 18).

Becoming a Member

Ethnomethodologists speak of "members" rather than people or subjects. The notion of membership is central (ten Have, 2002), and many ethnomethodologists insist that "researchers themselves become the phenomenon" and that one "must become a full-time member of the reality to be studied" (Mehan & Wood, 1975, pp. 225, 227). Garfinkel has defined the "unique adequacy requirement": "[F]or the analyst to recognize, or identify, or follow the development of, or describe phenomena of order in local production of coherent detail the analyst must be vulgarly competent in the local production and reflexively natural accountability of the phenomenon of order he is 'studying'" (Garfinkel & Wieder, 1992, p. 182).

From this point of view, having "vulgar competence" is necessary to gain the "membership knowledge" that enables the researcher to recognize the relevant phenomena. "Vulgar" is used here in the old sense of "belonging to the people." This knowledge is the "common sense" of membership, and to obtain it one needs "embodied presence as a competent participant in the field of action" (Pollner & Emerson, 2001, p. 127).

At first, Garfinkel proposed that EM required a "posture of indifference," a refusal to judge the value or validity of members' common sense. In this regard, the researcher clearly differs from those whose practical activity is being studied, who presumably hold their knowledge to be valid. But now Garfinkel emphasizes "hybrid" studies, "studies of work in which the analyst is uniquely and adequately competent to produce the phenomenon" (Garfinkel, 1996, p. 13), such as Sudnow's study (1974) of playing jazz piano. Garfinkel has gone so far as to suggest that the results of research should be presented not to other researchers but to members, using their vernacular. He has proposed that ethnomethodology is an "applied" kind of inquiry that offers its "expertise" in the form of a "remediation" for phenomena "whose local, endogenous production is troubled in ordered phenomenal details of structures" (Garfinkel, 1996, p. 8). Troubles are local, and their solutions will also be local, not abstract or general.

It should be clear that ethnomethodology doesn't try to produce over-arching theories or models. Garfinkel has suggested that the products of ethnomethodological studies have the form of "pedagogies" – methods and practices of teaching. Descriptions of how order is achieved can provide the basis for teaching how to achieve it. As Garfinkel puts it: "EM's findings are described with the questions 'What did we do? What did we learn? More to the point, what did we learn, but only in and as lived doings that we can teach? And how can we teach it?'" (Garfinkel, 1996, p. 9).

Garfinkel explains, "In endlessly many disciplines, as local occasion demands, practitioners are required to read descriptive accounts alternately as instructions" (Garfinkel, 1996, p. 19). This "praxeological reading" is done in practices "chained bodily and chiasmically to places, spaces, architectures, equipment, instruments, and timing" (p. 19). Diagrams, recipes, even free-way signs, are both instructions and descriptions of the work by which the instructions are to be applied. The instructions and instructions-in-use are related as "Lebenswelt pairs." Descriptions *are* instructions in how to produce the order described.

These "pedagogies" are not abstract formalizations but "tutorial problems" that are "learned in settings in which teaching and learning being done in concert with others were locally and endogenously witnessable" (Garfinkel, 1996, p. 9). Studies by ethnomethodologists of science, work, and professions have shown that "[t]he praxeological validity of instructed action *is* (i.e., 'exists as,' 'is identical with,' 'is the same as') the phenomenon" (p. 9, emphasis original). Activities of which instruction is a part offer opportunities for eth-nomethodology. Equally, ethnomethodology offers instruction to members.

Disrupting the Familiar

But ethnomethodology has used other ways to gain access to local phe-nomena. One strategy has been to employ "troublemakers" in the form of "Heideggerian uses" of inverting lenses, disability, and other kinds of break-downs to overcome the transparency and reveal what is "relevant to the parties" (Garfinkel, 1996, p. 12) among the details of "phenomenal fields." In such investigations, the concern has been with "practices that are chiasmically chained embodiedly to the environment of ongoingly ordered phenomenal details" (p. 13). By arranging "breaches" and "making trouble," the sociologist is able to "produce reflections through which the strangeness of an obstinately familiar world can be detected" (Garfinkel, 1964, p. 227). With this strategy of defamiliarizing the ordinary, Garfinkel has drawn on both Schutz and Heidegger. The echoes of Heidegger and Merleau-Ponty should be clear when Garfinkel writes of "reflexive body/world relations" and "the accomplished transparency and specifically unremarkable smoothness of concerted skills of 'equipmentally affiliated' shopwork and shoptalk" (Garfinkel, 1996, p. 12).

Accounts and Reflexivity

Garfinkel emphasizes the *reflexive* character of practical activity. "Reflexivity refers to the simultaneously embedded and constitutive character of actions, talk and understanding" (Pollner & Emerson, 2001, p. 121). Action is "bound up with the capacity of human agents for self-reflection, for the rational 'monitoring' of their own conduct." Members are continually monitoring their own actions and those of others and are able to provide "accounts" of these actions when called upon. Often this reflexivity is treated by social scientists as a nuisance, as Giddens noted. But ethnomethodology sees it as a central part of everyday life, another continuity between sociological activity and everyday activity.

For ethnomethodology, the location of action in place and time is of central significance. Formal models ignore something crucial, "the temporal 'succession' of here and now situations" (Garfinkel, 1967, pp. 67–69). In Chapter 3, we mentioned Garfinkel's interest in indexicality. Indexical expressions demonstrate their properties only in local settings. In context, they are able to achieve "coherent sense, reference, and correspondence to objects" (Garfinkel, 1996, p. 18). They do this not as cognitive functions or "transcendentalized intentionalities of analytic consciousness" but as practical activities with "procedural relevance" to people, settings, equipment, architecture, and so on. The exploration of these "rational properties of indexical expressions" is central to ethnomethodological inquiry. Occasionality, indexicality, "specific vagueness," "retrospective-prospective sense," and temporal sequencing of utterances are "sanctioned properties of common discourse" (Garfinkel, 1964, p. 229). They are conditions people use to be understood and to understand others in conversation, conditions that are usually "seen but unnoticed."

Garfinkel recommends that we notice that accounts are *part* of the actions that they make accountable. He has written that his "central recommendation is that the activities whereby members produce and manage settings of organized everyday affairs are identical with members' procedures for making these settings 'accountable'" (Garfinkel, 1967, p. 240). Accounting practices are not descriptions of a separate reality but are *constitutive* of the order they report. They are recommendations or instructions in how to see what is happening. Just as for Heidegger interpretations articulate the practical understanding of involved activity in order to inform that activity, for Garfinkel giving an account has an "'incarnate' character" (p. 240). It is in this respect that "knowledge" and "rationality" are themselves practical social accomplishments: people construct reality – not just moral order but all kinds of order – in and by means of their social interactions. At the same time, and as *part* of this work, they construct *accounts* that are taken as rational and objective by their fellow participants. Accounts are *part* of mundane reality, constructed and understood by people as they engage in concrete, practical tasks:

> When Garfinkel refers to behavior as being *accountable*, the word can be understood in two senses. First, members can be (and are) responsible for their actions and are accountable to their interlocutors for utterances and actions which may appear to be without reason or rationale. Second, and more obliquely, Garfinkel is contending that all behavior is designed in ways to give an account of the action as an instance of something or the other. (Koschmann, Stahl, & Zemel, 2004)

This means that every account is indexical: it has intrinsic links to its setting. And this in turn leads to the important insight that ordinary language and its "ambiguity" cannot be replaced by a scientific language that is "more precise," meaning less ambiguous or less context-bound.

This is yet another way in which "practical sociological reasoning" is placed by ethnomethodology on the same level as any other everyday practical activity. The scientist does not have a special status; sociological accounts are on a continuum with all the other kinds of accounts that are a continual accomplishment of everyday life. Giving accounts – accounting – is an endless process, too. There is no final point at which an exhaustive, objective accounting has been completed. Garfinkel rejects – or is indifferent to – attempts to translate the situated events of the social world into a neutral and objective scientific terminology. There is no valid basis for the notion that a researcher can, or should, adopt the stance of an *external* observer of a social world. In the multiplicity of life-worlds, the life-world of professional sociology is just one among many, with no special claim to objective knowledge. On the contrary, "knowledge" and "rationality" themselves are always practical social accomplishments.

A New Model of Language

Ethnomethodology pays attention to language as a dynamic, social phenomenon and to speech not as an inert vehicle – the expression of inner meanings – but as fundamental to the constitution of social life. Social reality is "talked into being" (Heritage, 1984, p. 290). Words are viewed as indexes, not as symbols or representations – or not necessarily these. Garfinkel (1967) pointed to the *multiple* ways in which language is used. Language "is conceived, not simply as a set of symbols or signs, as a mode of representing things, but as a 'medium of practical activity,' a mode of doing things." Ordinary language and its "ambiguity" cannot be ignored or replaced by a scientific language that is "more precise." "[T]o study a form of life involves grasping lay modes of talk which express that form of life"(Giddens, 1977, pp. 167–169).

We saw in Chapter 3 that Garfinkel rejects the typical model of language in research. This model amounts to a theory of signs: it assumes that language works by linking words and things through concepts. The word "tree" and the object tree are linked by a concept made up of features: trunk, leaves, and so on. It assumes that these features require no interpretation and that the concepts are the common property of all of us who speak a language, a shared background knowledge. This model skips over what has been said and tries to elaborate a "meaning" that is assumed to lie "within" the words, their hidden content.

Garfinkel proposes a different model of language (see Table 8.3), in which understanding what someone says is seen not as a matter of reconstructing the inner meaning of their words but of recognizing *how* they speak. Common understanding is something that must be achieved by the partic-ipants in a conversation, and there is no single way to do this. People speak in countless ways, and *multiple* sign functions can be accomplished by speaking – "marking, labeling, symbolizing . . . , analogies, anagrams, indicating . . . , imitating . . ." (Garfinkel, 1967, p. 258) – and many more. Understanding what someone says is a matter of recognizing which of these was done, the *method* of their speaking. Explaining what was talked about is a matter of describing this method, *how* they spoke jokingly, and so on. This description will provide instructions on how to see what was said, how to recognize what was done by the speaker. But these instructions can never be fully spelled out and will never be complete in themselves. They remain "organized artful practices" that we must study. The job of the researcher is not to explain what talk means but to describe how people can come to agree on what they mean (and can lose that agreement).

Embodied Know-How

Ethnomethodology places emphasis on *embodied* know-how. "It is Garfinkel's position that the knowledge of the practices he is trying to introduce is not a conceptual or cognitive knowledge but, rather, an embodied knowledge that comes only from engaging in practices in concerted co-presence with others"

TABLE 8.3. *Garfinkel's Model of Languages*

Recognizing That Someone Was Speaking	Recognizing How They Were Speaking	Explaining What Was Talked About
The words spoken by a specific person in specific circumstances	A *method* of speaking (e.g., metaphorically, euphemistically, jokingly)	An explanation of what was talked about consists of describing how someone was speaking. *How* they talked jokingly, etc.
	There are *multiple ways* of speaking. There are a multitude of sign functions: marking, labeling, analogies, anagrams, simulating. Common understanding must be *achieved* by participants. People are always "recognizing, using, and producing the orderly ways of cultural settings from 'within' those settings."	This description will provide *instructions* in how to see what was said. But these instructions can never be complete or transparent by themselves. The methods people use to achieve understanding are not formal procedures but "organizational phenomena." They are "organized artful practices." These are *topics* for ethnomethodology.

Note: Based on Garfinkel (1967).

(Rawls, 2006, p. 5). The distinction between embodied and cognitive knowledge is crucial, and:

[A]pproaches which reduce the detail of social life to concepts, typifications, or models lose the phenomena altogether. They end up focusing on the self as a carrier of concepts, instead of the situations in which they are given meaning. Learning to see differently sociologically means learning to see social orders in their details as they are achieved in real time by persons through the enactment of these details, instead of through conceptual glosses on those details after the fact. (Rawls, 2006, p. 6)

A contrast can be drawn between Garfinkel and sociologist Erving Goffman (1922–1982) that directly parallels the distinction I have made between ontological and epistemological constitution. Goffman conceived of interaction as dramaturgical, like a theatrical performance. He proposed that people engage in "impression management" and the "presentation of self." Rawls notes that:

For Goffman the world of action was essentially messy and lacking order. It was the actor's job to create the appearance of order – a thin veneer of consensus. For Garfinkel, by contrast, the world of embodied practice – created and lived in by groups of actors working in cooperation with one another – was ordered in and through their efforts and had coherence and meaning only in and through – or as – recognizable orders of practice. . . . To view things otherwise was to allow conceptual reduction to hide the achieved coherence of events: to render social order invisible, as Garfinkel would repeatedly say. (Rawls, 2006, p. 4)

In short, ethnomethodology undertakes "studies of shared enacted practices" using "a detailed qualitative approach" (Rawls, 1997, p. 5). Traditional social science has followed Kant in assuming that individuals have only cognitive knowledge and that researchers also must work principally with this kind of knowledge, conducting investigations with an attitude that is entirely theoretical. Ethnomethodology moves in a different direction, for "Garfinkel argues that the theoretical attitude is responsible for many of the problems with social research" (p. 4). Ethnomethodology focuses on constitution – on the problem, the apparent mystery, at least to Kant and Kantians, of the orderly character of society. It asks, "What more?" It is the study of the work that people do to produce epistemological and moral order. The promise of these investigations "is that they might shift the gestalt of theoretical perception such that we could be enabled to ask new questions about the world" (p. 6). Garfinkel has said that what ethnomethodological investigations can do is make evident a "territory of new organizational phenomena."

CONCLUSIONS

We have examined four explorations that reject the story about "constitution" that Kant told. Each proposes that there is more to human beings than was captured by Kant's model. Kant emphasized individual mental representation, a rational synthesis of perceptual data, as being the basis for valid knowledge and ethical action. The work in this chapter has emphasized practical understanding, embodied comportment with tools and equipment, our absorption in everyday social interaction, and our inescapable entanglement in the material world as the basis for epistemological and ethical order. For Hegel, Heidegger, Merleau-Ponty, and Garfinkel, theoretical knowledge is made possible by a more fundamental relationship between humans and our world. Representing the world is secondary, and in some ways distorting. What is more fundamental is the practical involvement – historical, embodied, and social – that is prior to the subject–object distinction but remains invisible to the traditional human sciences.

For Hegel, Kant failed to see how tensions *within* experience can propel it forward to achieve a grasp of things-in-themselves and the underlying relation between subject and object.

For Heidegger, experience is grounded in, and derivative of, a practical involvement in the world in which subject and object have not become distinct. When practice is suspended for practical circumspection, understanding is articulated as interpretation. Only in the complete detachment of philosophical reflection does "representation" seem primary. The central characteristic of practical involvement is its *temporality*: human beings are "thrown" into the public, social world of human affairs, grasping it and understanding themselves in terms of its history and projecting their practical activities into its future. This social world provides the ground for human beings and the entities we encounter. It defines the possible ways entities can be.

For Merleau-Ponty, the bodily character of human involvement in the world is primary. Consciousness is embodied perception, not representation. Our world is not constructed in representational thought but constituted through being "lived through." Bodily consciousness offers an infrastructure for thought; concepts are a way of seeing, a gaze of the mind, that is invited by the gestures of speech.

Finally, Garfinkel explored how the order of social reality is constituted in and through everyday social interactions. Rejecting formal programs of investigation, Garfinkel argued that we need to attend to the details of concerted activity, the work in which every kind of order is produced. Ethnomethodology shares with Heidegger and Merleau-Ponty the view that this order is assembled through embodied practice rather than conceptualization; that formal analysis does not adequately characterize this work; and that investigation requires a radical attitude:

> firstly, the idea that the experienced social world is composed not of discrete "variables" of one sort or another but of gestalt contextures that are assembled in and through actors' intrinsic ordering activities. This intrinsic ordering activity includes the lived way in which percipient bodies initially bring the world into being and only secondarily conceptualize it. Secondly, the ordering of the world does not occur through following rules or roles or other abstractly formulated proscriptions. Such proscriptions are themselves usable resources for "doing" nameable activities and providing for a visible, sensible social environment. Finally, the experience of an objective world, whether in everyday or scientific settings, depends upon practical adherence to a set of idealizations or presuppositions that require a radical investigative stance for proper inquiry. (Maynard & Clayman, 1991, p. 392)

The analyses in this chapter cut deeper than the studies of epistemic constitution in the previous chapter. For epistemic constructivists such as Husserl, Schutz, and Berger and Luckmann, our representations constitute what we *take* to be reality, but reality "in itself" is unknowable. Our capacity for representation is not questioned but is taken to be natural. We can call the approach of Heidegger, Merleau-Ponty, and Garfinkel an *ontological* constructivism because for them objects and subjects, not just ways of knowing, are formed in practical activity. This is a nondualistic "radical realism" (see Box 8.3).

Although this explicitly ontological approach to constitution (see Table 8.4) avoids the problems of the epistemic approach, Heidegger and Merleau-Ponty can be accused of failing to be specific enough. They went too quickly (Deleuze, 1986/1988, p. 112). For Heidegger, human understanding is based on a general and universal time rather than on the specific times of a particular society. Merleau-Ponty's focus is body conceived in general terms rather than the different kinds of bodies that are shaped in different circumstances (compare a weightlifter and a housewife). These analyses are empirical, but each rests on an abstraction, with the result that their efforts to overcome dualism and explain how humans can validly know our world lack concreteness. They value practical activity, but they don't foster it. Garfinkel has come closest to an exploration of the ethnomethods specific to particular forms of life and to seeing that inquiry can only produce accounts that have practical, local relevance.

Language plays a central role in these new analyses, which see that its role is not simply representing the world. As Merleau-Ponty put it: "[S]peech opens a field of the nameable and sayable" (1964/1968, p. 178). Thinking isn't a liberation from perception, from mere appearances, it is a *transformation* of perception, of visibility: "a *metamorphosis* of the flesh of the sensible into the flesh of language" (Carbone, 2004, p. 39). For Merleau-Ponty, thinking "shows by words"; concepts, generalities, and abstractions are transformations *of* the visible, not some separate realm. A concept is the *style* of a collection of things in general. Thought and reason can never completely possess the world intellectually, and no language ever rids itself of all sensory material.

Language is a multipicity of sign-relations; *how* language is used is multiple, and we instruct each other in how to see what was said. Accounts are features of the settings – the places and moments – in which they are given. They, too, have to be "seen" in the right way. And the objects, regions, and times to which they refer depend on the speaker's position as well as the hearer's relation to the speaker. Language doesn't describe the world from outside; it is implicated in the world, and it participates in the contingencies of the world. Language is not representation imposed on things from outside. What is reasonable (effective, clear, consistent, objective, etc.) is what is *accountable*, and making circumstances accountable is something people do all the time.

I hope that by now you are questioning your assumptions about what kinds of things exist in the world. This chapter has thrown cold water on the commonsense assumption that humans form representations of the world around them. For Kant, the "concept" was the way an individual intellectually

BOX 8.3. *Radical Realism*

One of the most important and exciting aspects of the work described in this chapter, though one of the least understood, is how it avoids dualism. It adopts what has been called a "radical realism," in which "there are not two worlds [mental and material] that must somehow be shown to be connected by the ingenuity of philosophers, but *one*: the subject is located in objective reality" (Bakhurst, 1991, pp. 115–116, emphasis original). Each human is part of the material world, albeit a rather special part. Our perception of this world is an *opening* to it, "an openness to reality itself" (p. 116).

In radical realism,

> The subject must be seen as having immediate or direct access to reality. None of this is to say, of course, that we have instant access to the truth. Our conception of the world can be, and often is, riddled with error. But we are able to be wrong about reality because our minds are capable of reaching right out to it. (Bakhurst, 1991, p. 116)

It would be better to say not that our minds "reach out" to reality but that mind, or better still consciousness, is always embedded and embodied *in* reality. To have a "conception" of the world is not to form a picture of it in the mind but to have an altered way of being embedded. To have a conception is to touch and see the world differently; recall Merleau-Ponty's proposal that concepts are the invisible *in* the visible.

This means that subject and object are identical, at least initially, as are thinking and being. Kant tried to locate reality in the experience of the individual subject. Radical realism locates the experiencing subject in reality, as an embodied person among other people. Objective reality is the world we inhabit.

One might ask if this is not reducing the world to the material universe. Does it not ignore culture: the fact that there are somehow multiple social worlds? That an object can have very different significance in different societies? The radical realist position does not ignore these objections, but its response to them is very different from Kant's and from that of the vast majority of social scientists. The standard view is that individuals in different cultures form different concepts, so they live in different subjective worlds. For radical realism, in contrast, an object becomes something of significance by virtue of its incorporation into human *practices*. It is incorporation of objects, both fabricated and natural, into human practical activity that gives them the significance with which they show up as objects of a certain kind. As Heidegger said, being – what something is – is an issue only for humans. What something is – that this object in front of me is a cup, for example – is an anthropocentric fact, dependent on the continued existence of human-kind, but it is a fact independent of any *individual* mind.

For radical realism, there is nothing "between" object and subject – no level of mental representations or linguistic representations. It is a *direct* realism. The human subject is immersed in a single natural, material environment that contains things of shared public significance. In this ontology, thinking and being do not stand in a relation of correspondence (or noncorrespondence when we think incorrectly) but in a relationship of *identity*. Of course, the immersion is *practical* before it is intellectual; we grasp things in the ongoing flow of activity.

In this radical realism, it is entirely reasonable to ask what an object – or the world as a whole – would be like in-itself. To pose such a question is to ask what difference our activity makes, to ask about the character and degree of influence of human activity on our planet, and in answering this question to learn about our place in nature. At the same time, radical realism suggests that what the sciences do is develop new kinds of social practices that continuously bring to light new ways for things to be. They do not – they cannot – study things as they are in-themselves.

Radical realism, then, rejects Kant's dualism, his individualism, his transcendental idealism, and his universal rationality. It offers a view of humans as embodied, social beings who are immersed and involved in a material world, who *constitute* the orders of this world in ongoing practical and collaborative production and reproduction and consequently know the objects of the world directly, in terms that are defined by their public social practices. Thinking itself is an embodied, practical, and social activity in which objects of the social world reveal new aspects but still within their being as public objects. The individual, in this account, is not the owner of a private mental space, is not "a self-contained, self-sufficient, and ready-made subject of 'inner states,' but . . . a socially formed being, essentially dependent on his or her ancestors and peers" (Bakhurst, 1991, p. 215).

grasps the essence of perception, the abstract and general character of what is seen. Thinking was how we actively make what is "real" from what is passively given to us. This chapter has explored a very different view, that the world is a place of activity in which each of us is but one small part. The sensible world is rich, complex, and baroque, and any way of talking about it grasps just a part, is just one way of participating in it, one style of perception, one way of being. The term *concept* originally meant being hollow and therefore able to accept and contain something (think of *conception*), and thinking can be seen as an activity of creating space for a thing to *be* something (Carbone, 2004). Language discloses how things can be. More broadly, speech can change the world and change the people in it; language has an ontological power. When I speak, I produce an utterance in order to invoke a way of seeing the world, to

TABLE 8.4. Conceptions of Constitution, Part 2

	Heidegger	Merleau-Ponty	Garfinkel
	Hermeneutic Phenomenology	Phenomenology of Embodiment	Ethnomethodology
Constitution	World is the upon-which of being. Cultural practices are the ground, the "meaning" of being.	All forms of knowing are constituted on the system of possibilities the world provides to the human body.	Social order is an ongoing, contingent accomplishment. A negotiated improvisation that includes successful accounting practices.
Subjectivity	Dasein: being-there; being-in-the-world. Thrown and projecting. Concerned and involved.	Perception is an embodied way of being.	A member, engaged in practical social activity in unique settings.
Methodology	Hermeneutics of everydayness; hermeneutics of suspicion.	Slacken the threads that attach us to the world.	Become a member.
Products of Investigation	Existential structures of human being.	Syncretic, immovable, and symbolic structures of behavior.	The methods that members use. Accounts of practical accomplishments.

pick out an entity in this sight, interpret that entity (e.g., make a claim about it), and act on other people, *move* them, and perhaps *change* them.

So although Garfinkel has recommended that researchers should talk primarily with members, it is clear that he, like Heidegger and Merleau-Ponty, is offering a new kind of discourse to social scientists that enables us to see old things in fresh ways and see new things that had previously been invisible. Escaping dualism, in particular, is largely a matter of seeing in new ways, and this in turn is facilitated by using language in new ways. Metamorphosing one's ontology involves changing how we talk and write – including how we talk and write about language.

The most important thing we have learned in this chapter is that constitution itself is *visible*. Embodied, practical, and concerted activity in the material world can be seen; it is not hidden away on some transcendental level of the mind. And if it can be seen, it can be studied. We can envision a form of qualitative inquiry that asks and answers questions that the "objective study of subjectivity" cannot frame, questions about the kind of subjects we become and the different subjects and objects of different places and times. But how, exactly? How can we best investigate the constitution – the *ontological work* – that has been pointed out in this chapter? What are needed are concrete and specific investigations of the actions of particular bodies, in specific times, as they interact together practically. Can ethnography – immersion in a way of life – do this, as Giddens, Taylor, and Geertz promised? In the following chapter, we will explore this issue.

9

The Crisis in Ethnography

Ethnography has, perhaps, never been so popular within the social sciences. At the same time, its rationales have never been more subject to critical scrutiny and revision.
Atkinson & Hammersley, 1994, p. 249

We can now return to those calls in the 1970s for a new approach to research in the social sciences, one that would escape dualism by adopting an ethnographic mode of investigation through "immersion" in a foreign form of life. We can see now that this approach has promise to the extent that it recognizes that practical activity operates prior to the separation of subject and object, of subjectivity and objectivity. The logic seems clear: if embodied practical activity is the locus of a constitution of both social order and the knowing subject, we need to study this activity in specific settings. The researcher ought to acquire a *practical* familiarity with the people and way of life being studied, and this can come only from *participating* in this way of life. Ethnographic fieldwork ought to be the way to go.

But over the past 20 years, cultural anthropologists have been debating and rethinking the character of ethnography, and right now there is little consensus about ethnographic investigation. The basic premise of anthropological ethnography has been that one lives among the people one wants to know, participates in their practices, observes what they do, and writes a summary report. But there are significant problems with this conception of ethnography, and these problems have led to a reexamination in cultural anthropology of the character of ethnographic fieldwork. In this chapter I will review this debate and see what conclusions we can draw about how ethnographic fieldwork should be conducted if it is to help us study constitution in a program of relevant qualitative research.

The central importance of ethnography comes from the recognition that our planet contains a rich variety of human forms of life and the conviction that to understand them one has to witness them firsthand, "actually taking part in the activity together with the people involved in it" (Winch, 1956, p. 31; see Winch, 1958):

> One of the chief tasks of the *social* scientist who wishes to understand a particular form of human activity in a given society will be ... to understand the concepts involved in that activity. It would appear then that he cannot acquire such an understanding by standing in the same relation to the participants in that form of activity as does the natural scientist to the phenomena which *he* is studying. His relation to them must rather be that between the natural scientist and his fellow natural scientists; i.e. that of a fellow participant. (Winch, 1956, p. 31)

We may not agree that what needs to be understood in a form of activity are its "concepts." But the argument that to understand a form of life one must participate in it seems convincing.

Ethnography has been defined as "a special scientific description of a people and the cultural basis of their peoplehood" (Vidich & Lyman, 2000), as a "snapshot" of the unified whole that comprises a culture (Erickson, 2002). Although in anthropology the pioneering studies were of foreign and exotic peoples, ethnography is now also conducted with Western cultures and sub-cultures. The term "ethnography" is applied both to firsthand inquiry in a culture and to the written report that is the product of such inquiry. In the first case, the ethnographer enters, often for an extended period of time, the place where people live and forms personal relationships with them. Ethnography is "fieldwork."

Ethnographic fieldwork has always been central – "at the heart of cultural anthropology for over seventy years" (Marcus, 1994, p. 42) – though it has been important in other social sciences, too (Robben & Sluka, 2007). It was a central part of the Enlightenment dream for knowledge of mankind gained through empirical study rather than religious belief or armchair speculation. Ethnographers would study "the natural, moral and political history" of the many societies covering our planet and "we ourselves would see a new world come from their pen, and we would thus learn to know our own" as Jean-Jacques Rousseau (1712–1778) wrote in 1755 (Giddens, 1987, p. 20). Voyages of expansion in Africa, Asia, and the Americas in the 18th century led to contact with indigenous people seemingly lacking civilization, and this motivated "the attempt on the part of Western man to discover the position of his own civilization and the nature of humanity by pitting his own against other cultures" (Gay, 1977, p. 319). The very definition of culture was built on the shifting sands of the distinction between savage and civilized. The philosophy of "the noble savage" (Dryden, 1672/1968) quickly changed to a view in which "savage" was equated with "primitive" (Hymes, 1972). Whether it was to justify colonial exploitation or to further tolerance among different peoples, ethnography was viewed as both scientific instrument and moral force. Only recently has reflection begun on these historical, political, and ethical conditions of ethnographic practice.

The founding figures of anthropology considered ethnography crucial. Franz Boas (1858–1942), Bronislaw Malinowski (1884–1942), and Alfred Radcliffe-Brown (1881–1955) all regarded firsthand contact as necessary for the scientific study of cultures (see Urry, 1972). Boas studied the Inuit people on Baffin Island and insisted on the importance of investigating the physical surroundings and history of a culture. Radcliffe-Brown conducted fieldwork in Australia and focused on how social customs preserve a society's structure. Malinowski worked in Papua (New Guinea) and the Trobriand Islands and emphasized the importance of studying the mundane details of everyday life. From these the anthropologist could construct a picture of the whole culture. Malinowski proposed that the "final goal of anthropology," "of which the Ethnographer should never lose sight" (Malinowski, 1922/1961, p. 25), was "to grasp the native's point of view, his relation to life, to realize *his* vision of *his* world" (p. 25, emphasis original). Achieving this goal might give us "a feeling of solidarity with the endeavours and ambitions of these natives.... Perhaps through realizing human nature in a shape very distinct and foreign to us, we shall have some light shed on our own" (p. 25).

Sociology, too, has embraced ethnography. The University of Chicago was home in the 1920s to the "Chicago School" of field research (Adler & Adler, 1987; Deegan, 2001). A number of studies directed by Robert Park (1864–1944) and Ernest Burgess (1886–1966) (e.g., Burgess, 1927; Park, 1915/1997) developed basic techniques of fieldwork. Park had worked as a journalist and studied with both pragmatist philosopher John Dewey (1859–1952) and German sociologist Georg Simmel (1858–1918). He and Burgess taught their students to view the city of Chicago as a natural laboratory. He advised students to "go and get the seat of your pants dirty in real research." Through active and empathic participation, they were to study the "human ecology" of Chicago and the subjective points of view of its inhabitants:

> The city is, rather, a state of mind, a body of customs and traditions, and of the organized attitudes and sentiments that inhere in these customs and are transmitted with this tradition. The city is not, in other words, merely a physical mechanism and an artificial construction. It is involved in the vital processes of the people who compose it; it is a product of nature and particularly of human nature. (Park, 1915/1997, p. 16)

But there have always been conflicting approaches to ethnography, and often there has been little systematic training (Shweder, 1996). Certainly today "it has become a site of debate and contestation within and across disciplinary boundaries" (Atkinson, Coffey, Delamont, Lofland, & Lofland, 2001, p. 1). Thirty years ago, Spradley (1980) felt an "urgent need to clarify the nature of ethnography," and others feel the same need today. But once again the "how to" books usually treat ethnography as straightforward. Fieldwork is a matter of collecting naturalistic data, data about what people actually do, not just what they say (in interviews). It calls for participant observation: one hangs out with the natives and then writes down what one has observed. The books offer techniques and tips on how to enter the field setting, the role of gate-keepers in gaining access, establishing a rapport with people, the kinds of roles to perform, the art of developing key informants, how long to observe, when to interview, and how to record fieldnotes. But recent debates in anthropology have shown that these matters are much more complex. These debates are relevant to our interest in deciding how to study constitution.

ETHNOGRAPHIC FIELDWORK

There are two components to ethnographic fieldwork: the *field* and the *work*. The *field* is the place where the ethnographer goes, the scene in which an ethnography is conducted, and the object of investigation. The *work* is what is done there. Ethnographic work itself has two sides: fieldwork has often been termed "participant observation," a "duality of role" as both "stranger and friend" (Rock, 2001, p. 32), requiring both "involvement and detachment" (Powdermaker, 1966, p. 9). *Participation* is involvement in the everyday practices of a group of people, whereas *observation* is systematic and regular recording of what goes on, usually in written form: "Since Malinowski's time, the 'method' of participant-observation has enacted a delicate balance of subjectivity and objectivity. The ethnographer's personal experiences, especially those of participation and empathy, are recognized as central to the research process, but they are firmly constrained by the impersonal standards of observation and 'objective' distance" (Clifford, 1986, p. 13).

But is this "a delicate balance" or an impossible contradiction? How can objective distance be combined with empathic participation? How can this "dynamic and contradictory synthesis of subjective insider and objective outsider" (Sluka & Robben, 2007, p. 2) be achieved? What sense can we make of the notion that ethnographic fieldwork involves, as Malinowski put it, "plunges into the lives of the natives" (Malinowski, 1922/1961, p. 22), after which the ethnographer climbs out, dries off, and takes notes? We need to examine both of these sides of ethnographic work more carefully. What we will discover, once again, is a troubling dualism.

The Work: Participation

The first aspect of the work of fieldwork, then, is the researcher's *participation* in an unfamiliar form of life. In the opening chapter of what has been called "probably the most famous, and certainly the most mythicized, stretch of fieldwork in the history of the discipline" (Geertz, 1988, p. 75), Malinowski provided a detailed account of the ethnographic method and how it achieves an "objective, scientific view of things" (Malinowski, 1922/1961, p. 6). Observation is central, and three types of phenomena need to be recorded. First is a complete survey of "the firm skeleton of the tribal life" (p. 11) using records of kinship terms, genealogies ("[n]othing else but a synoptic chart of a number of connected relations of kinship" [p. 14–15]), maps, plans, and diagrams. But all this is "dead material" that "lacks flesh and blood" (p. 17). Second, fieldworkers need "the realities of human life, the even flow of everyday events, the occasional ripples of excitement over a feast, or ceremony, or some singular occurrence" (p. 17). Fieldworkers need to grasp the *manner* in which customs are followed: the "imponderabilia of actual life and of typical behaviour" (p. 20). And so, third, they need to record "the spirit – the natives' views and opinions and utterances" (p. 22). In Malinowski's view, finding the flesh, blood, and spirit required "cutting oneself off from the company of other white men, and remaining in as close contact with the natives as possible, which really can only be achieved by camping right in their villages" (p. 6). He described how "I had to learn how to behave, and to a certain extent, I acquired 'the feeling' for native good and bad manners. . . . I began to feel that I was indeed in touch with the natives, and this is certainly the preliminary condition of being able to carry on successful field work" (p. 8). One needs to keep "[a]n ethnographic diary, carried on systematically throughout the course of one's work" (p. 21), including the normal and typical along with any deviations. But it is also necessary for the ethnographer "to put aside camera, note book and pencil, and to join in himself in what is going on" (p. 21). Malinowski insisted that the result of these "plunges into the lives of the natives" (p. 22) was that "their behaviour, their manner of being . . . became more transparent and easily understandable than it had been before" (p. 22).

Becoming "One of Them"

It has remained the central claim about fieldwork that the scientific investigation of a foreign culture requires "immersion." The justification for ethnographic participation has been that "it is only by attempting to enter the symbolic lifeworld of others that one can ascertain the subjective logic on which it is built and feel, hear and see a little of the social life as one's subjects do" (Rock, 2001, p. 32). The ethnographer seeks the "insiders' viewpoint" (Jorgensen, 1989, p. 14), and this can be achieved only by *becoming* an insider, or at least becoming as like one as possible by establishing rapport with members and "plunging" into their life.

If the aim of anthropology is truly "to grasp the native's point of view," it would follow that the ethnographer needs to achieve full membership. But anthropologists grant that in practice there is "a continuum of roles ranging from the empathic but less involved participant who establishes a peripheral membership within the group, to the fully committed convert or prior participant" (Adler & Adler, 1987, p. 8). As Bittner (1973) points out, most ethnographers experience none of the constraints and necessities that members of a culture live with. What are objective matters for members are seen by the ethnographer as subjective beliefs; what are experienced with nuance and subtlety by members can be described only superficially by the ethnographer. We must ask to what extent a Western ethnographer can become a full member of the culture of a tribal village and, more broadly, to what degree an ethnographer can become a full participant in any form of life they were not born into.

Geertz's ethnography of the Balinese cockfight provides an important illustration of the power of this central claim. We saw in Chapter 6 that the "guiding principle" of Geertz's interpretive anthropology is that "societies, like lives, contain their own interpretations. One has only to learn how to gain access to them" (Geertz, 1972, p. 29). How then did Geertz gain access to Balinese society? How could Geertz look through this "window" on the culture? How did he arrive at his interpretation of the significance of the cockfight *to the Balinese*?

Geertz is keen to tell us. His report begins with an entertaining anecdote that at first glance has no connection with his analysis of the cockfight. Geertz describes how he and his wife were treated when they first arrived in the 1950s in the small and isolated village that he intended to study. "We were intruders, professional ones, and the villagers dealt with us as Balinese seem always to deal with people not part of their life yet who press themselves upon them: as though we were not there. For them, and to a degree for ourselves, we were nonpersons, specters, invisible" (Geertz, 1972, p. 1). But a dramatic change occurred when they attended a cockfight, which, although illegal, was held in the central square of the village. The villagers apparently believed that they had paid adequate bribes to avoid police attention, but in fact "a truck full of policemen armed with machine guns roared up" (p. 3).

In the face of this police raid, the villagers turned and ran, as did the Geertzes. Without thinking, they ran in the direction opposite to the house where they were staying and followed one man, who ducked into a compound. Immediately, "his wife, who had apparently been through this sort of thing before, whipped out a table, a tablecloth, three chairs, and three cups of tea, and we all, without explicit communication whatsoever, sat down, commenced to sip tea, and sought to compose ourselves" (Geertz, 1972, p. 3). When a policeman arrived, the villager and his wife not only provided an alibi for the two anthropologists but also gave a detailed account of the legitimacy of their presence in the village, their official clearance, and their well-intended aims. Geertz reported that at this "it was my turn, having barely communicated with a living being save my landlord and the village chief for more than a week, to be astonished" (p. 3).

According to Geertz, "[t]he next morning the village was a completely different world for us. Not only were we no longer invisible, we were suddenly the center of attention, the object of a great outpouring of warmth, interest, and most especially, amusement" (Geertz, 1972, p. 4). Geertz was asked to tell the tale "fifty times by the end of the day." The villagers mimicked their flight, teased them about their actions, and were "extremely pleased and even more surprised that we had not simply ... asserted our Distinguished Visitor status, but had instead demonstrated our solidarity with what were now our villagers" (p. 4). Geertz and his wife were now "accepted." With this "turning point so far as our relationship to the community was concerned, ... we were quite literally 'in.' The whole village opened up to us" (p. 4). The lucky accident of getting caught up in the police raid "worked very well ... for achieving that mysterious necessity of anthropological field work, rapport" (p. 4). Geertz wrote that he now had "a sudden and complete acceptance into a society extremely difficult for outsiders to penetrate" and so was granted "the kind of immediate, inside-view grasp of an aspect of 'peasant mentality'" that most anthropologists don't get (p. 4).

The story is certainly entertaining. But it also serves a crucial function for Geertz's ethnography: it demonstrates how he gained entry, and that he became an insider. Yet when we examined his analysis in Chapter 6 it was clear that Geertz remained very different from the members of the culture he was studying. His writing displayed both his cultural background and his professional training. His interpretation of the cockfight made reference to work that no "peasant mentality" would have access to: Aristotle, Freud, Nietzsche, Northrop Frye, W. H. Auden, Jeremy Bentham, Schoenberg, and Wallace Stevens.

Although Geertz insisted that the fieldworker needs to read the texts of a culture *as though he or she is a member*, and although his story of "the raid" offered an account of how he was accepted as a member of village society, what he actually wrote about the cockfight is a complex and sophisticated text that includes dialogue with other anthropologists, shows broad familiarity with the Western humanities, and draws on game theory, utilitarian philosophy, literary criticism, and Aristotelian philosophy. It refers to reports of fieldwork, historical data, first-person reports, and statistical summaries, and regrets the lack of video-recording and microanalysis. None of this was available to the Balinese who attended the cockfight.

It is evidently not the case that Geertz's reading of the cockfight was based on an "inside-view grasp." The implication of his analysis of the Balinese cockfight – though Geertz himself recognized this only gradually – is that it belongs as much to the culture of anthropology as to that of the "native."

In "Deep Play" (Geertz, 1972) there was a central contradiction between Geertz's substantive claim about what a culture provides to its members and his account of the methodology with which he was able to study the cockfight.

If human subjectivity is shaped and educated by participation in a culture, presumably over an extended period of time, how can an anthropologist hope to be able to read the texts of a culture in which he or she was not raised "over the shoulders" of its participants? He or she will not have the same sensibilities as the participants, nor will they quickly step out of the way in which their subjectivity has been shaped by their own culture of origin.

Geertz soon (1976/1979) explicitly took up the issue of anthropological understanding and adopted a position quite different from what he called the "naïve culturalism" of his earlier work. No longer did he argue that an anthropologist goes native. Instead he focused on the *hermeneutics* involved in understanding a foreign culture. In 1967, Malinowski's private diary was published (Malinowski, 1967), and it contained many entries in which he expressed his distaste for and frustration with the people he studied. As Geertz pointed out, the publication of the diary demolished "[t]he myth of the chameleon fieldworker perfectly self-tuned to his exotic surroundings – a walking miracle of empathy, tact, patience, and cosmopolitanism" (Geertz, 1976/1979, p. 225). Malinowski's diary showed that he was impatient and even unsympathetic with "the native's point of view" he claimed to seek (but cf. Powdermaker, 1967).

Unlike most of the scandalized commentators, Geertz was not interested in Malinowski's moral character or the discovery that "he was not, to put it delicately, an unmitigated nice guy" (see Rapport, 1990). Geertz recognized that the real issue raised by the diaries "is not moral; it is epistemological" (Geertz, 1976/1979, p. 226):

> [I]f anthropological understanding does not stem, as we have been taught to believe, from some sort of extraordinary sensibility, an almost preternatural capacity to think, feel, and perceive like a native (a word, I should hurry to say, I use here "in the strict sense of the term"), then how is anthropological knowledge of the way natives think, feel, and perceive possible? (Geertz, 1976/1979, p. 226)

Malinowski clearly didn't enjoy joining in. He didn't relish his plunges into the life of the natives. He didn't show a great deal of tolerance. How, then, was he able to write what are still considered to be exemplary ethnographies?

Geertz continued to accept Malinowski's view that the goal of anthropology, and of ethnography in particular, is "to see things from the native's point of view." But he now reconsidered the question of how this goal can be achieved. If being a "chameleon" and attaining an "inside-view grasp" is not the basis for understanding a foreign culture, what is? Geertz now proposed that ethnography is a matter of "searching out and analyzing the symbolic forms – words, images, institutions, behaviors – in terms of which, in each case, people actually represent themselves to themselves and to one another" (Geertz, 1976/1979, p. 228). He insisted that because these forms are public they are "readily observable."

Geertz appealed to a distinction originally drawn in linguistics between *emic* and *etic* terms or concepts. Phonology distinguishes between *phonetic* analysis of the acoustic characteristics of speech and *phonemic* analysis of the functional sounds of a particular language. The phonemes of English are the 26 different sound categories that can be combined to produce words of the language. The anthropologist Kenneth Pike (1954) drew an analogous distinction between the "emic" concepts that are part of a culture, a particular way of life, and the "etic" concepts used by an anthropologist. Love, for example, is an emic concept in Western societies. Object cathexis is an etic concept, used in the explanatory framework of psychoanalytic specialists to describe the phenomena that in everyday life we call love.

Geertz argued that what is required in ethnography to understand and analyze the symbolic forms of a culture is interpretive work that moves constantly between emic (or "experience-near") and etic (or "experience-far") concepts, between local detail and global structure, between part and whole. What Malinowski's diary showed was that "you don't have to be one to know one" (Geertz, 1976/1979, p. 227). "In short, accounts of other peoples' subjectivities can be built up without recourse to pretensions to more-than-normal capacities for ego-effacement and fellow-feeling" (p. 240). Of course, if we are to intrude into strangers' lives, they are more likely to accept us if we are polite and sensitive. "But whatever accurate or half-accurate sense one gets of what one's informants are 'really like' comes not from the experience of that acceptance as such, which is part of one's own biography, not of theirs, but from the ability to construe their modes of expression, what I would call their symbol systems" (p. 241). Geertz suggested that in both his own ethnographic analyses and those of Malinowski there was a "characteristic intellectual movement," an "inward conceptual rhythm," "a continuous dialectical tacking between the most local of local detail and the most global of global structure in such a way as to bring both into view simultaneously": "[O]ne oscillates restlessly between the sort of exotic minutiae (lexical antitheses, categorical schemes, morphophonemic transformations) that makes even the best ethnographies a trial to read and the sort of sweeping characterizations ('quietism,' 'dramatism,' 'contextualism') that makes all but the most pedestrian of them somewhat implausible" (Geertz, 1976/1979, p. 239).

"Tacking," of course, is the zigzagging movement a sailboat makes as it changes direction to-and-fro to achieve the maximum benefit from the wind. Reading and interpreting a culture as an assemblage of texts involves "[h]opping back and forth between the whole conceived through the parts which actualize it and the parts conceived through the whole which motivate it, we seek to turn them, by a sort of intellectual perpetual motion, into explications of one another." And, as Geertz pointed out: "All this is, of course, is but the now familiar trajectory of what Dilthey called the hermeneutic circle, and my argument here is merely that it is as central to ethnographic interpretation, and thus to the penetration of other people's modes of thought, as it is to literary, historical, philological, psychoanalytic, or biblical interpretation" (Geertz, 1976/1979, pp. 239–240).

He concluded that "the ethnographer does not, and in my opinion, largely cannot, perceive what his informants perceive. What he perceives – and that uncertainly enough – is what they perceive 'with,' or 'by means of,' or 'through' or whatever word one may choose" (Geertz, 1976/1979, p. 228). These are the "symbolic forms – words, images, institutions, behaviors – in terms of which . . . people actually represent themselves to themselves and to one another" (p. 228).

But this "dialectical tacking" cannot take Geertz where he wants to go. This interpretation of symbolic forms cannot achieve Malinowski's goal of "reconstructing *his* vision of *his* world." We saw in Chapter 4 that Dilthey placed emphasis on empathic attunement or *Verstehen* and insisted that the goal of the human sciences was to penetrate a form of life, a manifestation of the life process. In linking his work with Dilthey, Geertz was attempting to achieve what Malinowski had identified as the goal of ethnography by adopting a hermeneutics that Gadamer had already shown was flawed. Although Geertz himself raised the crucial question, "What do we claim when we assert that we understand the semiotic means by which, in this case, persons are defined to one another? That we know words or that we know minds?" (Geertz, 1976/1979, p. 239), he didn't directly answer his own question. He claimed that through a hermeneutic process we can "understand the form and pressure of . . . natives' inner lives" (p. 241). But if culture is public, found in words and behaviors, how and why are its members' lives *inner*?

There is a circle here, but it is a vicious one, not the "hermeneutic circle" that Geertz claimed. As long as the aim of ethnography is "to grasp the native's point of view," to locate something "inner," its task is endless and the circle of interpretation will never be completed. Malinowski's definition of the fundamental goal of ethnographic fieldwork sets the field-worker an impossible task. The goal itself needs to be rethought, not merely the means to achieve it.

What exactly is ethnography trying to do? Anthropologists have begun to ask whether ethnographic claims about participation have enabled ethnographers to ignore the political contexts in which they work, in particular the violence and exploitation of colonialism. The notion that ethnographers become for all intents and purposes members of the cultures they study has served to hide the connections that many cultural anthropologists have had with the economic and political systems of colonialism. In the 1950s and 1960s, U.S. anthropology was part of a mission of development in new nation-states. Ethnography played a role in supporting colonial domination by providing useful information about dominated cultures or by legitimizing ideological models of social life. The fieldworker was taken – both by the colonial author-ities and often by the locals – to be a model of "liberal decency" (Marcus, 1997), but ethnography has frequently been a means to appropriate a local form of life for the purposes of the colonial powers.

Marcus has suggested that Geertz skirted this issue and "pulled back from looking too closely at the conditions of the production of anthropolog-ical knowledge" (Marcus, 1997, p. 91). In his account of the police raid, for example, Geertz presents himself as both shrewd and innocent. Renato Rosaldo (1989, 1993) has written of the "imperialist nostalgia" that permeated the interpretive ethnography of Geertz and others and its complicity with outside agents of change.

Acknowledging this colonial connection has drawn attention to the fact that when an ethnographer conducts fieldwork what occurs is never merely a contact between an individual and a culture, let alone the ethnographer becoming a member, but always the *meeting* of two cultures, two forms of life. Recognizing and rejecting these connections with colonialism has led to proposals for collaborative and participatory fieldwork in which ethnogra-phers would relinquish their authorial authority. But before we draw meth-odological conclusions such as these, we need to explore more closely the character of this meeting.

The Work: Observation

The second aspect of ethnographic work is *observation*. Writing fieldnotes has always been central to ethnography: as we've seen, Malinowski recom-mended keeping an ethnographic diary to record the details of everyday life. This "work of collecting and fixing impressions" is needed because "certain subtle peculiarities, which make an impression as long as they are novel, cease to be noticed as soon as they become familiar. Others can only be perceived with a better knowledge of the local conditions" (Malinowski, 1922/1961, p. 21).

What, exactly, does the ethnographer write? Fieldnotes are often treated as a straightforward record of what has been observed, "the written account of what the researcher hears, sees, experiences, and thinks in the course of collecting and reflecting on the data in a qualitative study" (Bogdan & Biklen, 1992, p. 107; see Bogdan, 1983). It seems common sense that ethnographers must record what they see and hear because memory is limited and fallible, because science requires that data be archived, and because analysis starts with such data. Students are advised, "Ideally, your notes would provide a literal record of everything transpiring in the field setting" (Jorgensen, 1989, p. 96). But the notion that "everything" be recorded is impractical, and the sugges-tion that a written record can be "literal" rests once again on the conduit model of language that we encountered in Part I.

Thick Description

One of the strengths of Geertz's ethnographic work was his recognition that fieldnotes are *not* a literal record but an interpretation. He recommended that to grasp what anthropology *is* we need to look at what anthropologists *do*. The heart of ethnography is not a matter of method, technique, or procedure (though it has its share of them, with diaries, genealogies, maps, fieldnotes, etc.) but what Geertz famously called "thick description" (Geertz, 1973). This notion has become part of the jargon of ethnography and of qualitative research in general, yet it is often misunderstood. Phrases such as "thick (empathic) description" (Johnson & Onwuegbuzie, 2004) miss Geertz's point because, as we have seen, ethnography has little to do with empathy, and thick description is not a matter of the amount of detail piled on. Thick description is how ethnographers "inscribe" the events they have witnessed and turn them into "accounts."

It is important to examine Geertz's proposal and understand both its strengths and its weaknesses. A culture may or may not be an assemblage of texts, but the researcher's fieldnotes certainly are. Geertz argued that an ethnographer must "fix" a culture to study it. Thick description is not merely description but "inscription." An ethnographer doesn't merely observe but *inscribes* and *interprets*. Ethnographers don't simply collect and record artifacts, fragments of a foreign way of life. They try to "clarify what goes on in such places, to reduce the puzzlement – what manner of men are these? – to which unfamiliar acts arising out of unknown backgrounds naturally give rise." Ethnography, as Geertz viewed it, studies "the flow of social discourse" and seeks to fix it "in perusable terms." Culture "is not a power, something to which social events, behaviors, institutions, or processes can be causally attributed; it is a context, something within which they can be intelligibly – that is, thickly – described" (Geertz, 1973, p. 14).

Geertz borrowed the phrase "thick description" from the Oxford philosopher Gilbert Ryle (1900–1976), together with the example Ryle used to illustrate it. Ryle (1968) invited us to imagine two boys, each of whom quickly contracts the eyelids of one eye. One, though, has an involuntary twitch, and the other is winking to signal to a friend. The same "thin description" could be given in both cases: "rapidly contracting his right eyelid." But the second boy is making a gesture – we have, as Geertz puts it, "a fleck of culture" – for which we need something more. A thick description brings out what Geertz calls the "pattern of life" to which the gesture belongs. Imagine a third boy, who gives us a parody of the wink. Here our thick description needs to be even more elaborate: "practicing a burlesque of a friend faking a wink to deceive an innocent into thinking a conspiracy is in action" (Geertz, 1973, p. 7).

Geertz proposed that the "object of ethnography" lies "between" the thin description and the thick description: it is "a stratified hierarchy of meaningful

structures in terms of which twitches, winks, fake-winks, parodies, rehearsals
of parodies are produced, perceived and interpreted, and without which they
would not ... in fact exist" (Geertz, 1973, p. 7). The anthropologist must
grasp "a multiplicity of complex conceptual structures, many of them super-
imposed upon or knotted into one another, which are at once strange,
irregular, and inexplicit, and which he must contrive somehow first to grasp
and then to render" (p. 10). Writing thick description is a central part of this
enterprise.

The notion of thick description has become popular, but it has often been
trivialized. Thick description is described as writing fieldnotes that are
"detailed, context-sensitive, and locally informed" (Emerson, Fretz, & Shaw,
1995, p. 10). But thick description requires more than this. It is tied up with
Geertz's ambition to discern what a culture *really* means. Although Geertz
insisted that thick description doesn't have to get into the minds of the actors
to identify their intentions, it does apparently need to identify *their* interpretations of their actions. Culture is public, but the ethnographer offers
anthropological interpretations of members' interpretations, "explicating
explications" (Geertz, 1973, p. 9).

This is once again the hermeneutics of Dilthey and Schleiermacher rather
than of Gadamer and Iser. Jean Bazin (2003) has criticized this view that an
ethnographer needs to interpret the meaning or significance *behind* an action
such as a wink, and suggests that Geertz misunderstood Ryle. Similarly,
Descombes (2002) argues that Geertz turned Ryle's notion of thick description
on its head (see Friedman, 2003). Bazin argues that only a *detached* observer –
one who views people as though they were mice, as he puts it – will have the
view that, unlike a twitch, a wink involves both an act and an interpretation.
Bazin suggests instead that a culture is not a text that can be read over the
shoulders of its members but "a multiplicity of actions structured by different
if simultaneously present logics" (Friedman, 2003, p. 417). Geertz's image
of text flattens out this multiplicity into something one-dimensional that
the ethnographer then interprets. Bazin argues that Geertz had accepted the
Kantian model of representation: meaning is distinct from action; on the one
hand, there is an action, and on the other hand, what it means. Bazin insists
that "we are not observing human behaviors of which it would be necessary, in
addition, to look for the meaning. *We are witnesses to actions*" (emphasis
original). He explains:

> It is sufficient that the winking scene be not only an example for an
> Oxford philosopher but also a real-life situation in order to see at what
> point the version which Geertz gives of it lacks plausibility. In all the
> cases where I would effectively be a witness to what is transpiring (and
> not just an observer of eyelid contractions), I would know or, at the
> very least, could know that a wink was exchanged without needing
> an "interpretation." For example, I would catch sight of B ostensibly

turning his head towards X and the latter would assume the knowing look of one who understood the message. For a wink (likewise a gesture or a word) is a relationship of B to X perceptible by a third party (me), and possibly seen if one finds oneself in a position to perceive it. This is not what is taking place in B's head. A wink is not the idea or the intention of winking. The intent to wink is not yet a wink. (Bazin, 2003, p. 422)

Bazin argues that to describe a cultural action we need to locate it in context, not look for a mental intention or even a member's interpretation:

In every situation where actions to which we are witnesses have to do with a world unfamiliar to us – a type of situation in which anthropologists often find themselves and into which they even ardently inquire – we tend to assume that what we are missing is not knowing what people are doing (of being in a position to describe it), but rather understanding the meaning of what they're doing. We liken what they do – their interactions – to a speech (or to a "text") to which we do not have the key and of which we would have to give an "interpretation." (Bazin, 2003, p. 426)

Certainly for Ryle, writing a thick description wasn't a matter of adding intentions to a thin description. Even a twitch may call for thick description – as a symptom of disease, for example. A thick description, as Ryle described it, has more than one layer. His view was that "thick description is a many-layered sandwich, of which only the bottom slice is catered for by that ... thinnest description." To closing the eyes we need to add "subordinate clauses" about what has been accomplished using an "already understood code." The winker has not done one thing that is interpreted as another (blinking interpreted as winking). He has winked and in doing so also signaled. He has (let's imagine) warned of the arrival of a third party. A thick description "has to indicate success-versus-failure conditions" of the action being performed. These are not in the head but in the cultural practices. To successfully answer the question "What is this person doing?" we need to add layers to the sandwich, not look underneath it: "A statesman signing his surname to a peace-treaty is doing much more than inscribe the seven letters of his surname, but he is not doing many or any more things. He is bringing a war to a close by inscribing the seven letters of his surname" (Ryle, 1968). In order to write a thick description of this action, a researcher needs to know what Ryle calls the "code," and the setting, the circumstances, and the history, not what is passing through the statesman's mind.

But Ryle was surely naive if he assumed that there is a "code" that enables all members of a culture to agree on the character of an action. When British Prime Minister Chamberlain signed a peace treaty with Hitler in 1938, was he ensuring peace for his citizens or making a gesture of appeasement? Was he to be cursed or blessed by future generations? In fact both; opinions differed at the time, and they have changed over time. Actions are retrospectively reidentified. Acts *in* context are also subject to *multiple* interpretations. But clearly Ryle was not suggesting that a thick description must capture the agent's intention, *their own* interpretation of their action.

Bazin argues that Geertz's use of Ryle's concept of thick description, just like his notion that one needs to read a culture as an ensemble of texts, assumes that meaning is hidden and unitary, whereas in fact it is public and multiple. The fact that Geertz needs to interpret shows that he is not a member at all. If cultural actions are public, they are describable by an ethnographer who knows what is appropriate in a given situation. But what exactly is required to know what is appropriate? Bazin proposes that it is a matter of learning to describe "which ways of acting are appropriate to the situation." In his view, the ethnographer must identify "the rules they follow" when people, for example, play soccer or attend a religious service. He adds that these rules are "not laws. They are not of the type, *if p, then q*" (Bazin, 2003, p. 433). This doesn't seem satisfactory. As Koschmann, Stahl, and Zemel (2004) point out, "Wittgenstein (1953) had already demonstrated the incoherence of treating social practices as a matter of following culturally defined rules. Tacit practices and group negotiations are necessary at some level to put rules into practice, if only because the idea of rules for implementing rules involves an impossible recourse."

For example, although there are certainly rules of the road in many societies, driving a car successfully through city traffic is not simply a matter of following these rules. It requires constant negotiation among drivers through signals, glances, positioning of vehicles, and so on. Bazin is surely right to insist that there is no single deeper significance, no unique "meaning," behind these driving practices. But nor are the traffic rules the whole story. Bazin's analysis leads him into what he grants is a contradiction. "Paradoxically, it may therefore be that, in the end, I know better than them about what they are actually doing, all the while being barely capable of acting as they do, even clumsily and with numerous mistakes" (Bazin, 2003, p. 430). His illustration of this is that "[i]t would be altogether imaginable for somebody to find his way around in a city perfectly well, that is to say for him to choose, with complete confidence, the shortest route leading from one point to another and yet be totally incapable of drawing a map of it. The moment he tried it, he would only produce a *completely inaccurate picture*" (p. 430, emphasis original).

At this point, Bazin claims that an outsider has the advantage: "Insofar as I manage to describe what they are doing by continuing to study them, it is precisely because I am not in their position." But why does a map provide "better" knowledge of what the participants "are actually doing" than their ability to find their way around? As Bazin himself notes, "They themselves have no need of a description or a map in order to get their bearings" (p. 430).

We can draw different conclusions from Bazin's criticism of Geertz. It is certainly true that Geertz was not a full participant, and never could be. As we have seen, he came to recognize this. And it doesn't make sense to aim for an interpretation, in the form of a thick description, that can penetrate, as Geertz hoped, to a unitary meaning hidden *behind* the public practices. But, at the same time, the goal of ethnography cannot be what Bazin wants, a maplike description of what people are "actually doing." No such description is possible; or rather, *many* such descriptions are possible. The ethnographer is not in a better position than a member to know what is "actually" going on.

The problem is that *both* Geertz and Bazin continue to assume that an ethnographic description – whether it is thick or thin, whether it is an interpretation or an interpretation of an interpretation – can be "literal" in the sense of being exact, straightfoward, accurate, undistorted. They believe that the ethnographer can record *the* truth about peoples' actions (or a whole culture).

But we have learned from Gadamer that any reading of a text is an application, an appropriation. We have learned from Heidegger that every articulation of practical understanding is oriented by a practical task. And we have learned from Garfinkel that any account is a *constituent* of the setting it describes and offers *instructions to see* that setting in a specific way. Remove the description from the setting, as the ethnographer generally does, and it withers.

Fixing Action

Let us turn to Geertz's other proposal about thick description, that it involves "inscribing" or "fixing" the events witnessed by an ethnographer. This is more convincing, though its implications go further than Geertz recognized. He was drawing on Ricoeur's (1971/1979) exploration of how "action itself, action as meaningful, may become an object of science, without losing its character of meaningfulness, through a kind of objectification similar to the fixation which occurs in writing" (Ricoeur, 1971/1979, p. 97). We considered in Chapter 5 how this was based not on Romanticist hermeneutics but on a hermeneutics of application. Ricoeur's central concern was how when action is fixed it is transformed. His main point was that, just like speech, human action involves *events* that are fleeting but are fixed in everyday life, in official records and other kinds of registration. Human action can be said to leave a trace, to have "*left its mark* on its time" (p. 100, emphasis original). Action has what amounts to a propositional structure that can take different objects. Consider a baseball player hitting a home run to great applause from the crowd. Here we can distinguish three different acts. The first is the action of swinging the bat to hit the ball; we might say that this is the act *of* hitting. The second is the act of hitting a home run, the identification of which requires knowledge and application of the rules of baseball. This second act is what is done *in* hitting the ball. And third, this action has an effect on those witnessing it: pleasing and exciting the crowd is the *result* of hitting the ball.

Ricoeur emphasized that once an action is fixed, its significance is expanded. Action that is fixed has an *"importance"* that "goes 'beyond' its *relevance* to its initial situation" (Ricoeur, 1971/1979, p. 102, emphasis original). "To say the same thing in different words, the meaning of an important event exceeds, overcomes, transcends, the social conditions of its production and may be reenacted in new social contexts." This is why "great works of culture" can have continued significance, and fresh relevance in new situations. It is as true of fixed action as it is of a written text that each is a work that "does not only mirror its time, but it opens up a world which it bears within itself" (p. 103). This means that, when fixed, "the meaning of human action is also something which is *addressed* to an indefinite range of possible 'readers'" (p. 103, emphasis original). "Human action is an open work."

Geertz accepted Ricoeur's proposal that what is fixed is not the "event" but the "said": "the meaning of the speech event, not the event as event" (Geertz, 1973, p. 19). But Geertz failed to grasp Ricoeur's point that this meaning is equivocal and indeterminate, always open to fresh interpretations. No matter how detailed, an ethnographic fieldnote can never capture *the* meaning of a cultural event. Just as with speech, a distance develops between the agent's intentions and the effects of her act. Perhaps the home run leads to a teammate being run off, and this to the loss of the game, but the hitter hardly intended this when he swung. This "autonomization" of action follows from its essentially public character: "An action is a social phenomenon not only because it is done by several agents in such a way that the role of each of them cannot be distinguished from the role of the others, but also because our deeds escape us and have effects which we did not intend" (Ricoeur, 1971/1979, p. 101).

Ricoeur's analysis of fixing shows that actions escape their agents and have multiple and changing interpretations even *within* the culture in which they occurred. His analysis is congruent not only with Gadamer's observations about interpretation but also with Garfinkel's comments about the indexical character of all accounts.

But ethnographic fieldnotes are fixed accounts of events that are *removed* from the field in which they have occurred. The people they describe seldom have the opportunity to read them, and this ruptures the "dialectic between explanation and understanding" (Ricoeur, 1973, p. 116) that fixing makes possible. This dialectic is what Heidegger called the hermeneutic circle between understanding and interpretation. It is, as Garfinkel pointed out, the way that an account instructs us how to see the events it describes. Ethnographic fieldwork has severed this connection, and the consequences have been disastrous.

Writing Culture: The Crisis of Representation

Writing is certainly how fieldnotes are created, and we have seen that the ethnographer is not merely an observer but an inscriber and interpreter. The raw data of ethnography are actually already cooked. But writing is also how the final published ethnographic account is prepared, ready to be served. In the 1980s, attention turned to the techniques and forms of written representation used in ethnographies. Much of this discussion has been in anthropology, but it has also occurred in other social sciences. It led to a critical reappraisal of classic ethnographies and to recommendations for new writing strategies and fresh "representational options – writing, scripting, dramatization, hypermedia, poetry, biography" (Atkinson et al., 2001, p. 322). Here, too, it has become clear that a polished ethnography is not simply an objective and literal report of the facts about a culture.

In 1986, James Clifford and George Marcus published *Writing Culture* (Clifford & Marcus, 1986), an influential collection of articles dealing with various aspects of ethnography as text. Rather than focusing on fieldwork or the definition of culture, *Writing Culture* began with what had typically been considered the unproblematic end point: *writing* the ethnography. This volume had "immediate extraordinary effect" (Spencer, 2001, p. 443), but it was part of a larger phenomenon. In the same year, Marcus and Michael Fischer published *Anthropology as Cultural Critique* (Marcus & Fischer, 1986a), Victor Turner and Edward Bruner edited *The Anthropology of Experience* (Turner & Bruner, 1986), and later Clifford published *The Cultural Predicament* (Clifford, 1988). Together, these books announced a "crisis of representation" in anthropology.

In his introduction to *Writing Culture*, Clifford wrote that "[w]e begin, not with participant-observation or with cultural texts (suitable for interpretation), but with writing, the making of texts" (Clifford, 1986, p. 2). In the past, writing had been "reduced to method," merely a matter of keeping good notes and "writing up" results. What was new was a "process of theorizing about the limits of representation itself" (p. 10). This amounted to "an ongoing critique of the West's most confident, characteristic discourses ... an overarching rejection of the institutionalized ways one large group of humanity has for millennia constructed its world" (p. 10).

Clifford argued that we should recognize that "what appears as 'real' in history, the social sciences, the arts, even in common sense, is always analyzable as a restrictive and expressive set of social codes and conventions" (Clifford, 1986, p. 11). In ethnography, anthropological ideas are "enmeshed in local practices and institutional constraints, ... contingent and often 'political' solutions to cultural problems" (p. 11). In particular:

> Ethnographic writing is determined in at least six ways: (1) contextually (it draws from and creates meaningful social milieux); (2) rhetorically (it uses and is used by expressing conventions); (3) institutionally (one writes within, and against, specific traditions, disciplines, audiences); (4) generically (an ethnography is usually distinguishable from a novel or a travel account); (5) politically (the authority to represent cultural

realities is unequally shared and at times contested); (6) historically (all the above conventions and constraints are changing). These determinations govern the inscription of coherent ethnographic fictions. (Clifford, 1986, p. 6)

The classic ethnographies permitted only a specific role for the voice of the author. It was "always manifest, but the conventions of textual presentation and reading forbade too close a connection between authorial style and the reality represented. ... The subjectivity of the author is separated from the objective referent of the text" (p. 13). Such texts had no place for "[s]tates of serious confusion, violent feelings or acts, censorships, important failures, changes of course, and excessive pleasures" (p. 13) on the part of the author.

Clifford suggested that the voices of ethnography should no longer be limited to the authoritative author, monophonic, and the informants, "to be quoted or paraphrased." A new sensitivity to language was needed in ethnography with the recognition "that every version of an 'other,' wherever found, is also the construction of a 'self,' and the making of ethnographic texts ... has always involved a process of 'self-fashioning'" (Clifford, 1986, pp. 23–24). Although the classic ethnographies included only very restricted self-representation by their authors, each author was nonetheless fashioning an image of him- or herself. Often their very absence from the descriptions of cultural practices served to establish their authority. Drawing on the work by Hayden White that we explored in Chapter 5, Clifford proposed that ethnographic writing is both poetical and political. "Cultural *poesis* – and politics – is the constant reconstitution of selves and others through specific exclusions, conventions, and discursive practices" (pp. 23–24). What the contributors to the book offered were "tools for the analysis of these processes, at home and abroad" (p. 24). Now "an implicit mark of interrogation [is] placed beside any overly confident and consistent ethnographic voice. What desires and confusions [is] it smoothing over? How [is] its 'objectivity' textually constructed?" (p. 14).

In a second coedited volume published that year, Marcus and Fischer (1986a) proposed that the crisis in anthropology was the core of a new skepticism about dominant paradigms of explanation and a shift of theoretical debate to "the level of method, to problems of epistemology, interpretation, and discursive forms of representation themselves" (Marcus & Fischer, 1986b, p. 9). It had become evident that "problems of description become problems of representation" (p. 9). The crisis was a result of the recognition of "problems of the interpretation of the details of a reality that eludes the ability of dominant paradigms to describe it, let alone explain it" (p. 12).

Drawing on White's (1973a) distinction between romance, tragedy, comedy, and satire (or irony) as what they called "shifting modes of writing" (Marcus & Fischer, 1986b, p. 12), Marcus and Fischer described "persistent oscillation between the more realist modes of description and irony" (p. 14), with Geertz's writing marking the romantic moment. They recommended use of the ironic mode, which White described as being used in narratives that are "cast in a self-consciously skeptical tone, or are 'relativizing' in their intention" (White, 1973a, p. 37). Such satirical narratives "gain their effects precisely by frustrating normal expectations about the kinds of resolutions provided by stories cast in other modes" (p. 8). Marcus and Fischer proposed that:

> The task . . . is not to escape the deeply suspicious and critical nature of the ironic mode of writing, but to embrace and utilize it in combination with other strategies for producing realist descriptions of society. . . . [B]ecause all perspectives and interpretations are subject to critical review, they must finally be left as multiple and open-ended alternatives. The only way to an accurate view and confident knowledge of the world is through a sophisticated epistemology that takes full account of intractable contradiction, paradox, irony, and uncertainty in the explanation of human activities. (Marcus & Fischer, 1986b, pp. 14–15)

How ethnographers represent themselves to establish their "ethnographic authority" was examined in detail by Clifford (1983). He distinguished "experiential," "interpretive," "dialogic," and "polyphonic" modes of authority. Anthropological writing generally has an "I was there" element. Indeed, "the predominant mode of modern fieldwork authority is signaled: 'You [the reader] are there, because I was there'" (Clifford, 1983, p. 118). But, at the same time, the *suppression* of this element establishes the "scientific" quality of the anthropology: "The actuality of discursive situations and individual interlocutors is filtered out. . . . The dialogical, situational aspects of ethnographic interpretation tend to be banished from the final representative text" (Clifford, 1983, p. 132).

For example, Malinowski's ethnography "is a complex narrative, simultaneously of Trobriand life and of ethnographic fieldwork" (p. 123). His text defined a genre both scientific and literary, "a synthetic cultural description." It included innovations designed to establish and validate the persona of the ethnographer (contrasted with amateurs, for example), to justify relatively short stays in the field, to emphasize the power of participation and observation rather than reliance on informants, to focus on "certain powerful theoretical abstractions" (such as kinship) as the way to penetrate a culture, to give a sense of the whole from the study of its parts, and to work in "the ethnographic present" and avoid the speculations of long-term historical inquiry.

The "realism" in conventional anthropological ethnographies such as Malinowski's hides the process by which such accounts are produced, and it defines the reality of the people studied from a Western point of view, which is represented as a view from nowhere. Clifford emphasized the constructed and negotiated character of ethnographic research and writing, along with its political context and role. He advocated collaborative ethnography and texts that are multivocal and open-ended.

But *Writing Culture* included no feminist contributions. Clifford attempted to explain this with the claim that "[f]eminism had not contributed much to the theoretical analysis of ethnographies as texts. Where women had made textual innovations ... they had not done so on feminist grounds" (Clifford, 1983, p. 20). In *Women Writing Culture*, Ruth Behar and Deborah Gordon pointed out that women such as Jean Briggs (e.g., Briggs, 1970) and Laura Bohannan (e.g., Bowen, 1964) had been crossing the line between anthropology and literature for many years (Behar & Gordon, 1995). Behar and Gordon applauded the effort by Clifford and Marcus to question "the politics of a poetics that depends on the words of (frequently less privileged) others for its existence and yet offers none of the benefits of authorship to those others," and to "[d]ecolonize the power relations inherent in the representation of the Other." They noted that "The *Writing Culture* agenda promised to renew anthropology's faltering sense of purpose" (Behar, 1995, p. 4). But they pointed out that female anthropologists have good reason to be familiar with the practices and politics of objectification: "When a woman sits down to write, all eyes are on her. The woman who is turning others into the object of her gaze is herself already an object of the gaze" (Behar, 1995, p. 2). And they pointed out the irony that "[a]s women we were being 'liberated' to write culture more creatively, more self-consciously, more engagingly by male colleagues who continued to operate within a gendered hierarchy that reproduced the usual structure of power relations within anthropology, the academy, and society in general" (p. 5). "Even the personal voice, undermined when used by women, was given the seal of approval in men's ethnographic accounts, reclassified in more academically favorable terms as 'reflexive' and 'experimental'" (p. 4).

Women Writing Culture set out to be a more ambitious text, with "its humor, its pathos, its democratizing politics, its attention to race and ethnicity as well as to culture, its engendered self-consciousness, its awareness of the academy as a knowledge-factory, its dreams" (Behar, 1995, p. 6). The book was "multivoiced and includes biographical, historical, and literary essays, fiction autobiography, theatre, poetry, life stories, travelogues, social criticism, fieldwork accounts, and blended texts of various kinds" (p. 7). It stemmed from a "double crisis – the crisis in anthropology and the crisis in feminism" (p. 3). And it raised perhaps the most important question, "Who has the right to write culture for whom?" (p. 7).

Not surprisingly, given his background in literature, Geertz, too, came to emphasize how "the writing of ethnography involves telling stories, making pictures, concocting symbolisms, and deploying tropes" (Geertz, 1988, p. 140). Indeed, he acknowledged (in words one could apply to his analysis of the cockfight) that "the pretense of looking at the world directly, as though through a one-way screen, seeing others as they really are when only God is looking . . . is itself a rhetorical strategy, a mode of persuasion; one that it may well be difficult wholly to abandon and still be read, or wholly to maintain and still be believed" (p. 141). He had already noted that "[s]elf-consciousness about modes of representation (not to speak of experiments with them) has been very lacking in anthropology" (Geertz, 1973, p. 19n) and proposed that anthropological writings are "fictions in the sense that they are 'something made,' 'something fashioned' – the original meaning of *fictio* – not that they are false, unfactual, or merely 'as if' thought experiments" (p. 15). He now coined the term "faction": "imaginative writing about real people in real places at real times," writing that is imaginative but not imaginary.

Geertz emphasized that "the burden of authorship" cannot be shrugged off. Yet he described this burden in terms that still showed his faith in the "exact," the "actual," and the "veridical":

All this is not to say that descriptions of how things look to one's subjects, efforts to get texts exact and translations veridical, concern with allowing to the people one writes about an imaginative existence in one's text corresponding to their actual one in society, explicit reflection upon what field work does or doesn't do to the field worker, and rigorous examination of one's assumptions are supremely worth doing for anyone who aspires to tell someone leading a French sort of life what leading an Ethiopian one is like. It is to say that doing so does not relieve one of the burden of authorship; it deepens it. (Geertz, 1988, pp. 145–146)

Geertz sounded weary with what he called "'author-saturated,' supersaturated even" texts (p. 97) – ethnographic accounts in which "the self the text creates and the self that creates the text are represented as being very near to identical" (p. 97). Such writing, he suggested, conveys a lack of confidence, even a "malaise." Its overall message is that "Being There is not just practically difficult. There is something corrupting about it altogether" (p. 97).

Not corrupting, perhaps, but certainly a form of constructing that should be acknowledged. The analyses in *Writing Culture* made it clear that ethnographies, both new and classical, construct the cultures they describe just as much as they construct the authors who write them: "If 'culture' is not an object to be described, neither is it a unified corpus of symbols and meanings that can be definitively interpreted. Culture is contested, temporal, and emergent. Representation and explanation – both by insiders and outsiders – is implicated in this emergence" (Clifford & Marcus, 1986, p. 19).

The reader of a published ethnography, whether it is classic or experimental, is being invited to imagine a world and in doing so see their own forms of life in a new way. At this point, culture and the ethnographer become *effects* of the text. Clifford and Marcus were touching on the fact that in a

globalized world these effects fold back to touch the lives of the people
described.

For example, a small town in Mexico named Tepoztlan was studied in
turn by two American ethnographers, Robert Redfield (1930) and Oscar Lewis
(1951). When I visited Tepoztlan in 2003, a local group had formed to read and
criticize these published ethnographies. At this point, the notion of
"veridicality," or what the inhabitants of this town are "actually doing,"
becomes moot: the ethnographies were *changing* what the Tepoztecos were
doing. This is not what their authors planned, I am sure. But to take this kind
of meeting of two forms of life into account is to begin to grasp what fieldwork
can become when its accounts have *local* accountability.

The Field

The preceding section was a long and complicated look at the *work* of fieldwork. We saw that anthropologists have reconsidered both the possibility and
the necessity of participating as a member in a foreign form of life and have
developed the notion of observation to acknowledge that writing is essential
both in inscribing fieldnotes and crafting a polished ethnographic account.
Both traditional aspects of the work of ethnography, *participation* and *observation*, are being reconceived. But before we draw conclusions from these
debates, we will turn to the third and final element of ethnography, the *field* in
which work is done and notes are written. This, too, has been called into
question in important ways.

Debates over the character of ethnography have also dealt with the field
and have led to fresh views of the *object* of ethnography. "Once one compli-
cates and historicizes the 'notes' in 'field/notes,' the boundaries of the first
term, 'field,' begin to blur" (Clifford, 1990, p. 64). Fieldnotes are notes *about*
the field made *in* the field. The very term fieldnotes reinforces the image that
what an ethnographer studies is found in a particular place and should be
written about by entering that place.

Ethnographic fieldwork has generally taken for granted a "traditional
mise-en-scène" (Marcus, 1997, p. 87). The term is French, literally "putting
into the scene." In filmwork, it refers to the arrangement of what appears in
front of the camera. In anthropology, the mise-en-scène of fieldwork has been
"a bounded culture" – a form of life that is "distinctively other." In this
traditional view, "the scene of fieldwork and the object of study are still
essentially coterminous, together establishing a culture situated in place and
to be learned about by one's presence *inside* it in sustained interaction" (p. 92).
This conception of a culture as a field that is situated in place as though laid
out in front of the fieldworker goes hand in hand with the traditional
conception of fieldwork as entry and participation, the effort to become an
insider.

To put it another way, cultural anthropology has accepted certain "ontological presumptions" (Faubion, 2001, p. 44). These include *boundedness* ("the presumption that each culture, if not literally confined to an island, could be approached as if it were"), *integration* ("even if a thing of shreds and patches, a culture had always also to be a thing of stitches and seams, a quilt or tapestry"), and *systematicity* ("each cultural 'part' . . . [was] the interpretive context at once for every other part and for the totality that comprised them") (p. 44). A culture has been assumed to be in a specific place with clear boundaries, an integrated whole largely free from conflicts and contradictions, with its parts working systematically together.

Marcus traces this ontology back to Gadamer, for whom as we saw in Chapter 4 one has to be inside a tradition to understand a text. As Terry Eagleton points out, we need to

> ask Gadamer whose and what "tradition" he actually has in mind. For his theory holds only on the enormous assumption that there is indeed a single "mainstream" tradition: that all "valid" works participate in it; that history forms an unbroken continuum, free of decisive rupture, conflict and contradiction; and that the prejudices which "we" (who?) have inherited from the "tradition" are to be cherished. . . . It is, in short, a grossly complacent theory of history. . . . It has little conception of history and tradition as oppressive as well as liberating forces, areas rent by conflict and domination. History for Gadamer is not a place of struggle, discontinuity and exclusion but a "continuing chain," an ever-flowing river, almost, one might say, a club of the like-minded. (Eagleton, 1983, p. 73)

Gadamer is of course aware that historical differences exist and that there is often a temporal gap between a text and its reader. But he believes that shared tradition enables the gap to be bridged, the horizons fused – bridged or, as Eagleton rather harshly puts it, "effectively liquidated" (p. 73). Gadamer's hermeneutics has no place for conflicting points of view in a tradition, no time for interpretations coming from outside.

Ethnography, too, has usually treated culture as something shared, in which everyone has the same values and the same practices. It has neglected the foreign context from which the fieldworker arrives, in particular the broader colonial context that arranged access for many anthropologists. It has ignored the complex organization – the discontinuity and multiplicity – of cultural processes today: "What is missing in the evocation of the ideal of collaboration is the much more complicated and contemporary sense of the broader context of anthropology operating in a so-called postmodern world of discontinuous cultural formations and multiple sites of cultural production" (Marcus, 1997, p. 93).

Perhaps anthropologists should never have assumed that they were studying something bounded, integrated, and systematic. The fieldworker has never been simply "there," alone on a tropical island, but has always been at the same time "here," in the world of academic research. What is certain is that *today* there is no justification for such an assumption. The location of contemporary ethnography is not single and simple, but multiple and complex. The work of the fieldworker must traverse this multiplicity:

> However far from the groves of academe anthropologists seek out their subjects – a shelved beach in Polynesia, a charred plateau in Amazonia; Akobo, Meknes, Panther Burn – they write their accounts with the world of lecterns, libraries, blackboards, and seminars all about them. This is the world that produces anthropologists, that licenses them to do the kind of work they do, and within which the kind of work they do must find a place if it is to count as worth attention. In itself, Being There is a postcard experience ("I've been to Katmandu – have you?"). It is Being Here, a scholar among scholars, that gets your anthropology read … published, reviewed, cited, taught. (Geertz, 1988, p. 130, ellipsis in original)

Let us again turn to Geertz, who has seen how "[t]he gap between engaging others where they are and representing them where they aren't, always immense but not much noticed, has suddenly become extremely visible. What once seemed only technically difficult, getting 'their' lives into 'our' works, has turned morally, politically, even epistemologically, delicate" (1988, p. 130). The distance between Here and There has become much smaller in a world linked by new technologies of travel and communication. It is now evident that an ethnography is conducted as much in the academy as it is on a foreign island, and is as much *about* the form of life of the fieldworker as it is about the culture ostensibly studied. The morality of Western, white researchers describing the lives of "exotic," "primitive" peoples has become suspect, just as the veridicality of their accounts has become dubious. The easy realism of classic ethnographic accounts – saying, implicitly if not explicitly, "Here's where they are, here's what they do, and here's why they do it" – has lost its compelling and unexamined persuasiveness. What, exactly, is an ethnography about, and who is it for?

IMPLICATIONS: RETHINKING FIELDWORK

The debates over ethnography have called into question each of its three elements: the *work* of *participation* and *observation* and the *field* that is studied. The arguments over fieldwork, in anthropology and other disciplines, have thrown into doubt the very object of anthropological inquiry. The concept of *culture* itself is up for grabs, its ontology brought into disrepute. Faubion believes that "no shared ontological alternative, no common replacement model of culture," has arrived (Faubion, 2001, p. 48). But without an acceptable ontology, the character of ethnographic investigation will remain unclear.

Can we propose a possible ontology in which the work of the ethnographer will neither dissolve into membership nor be reduced to that of observer and object?

Rethinking the Object of Ethnography

The components of ethnography – *work* and *field* – stand or fall together. If the object of ethnographic inquiry is a "field" that is clearly bounded, integrated, and systematic – in the paradigmatic case an island – then the work of the ethnographer is to enter, become an insider, and describe what can be observed from this point of view. But if the object is multiple, dispersed, and contested, the ethnographer must work to find it and discover how it is defined. Anthropology can

> no longer present fieldwork as an encounter between subject and object, nor even between one object and another. It would instead have to present it as the encounter between (at least) one intersubjective order and another – that which the anthropologist, as an enculturated being, brought to the field, and that (or those) with which her informants and interlocutors confronted her. (Faubion, 2001, p. 49)

Anthropology is recognizing that it is "embedded in a world system" and that as a consequence it "moves out from the single sites and local situations of conventional ethnographic research designs to examine the circulation of cultural meanings, objects, and identities in diffuse time-space" (Marcus, 1995, p. 96). This requires new research strategies and designs, informed by "macrotheoretical concepts and narratives." The fieldworker will need to trace "unexpected trajectories," "following the thread of the cultural process itself," and work "across and within multiple sites of activity." This is "a differently configured spatial canvas" (p. 98).

By a "world system" Marcus does not mean something macro and holistic that provides context to local settings but, like Garfinkel, something "integral to and embedded in discontinuous, multi-sited objects of study" (Marcus, 1995, p. 97). The point is not that the researcher needs to study the global as well as the local but that "there is no global in the local–global contrast now so frequently evoked. The global is an emergent dimension of arguing about the connection among sites in a multi-sited ethnography" (p. 99). The researcher needs to trace unexpected and unexplored linkages and connections, all of which are local (Latour, 1996). What is required is "a research design of juxtapositions in which the global is collapsed into and made an integral part of parallel, related local situations rather than something monolithic or external to them" (Marcus, 1995, p. 102).

When culture is not located in a specific place, the researcher "maps" the "fractured, discontinuous plane of movement and discovery among sites."

They must explore "logics of relationship, translation, and association among these sites" (Marcus, 1995, p. 102). The "contours, sites, and relationships are not known beforehand," for one is studying "different, complexly connected real-world sites of investigation." The ethnographer must write "accounts of cultures composed in a landscape for which there is as yet no developed theoretical conception or descriptive model" (p. 102).

Anthropology has been engaged in a "practical critique of ontology" in which cultures are now viewed "not as naturally bounded wholes but instead as artfully constructed differentia – sometimes found, sometimes invented, from one case to the next" and "not as spatial but rather as temporal and processual" (Faubion, 2001, p. 46). The traditional view that ethnographic fieldwork is the study of "a culture" located in a particular space and time (a There) is giving way to the view that it must be the study of "cultural processes" distributed dynamically through spaces and times that these processes themselves play a part in defining, and that link to the ethnographer in complex ways so that the distinction between There and Here dissolves. Let us start to call this new object a "form of life."

Rethinking Participation

Whereas Geertz struggled to grasp how fieldworkers can understand the insider's point of view if they cannot become insiders, Marcus (1997) has argued that it is both inevitable and necessary for the ethnographer to remain an *outsider* because outsider status is a condition for ethnographic knowledge. An outsider can raise questions that a member cannot, and in so doing learn new things about their own culture. In a position that is similar to Gadamer's, Marcus quotes Mikhail Bakhtin:

> In order to understand, it is immensely important for the person who understands to be located outside the object of his or her creative understanding – in time, in space, in culture. In the realm of culture, outsidedness is a most powerful factor in understanding. We raise new questions for a foreign culture, ones that it did not raise for itself; we seek answers to our own questions in it; and the foreign culture responds to us by revealing to us its new aspects and new semantic depths. (Bakhtin, *Speech Genres and Other Essays*, 1994, cited in Marcus, 1997, p. 95)

Marcus argues that when we examine "the central relationship of anthropologist to informant" (Marcus, 1997, p. 88) we see that fieldwork involves not rapport but "complicity," which he defines as "being an accomplice; partnership in an evil action. State of being complex or involved." The "fictions of fieldwork relations" have disguised what actually happens; the myth of rapport supports the illusion that the ethnographer must become an insider in a self-contained field. Marcus points out that Geertz's story about the police raid can actually be read as an account of complicity rather than rapport: "In the cockfight anecdote, complicity makes the outsider the desired anthropological insider. It is a circumstantial, fortuitous complicity that, by precipitating a momentary bond of solidarity, gains Geertz admission to the inside of Balinese relations" (Marcus, 1997, p. 89).

But, for Marcus, complicity is more than this "ironic means to a rapport that cements the working bond between fieldworker and informant" (Marcus, 1997, p. 94). It is "the defining element of the relationship between the anthropologist and the broader colonial context" (p. 94). It was a mistake to think of fieldwork as "crossing the border" to enter an alien form of life that has its own unique "cultural logic of enclosed difference" (p. 96). The fieldworker is always a stranger, a newcomer, and what happens in a local setting is inevitably in part a reaction to things happening elsewhere. "In any particular location certain practices, anxieties, and ambivalences are present as specific responses to the intimate functioning of nonlocal agencies and causes" (p. 96). What this calls for on the part of the researcher "is an awareness of existential doubleness on the part of *both* anthropologist and subject; this derives from having a sense of being *here* where major transformations are under way that are tied to things happening simultaneously *elsewhere*, but not having a certainty or authoritative representation of what those connections are" (p. 96).

The ethnographer is a stranger and a *representative*, a "marker," of the outside world to the people they study. "The idea of complicity forces the recognition of ethnographers as ever-present markers of 'outsideness'" (Marcus, 1997, p. 97). They are always on the boundary, never full participants in the cultural processes they study, but this makes possible "an affinity between ethnographer and subject" that the traditional concept of rapport cannot grasp. The fieldworker must continually renegotiate not *entry* but the fragile *access* they have to the phenomena of interest (Harrington, 2003).

The notion of complicity changes our understanding of the goal of fieldwork. What the ethnographer seeks "is not so much local knowledge as an articulation of the forms of anxiety that are generated by the awareness of being affected by what is elsewhere without knowing what the particular connections to that elsewhere might be. The ethnographer on the scene in this sense makes that elsewhere *present*" (Marcus, 1997, p. 97, emphasis original).

Because the ethnographic fieldworker is inevitably and irremediably a member of *another* form of life, ethnography is not an *individual* activity but a contact between forms of life: one here, the other elsewhere. The aim of ethnography is not to explore an internal cultural logic but to articulate knowledge "that arises from the anxieties of knowing that one is somehow tied into what is happening elsewhere." The members of a form of life are always "participating in discourses that are thoroughly localized but that are not their own" (Marcus, 1997, pp. 97–98).

Participation in the Academy

We have seen that ethnographies written *about* the field must be understood as having been written *for* the academy. They are inevitably shaped by what Clifford has called the "politics of knowledge." As much when writing notes in the field as when polishing the final draft of a monograph, the ethnographer has professional colleagues in mind as the audience. The successful ethnographer is one who skillfully anticipates the effects their written representations will have on their colleagues. To manage this effect ethically requires fieldwork with a *reflexive* element: "The fundamental problem here is in confronting the politics of knowledge that any project of fieldwork involves and the ethnographer's trying to gain position in relation to this politics by making this terrain itself part of the design of the fieldwork investigation" (Holmes & Marcus, 2005, p. 1101).

Paul Rabinow pointed out in his chapter in *Writing Culture* that Clifford's analysis of the politics of knowledge neglected academic politics. Anticolonialism is beside the point today: what is needed is more attention to the impact of *academic* power relations in the production of ethnographic texts. Clifford studied the texts but ignored the power: "[The] micropractices of the academy might well do with some scrutiny. . . . My wager is that looking at the conditions under which people are hired, given tenure, published, awarded grants, and feted would repay the effort" (Rabinow, 1986, p. 253). Rabinow noted that feminist anthropology has acknowledged the "initial and unassimilable fact of domination" (p. 255) and sought not simply to improve contemporary academic anthropology but to "shift the discourse" – to change the audience, the readership, who speaks, and the subject matter of conversation. The basic goals have been "resistance and nonassimilation" (p. 256).

Rabinow distinguished three positions in academic ethnography. Interpretive anthropologists, such as Geertz, understand themselves as being scientists seeking truth. For them, representations are the issue and "the greatest danger, seen from the inside, is the confusion of science and politics. The greatest weakness, seen from the outside, is the historical, political, and experiential cordon sanitaire drawn around interpretive science" (Rabinow, 1986, p. 257). Anthropological critics, such as Clifford, attend to textual rhetorical devices and see "domination, exclusion, and inequality," but only as subject matter, something given form by the writer (anthropologist or native). Political anthropologists, like the feminists, seek to constitute "a community-based political subjectivity" (p. 257). Rabinow described his own position as that of a "critical, cosmopolitan intellectual" (p. 258). This is an oppositional position, suspicious, guided first by ethical considerations, second by understanding. The cosmopolitan lives and thinks *between* the local and the universal, and aims to be attentive to difference without essentializing it: "What we share as a condition of our existence . . . is a specificity of historical experience and place, however complex and contestable they might be, and a worldwide macro-interdependency encompassing any local particularity. Whether we like it or not, we are all in the same situation" (p. 258).

Rethinking Writing

The most charged debates over ethnography have been about the role of writing. The challenges to traditional fieldwork "do not arise simply from the complexities of a postmodern or now globalizing world" but as much from a "diminution" in anthropology "of its distinctive documentary function amid many competing and overlapping forms of representation comparable to its own" (Holmes & Marcus, 2005, p. 1100). Anthropology has experienced a crisis that is both moral and epistemological. The moral aspect stemmed from its links to colonialism and extended into the politics of academia – *why* and *for whom* should ethnographers write? The epistemological aspect was the recognition that anthropological description can no longer be viewed as *mimesis*, as a straightforward realism. *How* should ethnographers write? The goals, relevance, procedures, and motives of academic anthropology now must "all be questioned" (Geertz, 1988, p. 139).

Some, including Geertz himself, had doubts about the value of these explorations of ethnographic rhetoric. But they undoubtedly have had an impact. The analysis of rhetorical style and the exploration of alternatives have spread from anthropology to other disciplines that use ethnography. One can now find a variety of realist, impressionist, and confessional texts (Van Maanen, 1988). And "[t]he strong objections to the use of the first person, any concern with the position of the fieldworker, any dwelling on the issue of style, have, as it were, melted away" (Spencer, 2001, p. 449).

Clifford has recommended "self-reflexive fieldwork accounts" that include new uses of the first-person singular, autobiography, and ironic self-portraits. In these, the ethnographer "is at center stage," allowed to be "a character in a fiction" (Clifford, 1986, p. 14), and "he or she can speak of previously 'irrelevant' topics: violence and desire, confusions, struggles and economic transactions with informants." These topics are now "seen as constitutive, inescapable." In such writing, "the rhetoric of experienced objectivity" is replaced by something more fallible, more human (p. 14). These new fieldwork accounts may even stage dialogues between the ethnographer and members of the culture, or narrate interpersonal confrontations. Clifford suggests that:

> [T]hese fictions of dialogue have the effect of transforming the "cultural" text (a ritual, an institution, a life history, or any unit of typical behavior to be described or interpreted) into a speaking subject, who sees as well as is seen, who evades, argues, probes back. In this view of ethnography, the proper referent of any account is not a represented "world"; now it is specific instances of discourse. (Clifford, 1986, p. 14)

Such an approach "obliges writers to find diverse ways of rendering negotiated realities as multisubjective, power-laden, and incongruent. In this view, "culture" is always relational, an inscription of communicative processes that exist, historically, *between* subjects in relations of power" (p. 15).

Marcus has proposed new strategies for fresh experimental writing that avoid the tendency to present particular places, events, and people as representing cultural wholes (Marcus, 1994, 1998). The first set of strategies involves "redesigning the observed" by first "problematizing the spatial" in order to "break with the trope of community" and acknowledging that people are not defined entirely by their neighbors. Second is "problematizing the temporal" to "break with the trope of history" and of historical determination. Third is "problematizing perspective or voice" by questioning the concept of structure either as social organization or as an underlying system of rules, recognizing that all ethnographic knowledge is fundamentally oral, transformed by writing, and employing voice in the form of embedded discourse.

A second set of strategies involves "redesigning the observer" by first appropriating dialogue through a text's "conceptual apparatus." Where realist ethnography pulls indigenous symbols or concepts from their discursive contexts and reinserts them into the ethnographer's discourse, we should make this exegesis explicit and turn it into a dialogue with our own concepts. Rather than explicating what an indigenous concept "means," we would initiate an exploration of anthropological concepts and Western concepts more broadly. The second strategy is "bifocality," looking at least two ways, exploring the history of ethnographic inquiry as much as the history of the cultural process under investigation. Finally, "critical juxtapositions and contemplation of alternative possibilities" turn ethnography and anthropology more generally into forms of cultural critique.

"Creative nonfiction" (Gerard, 1996), the use of literary techniques when writing about real events, is growing in popularity. One of my favorite examples, though it is not from an anthropologist, is John McPhee's (1969) description of the U.S. Open semifinal tennis match between Arthur Ashe and Clark Graebner. McPhee moves between a blow-by-blow account of the game and a detailed exploration of each player's background and personality. In another work (McPhee, 2000), he presents a discussion of the geology of the United States as a personal journey along Highway 66 from East Coast to West.

"Autoethnography" has become more accepted, though the meaning of this term has shifted. Originally autoethnography referred to those rare occasions when the ethnographer was a full member of the group studied (Hayano, 1979). Now it generally indicates a focus on the researcher's subjectivity. Clearly these are related: when the ethnographer is a full member, their subjectivity presumably exemplifies the culture of the group. But autoethnographic texts now focus on the ethnographer's experience even when they are not a member, and they are often written in a consciously performative mode ("evocative ethnography": Holman Jones, 2005). At one extreme, autoethnography becomes autobiography. But it has also been argued that autoethnography is entirely compatible with the aims of conventional "analytic ethnography" and that ethnography has always involved biographical engagement (Atkinson, 2006). It has always been the case that "[t]he labour of ethnography is emotional and potentially intimate. Identity work and the (re)construction of the self is part and parcel of the ethnographic endeavour" (Atkinson et al., 2001, p. 322).

But still some complain that "genuinely radical experiments with the mode of ethnographic representation remain as rare as ever" (Spencer, 2001, p. 450). And Gordon has raised important questions about this new experimental writing in ethnography. In particular, she questions the tendency to distinguish categorically "between [writing] which reflects on processes of representation and [writing] which does not" (Gordon, 1988, p. 9). She suggests that feminist ethnography already exhibited experimental explorations, such as community building that did not pit agency and structure against one another and in which there was "a form of subjectivity and rationality which is neither 'always already' positioned nor transcendental but actualized with an ongoing movement of political prioritizing" (p. 20). She emphasizes that feminism has always argued that seemingly unified cultures are sites of difference and contestation.

Certainly, the calls for experimental writing and "performative" ethnography (poetry, dramatic reconstruction, and so on) risk missing the point that *all* writing is performative. Malinowski's ethnographic accounts remain compelling even though they are traditional and realist in their style. Undoubtedly there is value in loosening the bonds that define acceptable writing, but this alone does not resolve the contradictions in ethnography. If experimental writing becomes merely a celebration of relativity, it will undercut the radicality of the debate over ethnography, which is not that there is no truth but that truths are products of the contact among distinct forms of life, conditioned by historical and cultural circumstances that are themselves results of struggles for power. The ethnographer cannot step into a culture and out of these struggles. Experimental writing cannot be a substitute for a clear sense of *why* one conducts ethnographic fieldwork and *who* one writes *for*. Who writes culture, and for whom?

Rethinking Observation

Moreover, there is a troubling gap between the reflections on published ethnographies and reflections on the writing in fieldnotes. It is now clear that fieldnotes are not an objective record but a fixing and inscribing. Yet the notion that fieldnotes should be "veridical" keeps slipping back in. Very little is known about the ways fieldnotes provide the basis for a published ethnography. Fieldnotes are treated as personal and private and, as a result, "[a] great deal of ethnographic writing carries little or no explicit reference to the ethnographic work on which it is based" (Spencer, 2001, p. 448).

Sanjek (1990) surveyed anthropologists' methods for analyzing fieldnotes. The most common approach was to index them, using either prior categories or ones apparently relevant in the field or afterward. Once indexed, the fieldnotes could be rearranged in topical order. Notes dealing with the same topic could be collated, selected, and summarized, or even coded, quantified, and sorted (Sanjek, 1990, p. 404). This approach is alarmingly similar to the analysis of interview transcripts by abstraction and generalization we questioned in Chapter 4, and it poses the same dangers of removing notes from context and focusing on categories instead of concrete detail. Once again, it presumes that chronology is unimportant, reinforcing the traditional conception of culture as timeless.

This approach to analysis also fails to pay attention to the "prefiguration" of the field in writing, which Clifford insisted is of crucial importance. A thick description pays attention to context, but Hayden White's work has shown us that "[a]ny image of a historical context is itself, taken as a whole, a prior interpretation chosen for a particular purpose and is in no way less problematic than the literary text [substitute "social act"] that constitutes its part" (Kellner, 2005). Clifford (1990) draws on White to observe that the field can be prefigured in diverse ways: as an image or pattern (with metaphors such as the body, as in Malinowski's text, or architecture, or landscape); as a collection of facts (metonymy); as a hierarchical, functional, or organizational whole (synecdoche); or as temporal, passing (irony). Analyzing fieldnotes by coding and indexing them fails to consider the way of seeing that provides the basis for what the ethnographer writes, or how this way of seeing is transformed by the practice of writing.

Even now the poetic dimension of published ethnographies has been granted, albeit without complete agreement on the implications, ethnographers have not applied the same insight to the writing of fieldnotes. We need to explore how ethnographers write fieldnotes to *learn* how to see the social practices they are studying, and how writing educates their professional vision. When ethnographers put pen to paper, they must struggle to articulate the understanding they have obtained as a stranger who witnesses a new form of life. The connection between the two aspects of the work of ethnography is not one-way: participation may seem the basis for understanding, but the ability to recognize what is a wink and what is a blink and articulate this in writing seems to be the *product* of ethnography, not its *prerequisite*. Ethnography needs to find room for an account of how *mis*understanding can turn into understanding (Fabian, 1995). It needs to consider how an ethnographer's vision develops as they gain familiarity with a form of life and how writing notes informs this development.

The Dualism of Participant Observation

Ethnographers who grant that their fieldnotes should be *thick* will often continue to maintain that these notes are indeed *descriptions* of a cultural order. By continuing to use the term "description," Geertz perpetuated the idea that the task of ethnography is to "describe," that is to say represent, a culture (Clifford, 1990, p. 68). This distinction between the object and its verbal description is dualism once more:

> The problems encountered in describing and explaining social action hinge in part on the notion that, on the one hand, there is an event in the world, a social action, and on the other, another event which is the description of that action and the method of producing that description. This dualism – which is an instance of the subject–object dualism – generates the issue of the veridicality of the description and, in the context of science, the necessity of literal description. (Wieder, 1974, p. 171)

If they are finally to escape from this dualism, ethnographers need to recognize that their accounts are always elements *within* the social situations that they write about. Ethnographic accounts – both fieldnotes and published ethnographies – function as instructions *to see* actions and events in a particular way. When traditional ethnographers set out to describe a social order that they assumed was a bounded object, a tropical island to be entered, leaving their own world behind, they practiced what Garfinkel (following Mannheim) calls "the documentary method of interpretation." We saw in Chapter 8 that in the documentary method an account is written or read as evidence for an underlying, hidden organization. In this case, anthropologists analyze their notes to discover properties of the order that "produced" them and "explains" them. But in fact ethnographers are involved in the same kind of work as the people they study: they are *making* order, *organizing* the complex details of everyday life in an orderly way that will be compelling to the people who read their accounts. Participating (to gain access to the order, albeit as a stranger) and accounting (representing that order) cannot be kept separate.

Returning one last time to the texts of Geertz, we find in a footnote his suggestion that the notion of participant observation "has led the anthropologist to block from his view the very special, culturally bracketed nature of his own role and to imagine himself something more than an interested (in both senses of that word) sojourner." Participant observation, Geertz proposes, "has been our most powerful source of bad faith" (Geertz, 1973, p. 20). Once again, Geertz's writing contains implications he himself didn't fully explore or perhaps even see. The notion of participant observation builds a dualism into ethnography from the start. "As an insider, the fieldworker learns what behavior means for the people themselves. As an outsider, the fieldworker observes, experiences, and makes comparisons in ways that insiders can or would not" (Sluka & Robben, 2007, p. 2). But fieldworkers are never simply members, nor do they describe objectively. Ethnography cannot describe "the native's point of view," nor is it a literal report of the observed facts. The ethnographer is a visitor, a stranger, a newcomer, who writes about a form of life to offer a way of making sense of it, a way of seeing it. Ethnography is an activity in which one culture takes account of another to learn just as much, surely, about itself as about the other. "The ethnography – in sociology and in anthropology – was born with this paradox, this asymmetry.... Ethnographers are conscious of the cultural conventions that are their subject-matter, but have all too often remained blissfully unaware of their own cultural conventions" (Atkinson, 1990, pp. 177–178).

CONCLUSIONS

> Fieldwork is a complex historical, political, intersubjective set of experiences which escapes the metaphors of participation, observation, initiation, rapport, induction, learning, and so forth, often deployed to account for it. (Clifford, 1990, p. 53)

In reviewing more than 20 years of debate in cultural anthropology, I have tried to maintain a focus on the ontological commitments that are embedded in research practice, and their implications for it. In ethnography, we have seen the distinction between a traditional ontology in which the object is taken to be bounded, systematic, and integrated and a new ontology, by no means fully accepted, in which a form of life is taken to be open, dynamic, and contested.

With this new way of thinking of the field, the object of ethnographic investigation, we can no longer think of the work as participant observation. Fieldwork is not simply a matter of gaining entry, participating as a member, and writing veridical reports of the world seen from this viewpoint. It requires tracing changing linkages, always as a stranger from another form of life, writing about what one sees in order to see it better, in order to make evident the order of a form of life. Let us borrow a phrase from Husserl and call this order the *regional ontology* of a form of life. It will be a "variable ontology," to borrow a phrase from Bruno Latour. It is how things and people show up. Writing ethnography is itself a matter of making things show up: writing to have an effect on one's readers, so they will see in a new and different way.

The calls in the 1970s for interpretive inquiry emphasized that it would require immersion in practical activities and focus on the mutual constitution of person and form of life. For most of its history, ethnography has not studied constitution as we came to understand it in the preceding chapters. Even Geertz, who was one of those who called for a new interpretive anthropology, reduced constitution to "a kind of sentimental education" (Geertz, 1972, p. 236) in which an event like the cockfight offers the Balinese a "metasocial commentary" (p. 234). Although he claimed that such events "generate and regenerate the very subjectivity they pretend only to display" (p. 237), that they are "positive agents in the creation and maintenance of ... a sensibility" (p. 238), this boiled down to no more than the Balinese "reading" the text of their culture to see "what it says."

Ethnographic studies of constitution need to turn the tables on Kant, and they have not yet done so. We know that far from mind constituting reality, practical activities constitute mind and world. This needs to be fleshed out in detailed studies of specific bodies in concrete interactions in distinct spaces and times, but only a few of these exist. Fieldwork is necessary to discover not only the order of a form of life but also the work that goes into producing and reproducing this order.

The new ontology, acknowledging that a form of life is vital, hybrid, and multisited, requires that the fieldworker step into the flux, trace links and connections as they are forged and then dissolved, and study unfolding practices and processes. Written accounts, both (relatively) raw fieldnotes and cooked published reports, will inevitably be partial and interested. They should be subject to local accountability as well as professional accountability. Fieldworkers need to take seriously the responsibilities of their rhetoric, the effects they want to have on their readers, their strategy in the games of academic politics, and their impact on the people they study.

Malinowski held two contradictory views of the "final goal" of ethnography. The first was to grasp the native's point of view. The second was to transform *our* view of the world. On the one hand, he believed that an ethnographer can "grasp the inner meaning" (Malinowski, 1922/1961, p. 517) of a foreign culture and by "letting the native speak for themselves" can provide the reader with "a vivid impression of the events enabling him to see them in their native perspective" (p. 516). On the other hand, he believed that the ethnologist can, in a "final synthesis," grasp the varieties of cultures and "turn such knowledge into wisdom" (p. 517):

> Though it may be given to us for a moment to enter into the soul of a savage and through his eyes to look at the outer world and feel ourselves what it must feel to *him* to be himself – yet our final goal is to enrich and deepen our own world's vision, to understand our own nature and to make it finer, intellectually and artistically. ... We cannot possibly reach the final Socratic vision of knowing ourselves if we never leave the narrow confinement of the customs, beliefs and prejudices into which each man is born. (Malinowski, 1922/1961, pp. 517–518)

This would be a "Science of Man," "based on understanding of other men's point of view" (p. 518), with the promise to increase "tolerance and generosity" and "become one of the most deeply philosophic, enlightening and elevating disciplines of scientific research" (p. 518). This is indeed a noble goal, but achieving it will mean relinquishing the fantasy that we can enter the soul of someone in a foreign form of life. Yet contact with other forms of life is undoubtedly needed if we are to learn and grow. Fieldwork, properly understood and practiced, remains crucial.

10

Studying Ontological Work

[T]hese theories presuppose the prior existence of the question that they are seeking to resolve: that there is a yawning gulf separating the agent from structure, the individual from society. Now if there is no gulf, then sociological theory would find itself in the rather odd situation of having tried to provide ever more refined solutions to a nonexistent problem.

Latour, 1996, p. 232

If constitution goes on in the everyday practices of a form of life, then it makes sense that to answer questions about constitution we need to study these practices. One way would seem to be firsthand contact with the form of life through ethnographic fieldwork, but, as we have seen, ethnographic accounts that claim to describe what is "behind" everyday actions and events such as the cockfight, which are written for outsiders and are immediately removed from the form of life they represent, face insuperable problems of veracity. When fieldwork is understood as participating as a member and then writing a disinterested description of the order of a form of life, it is inherently dualistic.

Are there other ways to study practices? It might be objected that in considering only ethnography I have been unduly narrow in my exploration of the ways in which we could study constitution. Certainly other methods and conceptual frameworks exist for the study of practical interaction. In this chapter, I want to consider two of them, both of which are, at least ostensibly, focused on exploring the relationship of constitution between individual and society, between member and form of life. These two are critical discourse analysis (CDA) and conversation analysis (CA).

Proponents of critical discourse analysis have summarized its main assumptions as follows: it addresses social problems; treats power relations as discursive; sees discourse as constituting society and culture, as doing ideological work, as historical, and as a form of social action; sees the link between text and society as mediated; and offers a form of analysis that is interpretive and explanatory (Fairclough & Wodak, 1997).

The focus on constitution is obvious. Similarly, Emanuel Schegloff, a central figure in conversation analysis, has proposed that it is in face-to-face conversational interaction that the work of constitution often attributed to macro institutions is done, so that CA's study of how these interactions are organized is fundamentally important: "Conversational interaction may then be thought of as a form of social organization through which the work of the constitutive institutions of societies gets done – institutions such as the economy, the polity, the family, socialization, etc. It is, so to speak, sociological bedrock" (Schegloff, 1995, p. 187). Both of these seem promising candidates for a kind of inquiry that would enable us to answer questions about constitution. How do they compare, and how do they differ?

CRITICAL DISCOURSE ANALYSIS

Critical discourse analysis is an influential approach to the analysis of the constitution accomplished through talk. It is one form of discourse analysis, a term that covers a wide variety of approaches to the analysis of spoken language. One can easily find handbooks of discourse analysis, one as large as four volumes (van Dijk, 1985) and another 850 pages (Schiffrin, Tannen, & Hamilton, 2001), as well as a two-volume collection of discourse studies (van Dijk, 1997a, 1997b) and a variety of journals devoted to studies of discursive phenomena. Approaches to the study of discourse may focus on structure (form), use (function), or both (Schiffrin, 1994). The term "discourse" has many definitions, but most fall into three main categories: "(1) anything beyond the sentence, (2) language use, and (3) a broader range of social practices that includes nonlinguistic and nonspecific instances of language" (Schiffrin, Tannen, & Hamilton, 2001, p. 1).

What is generally referred to as critical discourse analysis falls into the third category. But CDA itself takes various forms, three of which I will examine here.

Preconstituted Objects and Subjects

Norman Fairclough gave linguistic analysis a political spin with his study of British Prime Minister Margaret Thatcher's rhetoric (Fairclough, 1989). Fairclough has proposed a methodology for "TODA" (textually oriented discourse analysis) that attends to three dimensions of discourse (Fairclough, 1992). The first is its linguistic organization (vocabulary, grammar, cohesion, and text structure), and Fairclough argues that attention to this kind of concrete detail is missing from much research on the effects of discourse (p. 37). Second is how discourse is produced, how it circulates, and how it is consumed. Forms of "intertextuality," the explicit and implicit links that discourse (and here Fairclough includes texts and text-analogues as well as spoken language) has to other texts, are relevant here. The third dimension is the way discourse is employed in social practices of ideology and hegemony.

Fairclough describes his principal aim as being to uncover the ways that social structures have an impact on discourse and vice versa. He argues that new forms of social organization – in particular, the "flexible production" of modern factories – have led to new uses of language and to new language forms that increasingly are designed deliberately and commodified. With "spinning," sound bites, and televised interviews, political discourse in particular has become, paradoxically, more informal and conversational, at least in its appearance. At the same time, processes of democratization, commodification, and what Fairclough calls "technologization" – efforts to "engineer" discursive practices – have changed power relationships and blurred the distinction between information and persuasion. Fairclough is particularly interested in the way that certain "genres" such as the examination (which objectifies people), the confession (which subjectifies them), management, and advertising have "colonized" an expanding territory of institutions and organizations. He is interested in the effects of technologization of discourse, such as techniques for putting interviewees at ease. This all sounds very compatible with the kind of constitution we explored in Chapter 8. But Fairclough is actually very skeptical about the suggestion that discourse can be constitutive (Fairclough, 1992, p. 56). He is willing to accept that both objects and subjects are "shaped" by discursive practices, but he is unwilling to accept that they are constituted in these practices: "I would wish to insist that these practices are constrained by the fact that they inevitably take place within a constituted, material reality, with preconstituted 'objects' and preconstituted social subjects" (p. 60).

For Fairclough, both people and the objects known and talked about exist outside and prior to that talk. He recommends that instead of using the term "constitution" we should instead "use both the terms 'referring' and 'signifying': discourse includes reference to preconstituted objects, as well as the creative and constitutive signification of objects" (p. 60). That is to say, although discourse can constitute *representations* of objects, it does not constitute the objects themselves. These exist as a material reality outside and prior to discourse.

This is, of course, to see discourse as having a solely representational function. The consequence is that when Fairclough focuses on phenomena such as "engineering of semantic change," it is solely in terms of the "ideational" function of language – and it is this aspect of discursive practice alone that he is willing to consider as "constructing social reality" (Fairclough, 1992, p. 169). At times, Fairclough is completely explicit that his focus is limited to "the role of discourse in constituting, reproducing, challenging and restructuring systems of knowledge and belief" (p. 169). This emphasis on knowledge and beliefs shows quite clearly that Fairclough considers constitution only in representational terms and that his treatment of constitution draws from Kant and Husserl, not from Hegel, Heidegger, and Merleau-Ponty.

As a result, Fairclough's critical discourse analysis tries to explore a relationship between two separate phenomena – talk and texts on the one hand and social structure on the other – which he assumes exist independently. He writes of them in dualistic terms, as "micro" and "macro," respectively (Fairclough, 1992, pp. 231–238). As a consequence, Fairclough's analyses move back and forth among "discourse practice," "text," and "social practice," from "macro" to "micro" and back again (pp. 231–238), without ever successfully clarifying how discourse is able to do the constitutive work he attributes to it.

How Discourse Structures Influence Mental Representations

Critical discourse analysis is also practiced by Teun van Dijk, who describes his focus as "the way social power abuse, dominance, and inequality are enacted, reproduced, and resisted by text and talk in the social and political context" (van Dijk, 2001, p. 352). But van Dijk, too, immediately introduces a dualism when he argues that "CDA has to theoretically bridge the well-known 'gap' between micro and macro approaches" (p. 354). In his view, "critical discourse analyses deal with the relationship between discourse and power" (p. 363), and this would be fine except that he considers them to be two distinct realms.

In fact, van Dijk seems to assume that macro-structures, mind, and discourse all exist separately, and he then sets out to explore the connections among them. For example, he takes for granted the existence of "mind" and mental representations rather than exploring how mind is constituted. He writes that we need to have at least "elementary insight into some of the structures of the mind, and what it means to control it" (van Dijk, 2001, p. 357), and adds that mind is a system of "beliefs, knowledge, and opinions." His interest is in how "power and dominance are involved in mind control" (p. 357) because "CDA also focuses on how *discourse structures* influence mental representations" (p. 358, emphasis original). In his view, "the crucial question is how discourse and its structures are able to exercise such control" on the mind (pp. 357–358), which again assumes that the two are distinct. When van Dijk describes this influence, it is by looking at such phenomena as the "topics" in a news headline, for example, or the "opinions that are 'hidden' in its implicit premises" (p. 358). That is to say, like Fairclough he considers discourse structures only as forms of representation.

Van Dijk acknowledges that "the details of the multidisciplinary *theory* of CDA that should relate discourse and action with cognition [mind] and society [context] are still on the agenda" (van Dijk, 2001, p. 363, emphasis original), and he adds that "[t]he relations between the social power of groups and institutions, on the one hand, and discourse on the other, as well as between discourse and cognition, and cognition and society, are vastly more complex" than the "schematic" picture he has sketched (p. 364). In such statements, the assumption is that the task of discourse analysis is to explicate connections among phenomena that are in distinct realms, construed in dualistic terms.

In a typical statement of the aims of critical discourse analysis, van Dijk writes that the "central task ... is [to] ... reconnect instances of local discourse with salient political, economic, and cultural formations (Luke, 1995, p. 11). Here again, he, like Fairclough, assumes a macro level of social formations on the one hand and a local level of discourse on the other. Given this kind of dualism, constitution is once again reduced to no more than a process of representation.

Hegemonic Practices

Finally, Ernesto Laclau and Chantal Mouffe have developed an approach to the study of discourse that draws from multiple sources. Laclau and Mouffe (1985) have argued that too often the "social" is thought of as a unitary totality when it is actually diverse, multiple, and fragmented. For example, Marxist theory (which we will explore in Chapter 11) became stuck in an economic determinism that focused on class antagonism, which Laclau and Mouffe argue is only one small part of what goes on in "political space." A major focus of this approach to discourse analysis is the way that language is used to *impose* a sense of totality on a reality that is actually more complex.

Laclau and Mouffe propose that all social phenomena should be viewed as discursive phenomena. Discourse, for them, is a way of organizing elements and delimiting them against what is excluded in a "field of discursivity." Elements are merely "empty signifiers" until they are brought into "chains of equivalence." Social practices bring elements into relationships and in doing so modify their identities. Practices "articulate" these elements, and discourse is the "structured totality" that results from such articulation. Laclau and Mouffe develop the concept of the "constitutive outside," arguing that identities are made possible by what *differs* from them. Only temporary closure against what is outside and different can ever be achieved, they argue, and in addition there will always be struggle among competing discourses to define the totality and what is outside it.

Laclau draws on Antonio Gramsci's concept of "hegemony" to emphasize that political struggle is at the heart of modern society, in its economic and political institutions and practices. Gramsci (1891–1937) was an Italian politician who proposed that the bourgeoisie maintains its dominance over the working class through a "cultural hegemony" in which a combination of institutions and social relations presents the values of one class as "common sense" for all. Laclau and Mouffe view hegemony as a strategy of combining distinct discourses into a totalizing ideology: "society" never actually exists as a complete whole. A discourse cannot organize itself; it is shaped by these "hegemonic practices."

Laclau and Mouffe propose, then, that what we experience as a complete and autonomous social reality is a myth that disguises the constituted character of any objectivity and the constitutive role of the nonidentical. This totality is in fact a hegemonic political vision that a majority of citizens have bought into.

Laclau and Mouffe propose that society should not be viewed as made up of individuals or political parties but should be seen as a set of unstable and ephemeral "subject positions" and contingent subjectivities. One of the effects of discourse is to assign subject positions to its participants. No position offers a privileged viewpoint, so actions and decisions are always contingent, will exclude some interests, and are in a sense blind. Even a consensus will exclude some interests. That is to say, as political agents change position and alignments and antagonisms emerge, what we witness is a hegemonic logic, an "agonistic pluralism."

Laclau (2000, p. xi) has described how this analytic approach tries "to move the center of social and political research from 'ontically' given objects of investigation to their 'ontological' conditions of possibility." Documents, advertisements, and television broadcasts are all treated as texts, as cultural artifacts that can be read as significant. The focus of this kind of CDA is the "mechanisms" by which meaning is produced and challenged. The researcher attends not so much to what people say as to which signs (elements) are privileged and how they are articulated (positioned and related) with respect to one another. In this way, one can "map the partial structuring by the discourses of specific domains" (Phillips & Jörgensen, 2002, p. 30). Researchers will explore antagonisms among social spaces, for these are "limits to the constitution of any objectivity" (Laclau, 2000, p. xi).

This is all promising, but nonetheless Laclau's ontological approach is described principally in representational terms. One introduction to this approach to critical discourse analysis, for example, illustrates its conceptions of meaning, discourse, and identity in the following way:

> [A] forest in the path of a proposed motorway . . . may simply represent an inconvenient obstacle impeding the rapid implementation of a new road system, or might be viewed as a site of special interest for scientists and naturalists, or a symbol of the nation's threatened natural heritage. Whatever the case, its meaning depends on the orders of discourse that constitute its identity and significance. (Howarth & Stavrakakis, 2000, pp. 2–3)

This talk of "identity" and "significance," of "representation," "meaning," and "symbol," is muddled. Howarth and Stavrakakis explain that "empirical data are viewed as sets of signifying practices that constitute a 'discourse' and its 'reality,' thus providing the conditions which enable subjects to experience the world of objects, words and practices" (p. 4). Their emphasis on signification and experience, and the quotes around the word "reality," suggest that Laclau and Mouffe, too, are taking a representational view of constitution.

CONVERSATION ANALYSIS

A very different approach to the investigation of everyday practices, one that steps beyond the treatment of discourse as representation, is "conversation analysis" (CA). This is a way of analyzing the pragmatics of ordinary conversation, focusing on the practical and interactive construction of these everyday interchanges. It was strongly influenced by ethnomethodology and has become a widely accepted approach to the study of everyday interaction. Conversation analysis, I will suggest, offers a way to include in our descriptions of a form of life an empirical warrant concerning the way its participants understand what they and others are doing. It does this without claiming to penetrate beneath the surface of events to some kind of deeper meaning. It views "social interaction as a dynamic interface between individual and social cognition on the one hand, and culture and social reproduction on the other" (Goodwin & Heritage, 1990, p. 284). It attends to the constitutive relation between person and form of life on the level of practical activity, not that of representations. In short, it is the study of what talk *does*, not what talk *means*.

Conversation analysis is an approach that views what people say in interaction together as first and foremost a kind of *action*. Working with audio or video recordings of naturally occurring interactions, it focuses on the "pragmatics" of speech rather than on the syntactics (grammar) or semantics (meaning) (see Box 10.1). These other aspects are not considered irrelevant, but the general approach has been to assume that the pragmatics is primary and the other aspects follow from it, rather than vice versa. Each utterance is considered as a speech act rather than as a statement or a grammatical string of words, and the sequence of utterances that make up a conversation is taken to be dynamically assembled. The successes of this approach have tended to confirm the validity of these assumptions.

Display of Understanding

Conversation analysis aims to describe conversation in a way that builds on the way it is understood by the people participating. It does this by treating each utterance as *displaying an interpretation* of the previous utterance. It is

BOX 10.1. *Speech as Action: The Pragmatics of Language*

Pragmatics is the study of what is done with language (see Levinson, 1983, p. 5ff, for a discussion of the difficulties in defining the term). It deals with *utterances*, what is said by particular people in specific times and places. Pragmatics looks at how talk is used to do things, how utterances are actions. The term was coined by Charles Morris (1938) as part of his general science of signs. Morris distinguished pragmatics from syntactics (the study of grammar) and semantics (the study of reference).

Ludwig Wittgenstein argued in his *Philosophical Investigations* that "meaning is use," that utterances can be understood only within the "language games" in which they play a part. The utterance "block," for example, is understood when it is used within the activity of building a wall.

The philosopher John Austin (1911–1960) explored "how to do things with words" (Austin, 1975) and drew a distinction between *performative* utterances and *constative* utterances. A performative utterance performs an action; for example, making a promise or issuing a threat. A constative utterance is saying what is true or false. But Austin pointed out that a single utterance can do both these things. He recommended that we distinguish, in any utterance, between the *locutionary* act, the act of saying something, the *illocutionary* act, the performance of an act *in* saying something, and the *perlocutionary* act, the act of producing certain effects in the hearer.

Herbert Grice (1975) emphasized the difference between *conveyed meaning* and *literal meaning* and studied the *conversational implicatures* by means of which people understand what is meant on the basis of what is said and what is left unsaid. Contextual processes of disambiguation enable us to use and understand expressions as acts of politeness, irony, and indirection. Speakers and hearers employ *conversational maxims* as guidelines; when a speaker appears not to follow these maxims, hearers infer that their utterance has a nonliteral meaning.

Philosopher John Searle, a student of Austin, developed the theory of "speech acts" (Searle, 1969), focusing on the "constitutive rules" for performing an utterance of a particular kind. Searle proposed that there are several kinds of felicity conditions that define a speech act: rules delimiting the propositional content of the illocutionary act, preparatory rules that specify what the act implies, sincerity rules that specify the speaker's commitments, and essential rules that define what kind of action the utterance counts as.

Searle further suggested that just five kinds of speech action are possible: representatives, directives, commissives, expressives, and declarations. *Representatives* commit the speaker to the truth of the propositional content ("It is hot"). *Directives* are attempts to have the hearer perform an action

("Please close the door"). *Commissives* commit the speaker to a future action ("I'll put out the trash later"). *Expressives* express the speaker's state of mind ("Thank you for the gift"). *Declarations* change the institutional state of affairs ("I declare this building open").

A different approach to pragmatics is the *ethnography of speaking*, which focuses on those *speech events* that are recognized within a particular culturally recognized social activity (Gumperz, 1982; Hymes, 1964). *Conversation analysis* (see this chapter) differs from speech act theory in its emphasis on the way hearers make central use of context (both conversational context and nonlinguistic setting) in order to understand a speaker's utterance (Duranti & Goodwin, 1992).

here that we find the empirical warrant, in the attention paid in CA to the understanding displayed by the participants themselves:

> The methodology employed in CA requires evidence not only that some aspect of conversation *can* be viewed in the way suggested, but that it actually is so conceived by the participants producing it. That is, what conversation analysts are trying to model are the procedures and expectations actually employed by participants in producing and understanding conversation. . . . We may start with the problem of demonstrating that some conversational organization is actually oriented to (i.e. implicitly recognized by) participants, rather than being an artifact of analysis. (Levinson, 1983, p. 319)

Conversational Pairs

The display of understanding by the participants in a conversation is derived from a fundamental organizational feature of conversation, namely that it is composed not of individual utterances but of *pairs* of utterances that are, of course, produced not by a single speaker but by two (or more) speakers.

We frequently speak of an "exchange of opinions" or "an exchange of greetings" because many conversational actions call for a particular kind of conversational response in return. Greetings and farewells typically call for another utterance of the *same* type. Other actions call for a *different* type of action: invitations with acceptances (or rejections); congratulations with thanks; offers with acceptances (or refusals). Such pairs of conventionally linked conversational actions are said to have two "parts": a "first part" and a "second part." The pairs can be said to have "conditional relevance"; that is to say, the production of one is a relevant contribution to a conversation only when and because the other is also produced.

More formally stated, adjacency pairs are sequences of two utterances that are: (1) adjacent (unless separated by an "insertion sequence"; see discussion on p. 257); (2) produced by different speakers; (3) ordered as a "first part" and a "second part"; (4) typed, so that a particular first part requires a particular second part (or range of second parts) – for example, offers require acceptances or rejections, greetings require greetings, and so on. There is a general expectation regarding the use of adjacency pairs, namely that having produced a first part of some pair, the current speaker should stop speaking and the next speaker should produce at that point a second part to the same pair. This is not to say that this is always what happens, but when it does not happen this is a noteworthy event that the participants will attend to and understand in the light of the expectation that has not been fulfilled.

The existence of adjacency pairs in conversation is the basis for a fundamental methodological point in conversation analysis. It is in producing a second pair-part as their response to a conversational action that participants display their understanding of that conversational action. This means that: "Conversation, as opposed to monologue, offers the analyst an invaluable analytical resource: as each turn is responded to by a second, we find displayed in that second an *analysis* of the first by its recipient. Such an analysis is thus provided by participants not only for each other but for analysts too" (Levinson, 1983, p. 321).

Taking Turns

The organization of conversation into pairs of actions is linked to a second aspect of conversational structure, namely *turn-taking*. The organization of pairs of conversational actions depends on this fundamental feature of everyday interaction, that the participants in a conversation take "turns" to speak. The pair-part produced by a speaker during their turn can be considered a conversational "move" of some kind or another. At the very least, the move will often serve to select who speaks next, but usually more than this is accomplished. We see here for the first time the notion that social interaction is metaphorically a game, a metaphor that we will explore further in Part III (see Box 13.1):

> Conversation is a process in which people interact on a moment-by-moment, turn-by-turn basis. During a sequence of turns participants exchange talk with each other, but, more important, they exchange social or communicative actions. These actions are the "moves" of conversation considered as a collection of games. Indeed, conversational actions are some of the most important moves of the broader "game of everyday life." (Nofsinger, 1991, p. 10)

Not all kinds of verbal exchange operate this way: a formal speech is planned in advance and is managed primarily by the person giving it (though responses from the audience play a part, too). Speech in settings such as a classroom or a courtroom will have a distinct organization. However, these other kinds of interactions can be and have been analyzed using the approach of conversation analysis (see Heritage, 1998).

Turns can be made up of a single word, a phrase, a clause, or a full sentence. They are defined not syntactically or semantically but in terms of their pragmatic force. One consequence is that what counts as a turn is itself contingent, and the participants may retrospectively redefine what they recognize as a turn as their conversation continues.

It is an evident fact about conversation, then, that participants alternate turns to speak, but how does this happen? How does a participant "get the floor"? It may seem that people simply wait for the speaker to stop and then talk, but the gaps between turns are generally too short for this to be the case. Sometimes they are just microseconds in length, and on average they are no longer than a few tenths of a second. The recognizable potential end of a turn is called in CA a "transition relevant place" (TRP). A TRP may be identified by "a change in the pitch or volume of the voice, the end of a syntactic unit of language, a momentary silence, or some sort of body motion" (Nofsinger, 1991, p. 81). Transition between speakers usually occurs at such a point, and in an early and important article, Harvey Sacks, Emanuel Schegloff, and Gail Jefferson (Sacks, Schegloff, & Jefferson, 1974) demonstrated that a small set of methods is generally used to organize the taking of turns in a conversation. In simplified form, these methods are as follows:

1. The current speaker (C) can select the next speaker (N) while still talking, but must then stop talking at the next transition-relevant place (TRP). (*Current speaker selects the next speaker.*)
2. If N is not selected, anyone can jump in, and the first to do so gains rights to the floor. (*Self -selection by the next speaker.*)
3. If neither (1) nor (2) occurs, C may (but need not) continue talking. (*Current speaker continues speaking.*)
4. If (3) happens, rules (1) to (3) apply again at the next TRP.

These methods assign the rights and responsibilities of the participants in a conversation. As Sacks et al. noted, "It is a systematic consequence of the turn-taking organization of conversation that it obliges its participants to display to each other, in a turn's talk, their understanding of other turns' talk" (p. 728).

In general, conversation analysts adopt the view that when people conduct a conversation it is an "interactionally managed" and "locally managed" phenomenon. That is to say, people organize the construction of a conversation together, cooperatively, and they deal with the organization at a "local" level, one utterance at a time. Conversation is treated as a special case of the

general view in ethnomethodology that order is the product of skilled improvisatory social activity.

Recipient Design

When people speak in an ongoing conversation, they do so in the light of what
has been said and in anticipation of what might be said in the future. We can
say that they "design" or "construct" their speech, and understand the talk of
other people, with attention to this ongoing organization of the interaction.
This construction of utterances to suit what has been said and what will be said
is called *recipient design*.

It is a central insight of conversation analysis that the way participants
understand an utterance will depend not just on its linguistic form but also on
its location in the conversational sequence, as well as the context, the identity
of the speaker, and other considerations. Consequently, when we are trying to
understand a particular utterance or conversational action, it is crucial to
consider where and how that action is located in the ongoing sequence of
conversational actions and in the extralinguistic context.

If the way an utterance is interpreted and the way it was designed depend
on its context, both verbal and nonverbal, this implies that an utterance's
interpretation is not fixed: it can change as the conversation develops.
Participants can and often do engage in an ongoing reappraisal of prior
utterances. "This ongoing judgment of each utterance against those immediately adjacent to it provides participants with a continually updated (and, if
need be, corrected) understanding of the conversation" (Nofsinger, 1991,
p. 66). Each person involved in a conversation has their own interpretation
of what is going on, but these interpretations are intersubjective in the sense
that every person treats the adjacent utterances in similar ways. People share
an understanding of the "game" they are engaged in and its "order."

Recipient design includes the way that participants shape their utterances
to take account of the identities of themselves and the other speakers, and what
they take their interests to be. Conversation analysis draws a distinction between
"speaker" and "hearer" as positions in the discourse and the incumbency of these
positions. For example, it is obvious that in a conversation between two persons,
the positions of speaker and hearer alternate. In addition, a "recipient" is someone to whom speech is not addressed but who nonetheless hears it. For instance,
"a speaker can focus on a subset of those present (for example, through the use of
a restricted gaze or an address term) while still designing aspects of his talk for
those who are not explicitly addressed. The notion of recipient, encompassing
but not restricted to explicit addressees, is thus also necessary" (Goodwin &
Heritage, 1990, p. 292). Speech is often designed in part for a recipient, even
though they are not the designated or selected hearer. This has implications for
the identity or position of the researcher, which I will return to in a moment.

Alignment

We have seen that the basic unit of the conversation is not the individual utterance but the adjacency pair. This is not to say, however, that the organization of a conversation extends no further than a couple of utterances. Adjacency pairs are generally linked together in closely integrated ways. One pair may *follow* another (question, answer; question, answer) or may be *embedded* inside another pair. A "presequence" is an example of the former, an "insertion sequence" an example of the latter.

A presequence occurs when some preliminary action is taken before initiating the first part of an adjacency pair, and the preliminary action itself involves an adjacency pair. Before making a request, for instance, it often makes sense to check whether the other person has the item one wants. Here a question–answer pair (turns 1 and 2) prepares for a request-agreement (or request–rejection) pair (initiated in turn 3).

1 A: Do you have the spanner?
2 B: Yes.
3 C: Can I have it please?
4 B: [...]

Here is a second example of a presequence:

5 TEACHER: Mike, do you think you know the answer to question four?
6 MIKE: Yes.
7 TEACHER: Can you tell the class, then, please?
8 MIKE: [...]

Another kind of linkage between adjacency pairs occurs when the person to whom the first part of an adjacency pair has been directed undertakes some preliminary action before responding with the second part. A request for clarification by the recipient will take place *after* the first pair part but *before* the second pair part. This is an insertion sequence. In the following example, turns 1 and 4 make up one adjacency pair, and turns 2 and 3 make up a second adjacency pair inserted between the two parts of the first pair:

9 P: Martin, would you like to dance?
10 M: Is the floor slippery?
11 P: No, it's fine.
12 M: Then I'd be happy to.

Here is another example of an insertion sequence:

13 TEACHER: Will you tell us the answer to question four?
14 MIKE: Is that on page six or seven?
15 TEACHER: Six.
16 MIKE: Oh, okay. The answer is factorial two.

For a conversation to run smoothly and effectively, the taking of turns must be organized and the pairs linked, but this is only one aspect of how a conversation must be kept on track. The understanding of prior talk that is displayed in each conversational move indicates what is called "alignment." In particular, specialized conversational actions serve to provide evidence to the speaker of how their talk is being understood. These include actions such as "assessments" ("That's good"), "newsmarks" ("Oh, wow!"), "continuers" ("uh huh"), "formulations" (giving the gist of what has been said), and "collaborative completions" (finishing the speaker's utterance).

Alignment is displayed and adjusted not only in responses to an utterance but also in advance. "Preventatives" such as "disclaimers" ("I really don't know much about this, but . . .") are examples of such "pre-positioned alignment devices." "Pre-sequences" (described earlier) do this, too. Alignment is especially important at the "openings" and "closings" of conversation.

At times, actions need to be initiated to "repair" a conversational breakdown and restore alignment. Breakdowns can be misunderstandings ("What did you say?"; "What do you mean?"), but disagreements ("I think you're wrong"), rejections ("No, I won't"), and other difficulties are also occasions of breakdown and repair. "Revisions" may occur when a speaker anticipates that trouble has occurred and reformulates their talk accordingly.

Such breakdowns are important because they provide powerful evidence to the researcher of how the participants are understanding their interaction and the ways such understanding can at times be discordant or contested:

> One key source of verification here is what happens when some "hitch" occurs – i.e. when the hypothesized organization does not operate in the predicted way – since then participants (like the analyst) should address themselves to the problem thus produced. Specifically, we may expect them either to try to repair the hitch, or alternatively, to draw strong inferences of a quite specific kind from the absence of the expected behavior, and to act accordingly. (Levinson, 1983, p. 319)

For example, the following exchange between student (S) and teacher (T) shows several of the features we have discussed. (This is a much abbreviated analysis; see Levinson, pp. 320–321, for more details.) First, each turn can be seen as displaying an interpretation of the preceding turn. Second, this includes the silence in line 2, which is taken as a negative reply. Third, the silence can be read as a lack of alignment, and S apparently reads it as such. Fourth, line 3 can be considered a repair to S's original question:

17 S: So I was wondering would you be in your office after class this week?
18   (2.0)
19 S: Probably not.
20 T: Hmm, no.

Here the two-second pause after the student's question – a "hitch" in the conversation – is interpreted as a negative answer to the question. It is a striking and important observation that although a silence has no features on its own, conversational significance is attributed to it, just like any other turn of talk, on the basis of the expectations that arise from its location in the surrounding talk. Depending on where silence occurs in a conversation – its location in the conversational structure – it will be interpreted as a "gap" between turns, a "lapse" in the conversation, or a "pause" that is attributed to the designated speaker. Furthermore, such interpretations are revisable. This means that any attempt to describe, let alone code, the contribution made to the conversation by a silence without attending to sequence and context is bound to fail. A gap is silence at the TRP when the current speaker has stopped talking without selecting the next speaker and there is a brief silence before the next speaker self-selects. A gap does not "belong" to anyone. A lapse is silence when no next speaker is selected and no one self-selects: the conversation comes to an end for at least a moment (which has the consequence that a gap and a lapse can be distinguished from one another only in retrospect). A pause is silence when the current speaker has selected the next speaker and stopped talking but the next speaker is silent.

PREFERENCE

When speakers have a choice between conversational actions, one will generally be considered more usual, more typical, than the others. This phenomenon is known as "preference," and the more usual action is called the "preferred response." The term refers not to the personal desires of the participants but to the norms of the shared conversational system. The existence of these shared norms has the consequence that "any of the conversational tendencies and orientations that we commonly attribute to participants' personalities or interpersonal relationships derive (at least in part) from the turn system" (Nofsinger, 1991, p. 89).

For example, in response to the first part of an adjacency pair, some second-part responses are preferred, whereas others are dispreferred. Refusals of requests or invitations are nearly always dispreferred, whereas acceptances are preferred (see Table 10.1).

Participants' grasp of preference will strongly influence their interpretation of conversational actions. For instance, a silence in response to a request may be taken as evidence of a likely upcoming dispreferred response (a refusal in the case of the preceding example), so that further inducements may be added. Conversational actions that are not the preferred response are often conducted in a manner that displays this: "dispreferred seconds are typically delivered: (a) after some significant delay; (b) with some preface marking

TABLE 10.1. *Conversational Preference*

| | | Preferred and Dispreferred Second Parts to Various First Parts | | | |
| | First Parts | | | | |
	Request	Offer/Invite	Assessment	Question	Blame
Preferred Second Part:	acceptance	acceptance	agreement	expected answer	denial
Dispreferred Second Part:	refusal	refusal	disagreement	unexpected answer or nonanswer	admission

Source: Modified from Levinson (1983, p. 336).

their dispreferred status, often the particle *well*; (c) with some account of why the preferred second cannot be performed" (Levinson, 1983, p. 307).

Extended Discourse

The previous examples have been excerpts from conversations in which the turns are brief, but the lengthy stories and substantial arguments that occur in conversations can also be examined with the focus and methods of conversation analysis. It turns out, counter to what one might expect, that "[s]tories and other extended-turn structures in conversation are not simply produced by a primary speaker, but are jointly or interactively produced by a primary speaker together with other cooperating participants" (Nofsinger, 1991, p. 94).

The same is true of argument. Researchers distinguish "making" an argument from "having" an argument: "Roughly, the first involves using reasons, evidence, claims, and the like to 'make a case.' The latter involves interactive disagreement (for example, 'You can't,' 'I can,' 'Oh no you can't,' 'Well I certainly can'). Conversational argument often consists of participants making arguments in the process of having one" (Nofsinger, 1991, p. 146). The organization of both narratives and arguments has been successfully described in terms of the methods evident in more typical conversations, including the turn-taking system, preference, sequences of adjacency pairs, and repair.

CDA AND CA COMPARED

Critical discourse analysis and conversation analysis both aim to show constitution at work, but they approach this task in very different ways. Most important, they have very different preconceptions about constitution. Critical discourse analysis assumes that there is an ontological distinction, a gap, between discourse and reality, and tries to bridge the gap by appealing to representation. Critical discourse analysts maintain that "discourses produce a perception and representation of social reality. This representation forms part of hegemonic strategies of establishing dominant interpretations of 'reality'" (Bührmann et al., 2007). In other words, discursive practices are thought to constitute reality by offering compelling representations of it, which the researcher can reconstruct without consulting other people. Critical discourse analysts "consider discourse to refer to a set of meanings, representations, images, stories and statements" (Hook, 2001, p. 530). It is evident that CDA employs the epistemological view of constitution that we explored in Chapter 7, with all its problems and limitations.

Ironically, critical discourse analysis short-circuits investigation by assuming that discourse has an effect solely because of its meaning and then seeing its task as identifying what a sample of discourse "really" means. This kind of approach "cannot rescue itself from claims that it functions as an interpretative activity that reifies the text, recuperates the author-principle (in the figure of the interpreter) and restores a central anchoring-point, not this time in the form of truth, but in the authoritative interpretation, which performs much of the same function" (Hook, 2001, p. 541). When this approach attends to what a particular piece of discourse *does*, it is without reference to who speaks, when, in what setting, and to whom.

Conversation analysis, in contrast, assumes that the order that people recognize as social structure and societal institutions is an ongoing product of their practical activity. It thereby avoids the dualism that CDA struggles with. For CA, discourse constitutes objects (and subjects) not by providing a representation of them that is somehow compelling but, to put it briefly, as a form of practical activity that establishes a space in which objects and subjects become visible, as *effects* of discourse.

The ontological presuppositions of CA are similar to those we explored in Chapter 8. Not surprisingly, given the influence of ethnomethodology, we can hear echoes of Garfinkel, but there are also points of contact with Heidegger and Merleau-Ponty. For conversation analysis, an utterance is like a tool, one that is made on the spot, offhand, in the moment. An utterance is an artifact that has been *improvised* (literally sudden, unforeseen), a "conversational device" that *may* provide a representation (but what kind? Remember Garfinkel's point that there are many) but equally may have quite another kind of pragmatic force. The focus of CA is what Schegloff (2001) has called the "action import" of utterances: "not just what they are about or what they impart" but the way they have an effect, "not as a syntactic code, or as a medium that reports events in some external world, but rather as a mode of action embedded within human interaction" (Goodwin & Duranti, 1992,

p. 29). These tools and the practical know-how to employ them are resources for the participants in a conversation, and they are also resources for the researcher, assuming that they know the language.

The production of an utterance makes a point, provides the speaker with a feel for the other person (through their response), and gives the speaker a sense of him- or herself (we discover ourselves in our words) as well as a way of accomplishing something socially. To understand an entity like an utterance is to be aware of its point by grasping the utterance and projecting it. The way the utterance is projected depends on (1) the ongoing conversation of which it is a part; (2) the context, the here and now; and (3) familiarity with the public conventions of language. To understand a speech act is to recognize its point – where the speaker is coming from; where they are going; what they are getting at – what their concern is, not what their beliefs are.

We can think of alignment among participants in a conversation in Heidegger's terms. When participants project an utterance in ways that line up, they are aligned. Their projections will generally not be *identical* but *reciprocal*: a speaker's good news may be a hearer's bad news.

Everyday conversation depends on and works with context in complex and critical ways; indeed, speech acts can be treated as first and foremost operations on context (Levinson, 1983, p. 276). Settings are continually reconstructed at the same time as they are used as grounds for ongoing activity. Nofsinger points out how conversational utterances are both context-shaped and context-renewing: "Context, in this immediate and narrow sense, is composed not just of what people know, but of what participants *do* to show each other which items of their shared knowledge should be used in making interpretations. The conversational actions produced by participants create an interpretive resource that is used to align conversational understanding" (Nofsinger, 1991, p. 143).

CONCLUSIONS

As a form of investigation, conversation analysis proceeds in the manner that Ricoeur described (Chapter 5): it moves from "understanding" to "comprehension"; it starts with "a naive grasping of the meaning of the text as a whole" and progresses toward "a sophisticated mode of understanding, supported by explanatory procedures" (Ricoeur, 1976, p. 74). Drawing on the researcher's intuitive recognition of the conversational actions that have been recorded, CA is a way of developing a detailed, step-by-step articulation of the work that people accomplish together in practical activity.

Conversation analysis, unlike CDA, focuses on the ontological work accomplished in practical activity. Its focus on pragmatics rather than representations enables it to avoid claiming to penetrate beneath the surface of phenomena and instead to pay attention to the ongoing activity in which phenomena are produced. Attending to what talk does, it is guided by participants' displays of their understanding in their ongoing interactions.

Some critics have suggested that conversation analysis suffers from a "disinterest in [the] question of external social or natural causes" (Wetherell, 1998, p. 391) and that it neglects the "broader interpretative repertoires" (p. 401) within which particular conversations are located, selecting small episodes for analysis without attention to the larger fabric. It has been accused of hiding its "a priori assumptions" (Billig, 1999a, p. 543) and of being based on "a naive epistemology and methodology" (Billig, 1999b, p. 573).

It is understandable that the careful attention given to the details of conversational interaction might create the impression that CA studies the "micro" and ignores the "macro." But Schegloff (1998, 1999) has responded convincingly to these criticisms. He acknowledged that CA is undoubtedly not free from presuppositions, and its technical vocabulary will inevitably differ from the vernacular that is studied. People *use* two-part pairs, for example; they don't *talk* about them. But Schegloff insisted that there is no incompatibility between CA and a concern with larger issues, such as political concerns. He recommended, however, that attention to such issues should not neglect the details of interaction, for it is in these details that participants demonstrate their own concern with a "here and now" that "can range across various orders of granularity" (Schegloff, 1998, p. 413). That is to say, a seemingly local interaction will have concerns that may be either large or small. Furthermore, Schegloff argued that any reference to context ought to refer to what is salient to the participants, not merely what is pertinent to the researcher. It is true that CA will offer an "internally analyzed rendering" (Schegloff, 1997, p. 174), but this means it pays attention to what is internal to the *event* (including the context, close at hand or far away), not only what is internal to the *discourse*. The critics have failed to see this distinction. This means that CA (like EM) is a "local" analysis in which the local can extend enormous distances.

These criticisms of CA replay many of the mistaken objections to ethnomethodology: that it is sociology without society, a microsociology that neglects larger social structures. We have seen that Garfinkel aimed precisely to *avoid* the divisions between micro and macro, local and global, agent and structure. Schegloff pointed out that CA does the same: analysis is based not on the assumption that there are "connections" between the local and the global that must be identified but on the contrary that the distinction between local and global is inappropriate and misleading.

Here, as we have seen, the difference between CA and CDA is very clear. It also is clear that any attempt to study constitution is hopeless once one begins with a dualist distinction between person and society, micro and macro. Once that dualism has been accepted, the only way that constitution seems to be possible is as a *representational* phenomenon, but then we are studying not reality but only "'reality."

This recourse to representation is at the root of the ironic fact that although the notion of social construction has become so popular that it is applied to virtually everything, at the same time it is taken to say nothing about reality! "Constructed" has become equivalent to "insubstantial," "conventional," and easily changed.

Bruno Latour is one person who has strongly opposed this interpretation of construction. He insists that in his own work the term does not have these connotations, and it should be thought of quite differently: "In plain English, to say something is constructed means that it's not a mystery that has popped out of nowhere, or that it has a more humble but also more visible and more interesting origin" (Latour, 2005, p. 88).

It is, after all, in a construction site that we can see people engaged in the concrete and practical work of making things: "films, skyscrapers, facts, political meeting, initiation rituals, haute couture, cooking" (Latour, 2005, p. 89). Visiting such a site provides the opportunity to see "the skills and knacks of practitioners" and "also provides a rare glimpse of what it is for a thing to emerge out of inexistence by adding to any existing entity its time dimension" (p. 89). One can see immediately that "things *could be different*" and "that *they could still fail.*" But this is not to say that things are not real or that change would be easy.

People have come to assume that to describe something as constructed is to say that it is not real, not true, that it is invented, made up. On the contrary, Latour points out that "to say that something is constructed has always been associated with an appreciation of its robustness, quality, style, durability, worth, etc. So much so that no one would bother to say that a skyscraper, a nuclear plant, a sculpture, or an automobile is 'constructed.' This is too obvious to be pointed out" (Latour, 2005, p. 89).

Constructed (or in our terms *constituted*) should generally be considered a synonym for *real*. To say something is *artificial* is not to deny that it is *real* and *objective*. In the study of science laboratories (Latour & Woolgar, 1979/1986), for example, it became clear that "to be contrived and to be objective went together" (Latour, 2005, p. 90). "Facts were facts – meaning exact – *because* they were fabricated." The important question is whether they were constructed badly or well.

At the same time, Latour is critical of the phrase "social constructivism" when it is used in the kind of inquiry that tries to "*replace* what this reality is made of with some *other stuff*, the social in which it is 'really' built" (Latour, 2005, p. 91, emphasis original). The study of construction must be the study of the ways in which humans and nonhumans are "fused together" in the real work of practical activity. What is necessary is to use the word "social" not "to replace one kind of stuff by another" but to explore "the associations that have rendered some state of affairs solid and durable" (p. 93).

How then can we use an approach such as conversation analysis to pay attention to the big picture while avoiding dualism? Latour suggests that face-to-face interactions appear to be purely local only as a consequence of what he calls "framing." When we observe primates, for example, we find that a whole troop of apes will be in continuous interaction. Humans, in contrast, arrange matters – walls, barriers, fences – to shut out most of their social network (family, colleagues, strangers). The result is a kind of interaction that is not "complex" (like the primates') but "complicated," together with the illusion that what has been shut out forms a larger whole, the "context" or "structure" to the local interchange: "For humans, an abyss seems to separate individual action from the weight of a transcendent society. But this is not an original separation that some social theory concept could span and which might serve to distinguish us radically from other primates. It is an artifact created by the forgetting of all practical activities for localizing and globalizing" (Latour, 1996, p. 234).

Latour recommends that in order to understand how face-to-face inter-actions are organized and how they are linked together over space and time, we need to pay attention to the complex material arrangements in which these interactions occur. What situates interactions, Latour suggests, is not inter-subjectivity but "interobjectivity." Latour's recommendation is "following practices, objects and instruments." When we do this, we "never again cross that abrupt threshold that should appear, according to earlier theory, between the level of 'face-to-face' interaction and that of the social structure; between the 'micro' and the 'macro'" (Latour, 1996, p. 240). Latour sees that we don't need to solve Kant's problem of the relationship between local, linguistic representations on the one hand and some larger reality as it actually is on the other. We simply need to dissolve the problem by recognizing that the gap apparently separating agent from structure, the gulf apparently between individual and society, is illusory. What we need is not an analytic strategy that will show how the gap is "bridged" but one that follows the practices wherever they lead: "The interactionists are right when they say that we should never leave interactions – but if one follows human interactions then one never stays in the same place, nor ever in the presence of the same actors; and never in the same temporal sequence" (pp. 238–239). The complex totality that we call Paris, for example, is structured each day by thousands of Parisians, locally and with their own methods. "This is the profound truth of ethnomethodology," Latour adds. "All that remains is to restore to it [sociology] what it had itself forgotten: the *means* of constructing the social world" (p. 240), including "the multiple panoptica that strive each day to sum up Parisian life" (p. 239).

Conversation analysis adds something very important to our program for investigating constitution: a way of studying practical activity that articulates the work that is done and that pays attention to the vernacular understanding displayed by the participants. Whereas ethnographic field-work has been concerned with describing a cultural order "as if" a member, an order that paradoxically lies "behind" what is evident, CA pays attention to the details of the work of ordering that constitute the order of a form of life.

I am not proposing that we are faced with a choice between ethnographic fieldwork and conversation analysis. Rather, it seems that each complements the other. Far from being disinterested descriptions, ethnographies are inter-pretations, and as such are occasions of *application*, in Gadamer's sense. This suggests that they make claims to *phronesis*, claims that could be tested in terms of their local accountability. I do not mean by this that we should ask participants to read our ethnographies and tell us whether our interpre-tations are correct or not. Such "member checks" confuse explicit know-that with practical know-how and ignore the complex dynamics of the research relationship.

What I mean is that a fieldwork account would benefit immeasurably from having some empirical warrant for the claims it makes about participants' understanding of events and actions. This is lacking when these claims take the form of statements about the meanings and intentions "behind" events. But the warrant is also lacking, or at least unconvincing, when the researcher claims to have "become" a member and thus able to understand the cultural order as a member does.

Conversation analysis provides a convincing way for a fieldwork account to have this kind of warrant. It shows how the study of a form of life by an outsider, a stranger, can produce accounts that are accountable not merely to the academy, to other scientists, but to the participants themselves.

In addition, conversation analysis offers fresh ways of thinking about the fieldworker's participation in a form of life, and the evidence that is available concerning that participation. First, CA helps us clarify the issue of what the researcher needs to be capable of and whether they can or should be a member. To study a conversation, the researcher needs to be able to *recognize* what is being said but not necessarily be able to *produce* a contribution to the conversation. Second, as we saw in Chapter 9, it was suggested by Bazin that the fieldworker is not an observer but a "witness," and in the terms of CA a witness is a recipient but not a hearer, someone to whom speech is not directly addressed but who nonetheless hears it and will be acknowledged by those who are talking in the design of their talk because "[t]he identity assumed by one party is ratified, not by her own actions, but by the actions of another who assumes a complementary identity toward her" (Goodwin & Heritage, 1990, p. 292).

It seems likely that fieldworkers will frequently be in this kind of position. In such cases, the way speech is designed will be evidence that enables them to infer how their presence was understood by the participants.

Geertz argued that the most important component of fieldwork is the *fixing* of events in writing. Conversation analysis employs fixing, too, but in the form of audiotaping and videorecording. Here, too, the transformations that Ricoeur identified play an important role, and here is another way in which the detailed analysis of interaction is an important complement to fieldwork. The latter articulates the order of a form of life; the former articulates the work of ordering.

The aim of ethnography, as we have seen, has generally been defined as describing a form of life as a member understands it. But what if participation in a form of life leads its members to *mis*understand it? There are reasons for thinking that the members of a form of life do not, perhaps *cannot*, grasp it correctly. To a great degree, this is the motivation for CDA: critical discourse analysts assume that people are mistaken about the character of the world around them and that the purpose of analysis is to discover how things *really* are.

The proposal that people necessarily misunderstand their form of life will be the central topic of the third part of this book. We will explore its implications for research, including the implication that participation, even as an outsider in a relation of complicity, cannot be a sufficient basis for inquiry. What I want to emphasize here is that even if we suspect that participants misunderstand what they and others are doing, we still need to take their understanding into account. We do not need to accept the understanding that participants display in an interaction, and our analysis does not need to stop there. But it does need to *start* there; we should not try to bypass the way participants grasp events and jump directly to claims about what is "really" going on. Indeed, we cannot assert that they are mistaken if we do not in fact know how they understand events. We cannot *critique* participants' understanding unless we first figure out what it is. How we should engage in the investigation of constitution with a critical dimension is the question we turn to next.

PART III

INQUIRY WITH AN EMANCIPATORY INTEREST

Kant worked hard to resolve a contradiction that emerged from the Enlightenment emphasis on rationality: the contradiction between the new view of man as an individual, natural creature and the need for valid scientific knowledge and ethical conduct. How could objective knowledge and binding ethics be built from subjective experiences and preferences? Kant's proposal was that ideas are representations that the mind actively constructs using its innate capacity for universal reason. Mind constitutes reality as we know it. As a result, individuals can have indubitable knowledge of the world and be clear on their moral duties.

The empirical-analytic social sciences accepted this solution and defined their job as the objective study both of people's objective behavior and their subjective beliefs, opinions, and desires. Qualitative research, when it is defined as the objective study of subjectivity, also accepts Kant's solution. But this solution operates at the *epistemological* level, while it presumes an ontological divide. It is torn in two by the dualisms of appearance and reality, subjective opinion and objective knowledge, for-itself and in-itself.

In the face of these problems, some people began to ask, "What more?" Their answers change our understanding of both the character and the location of constitution. They emphasize that practical activity is the origin of order and of subjects and objects. Constitution takes place in practical know-how before any theoretical or objective-seeming formal know-that. Knowledge – reason, logic, and so on – is a local accomplishment. Objects and subjects are formed in practical activity.

We can now understand better the calls for a new kind of interpretive research in the 1970s. These researchers wanted to study constitution, avoid dualism, focus on practices through immersion, and use hermeneutics. We have seen that the ontological approach to constitution does indeed offer a nondualistic account of humans and the world and emphasizes the primacy and importance of social practices. These practices provide a tacit know-how on which all formal knowledge is based, and they involve ways of being. It

seems reasonable that researchers have to obtain familiarity with social practices by participating in them, by becoming "immersed."

Parts I and II of this book have explored two central ways of defining
qualitative research, first as the objective study of subjectivity, centered
around interviewing, and then as the participatory study of intersubjectivity,
centered around ethnographic fieldwork. Both were initially conceived as
responses to the inadequacy and distortions of traditional research. Yet both
were infected by the disease they tried to cure. The first responded to objectifying social research by insisting that people are more than objects for
manipulation and measurement because each individual has a unique subjectivity to which research ought to attend. But this new human science was
caught in a contradiction: by emphasizing that each person has a unique way
of making meaning, living in their "own world," it could not explain how
researchers, who after all are people like everyone else, could possibly overcome the limitations of their own subjective experience to achieve objective
knowledge. Trying to be both subjective and objective at the same time proved
impossible.

In the second definition, the focus was not subjectivity but intersubjectivity. In anthropology, sociology, and political science, the predominant view
has been that a society or a culture is an objective structure that determines
human behavior and even human thinking. The objection to this was that it
left no place for human agency and treated people as no more than "cultural
dopes." What was needed was a fresh approach that would avoid or overcome
the dualism of agency and structure. Geertz rejected both subjectivism
and objectivism: "The cognitivist fallacy – that culture consists of 'mental
phenomena which can be analyzed by formal methods similar to those of
mathematics and logic' – is as destructive of an effective use of the concept as
are the behaviorist and idealist fallacies to which it is a misdrawn correction"
(Geertz, 1973, p. 12).

Consequently, in Part II we explored the proposal that practical activity is
more fundamental, both epistemologically and ontologically, than subjectivity and objectivity. It is in practical activity that the knowing subject and the
known object are *constituted*. Practices and processes, not objects and subjects,
are ontologically fundamental. Epistemologically, practical activity provides
know-how, a hands-on understanding, that is the basis for any kind of
articulated and formalized know-that. Practical activity offers a way of grasping, a way of seeing and feeling, a way of thinking, that is embodied, social, and
materially situated in time and place, in history and culture. From this point of
view, objective knowledge – if by this we mean knowledge that is disinterested
and general, abstract and decontextualized – is an illusion. But knowledge that
is interested and situated is not therefore subjective.

The site of *constitution* is not the individual mind but the material world.
In activity, we become specific kinds of persons in a world of particular things

and events. We grasp what is true and good in participation, not in solitary reflection, and so Kant's attempt to find the origin of valid knowledge and ethical action in the individual mind was bound to fail.

As we have seen, calls for a new kind of inquiry proposed that it would be based on immersion in the practices of a form of life. From ethnography to ethnomethodology, practical activity has been seen as where the action is, and participation and membership as the keys to access. Such inquiry aims to articulate and make explicit the implicit knowledge used by members of a culture or community as they act and interact; this knowledge is considered crucial for the skillful production and reproduction of social order. Inquiry requires participation, and the articulation of the understanding gained through participation, in a hermeneutics of application.

But fieldwork has often been defined as an approach in which the researcher *both* participates as a member of the intersubjective community being studied *and* writes an objective description of that community. The contradiction between participation and observation is just as problematic as that between subjectivity and objectivity, and in Chapter 8 we considered some of the problems that result. The proposal that the researcher becomes a full member of a bounded form of life and describes what they can then see is unrealistic. It trivializes the role of language, neglects the diversities and contradictions of any culture, and ignores the fact that the ethnographer writes for, and will return to, the cozy academy. Cultural anthropology has begun to reconceptualize the aims and methods of ethnography, though consensus has not yet been reached.

In Chapter 10, we saw that conversation analysis offers a way to study practical activity that attends to its pragmatics and, rather than claiming to penetrate beneath the surface, describes the ontological work that is accomplished, guided by participant displays of their understanding. Conversation analysis avoids presuming a gap between agency and structure, between micro and macro, local and global.

There is another problem with research that emphasizes immersion in practices. What if participation in practical activity provides a *mis*understanding of the world, of who we are, and of what is true and right? What if a member's practical know-how is *distorted*? What if the members of a form of life do not, perhaps *cannot*, understand their own circumstances accurately? What if social practices themselves are in some sense *malformed*? If this is the case, then participation, even as an outsider in a relation of complicity, is not a sufficient basis for inquiry. This is the troubling suggestion made most powerfully by Karl Marx. With Marx, a new dimension was added to social inquiry and a new aim defined. The dimension was that of *critique*, and the aim nothing less than *emancipation*. In the chapters that follow in the third part of this book I will consider first Marx and the origins of critique and then three distinct visions of how critical inquiry might be carried out. In each case, fieldwork and interviews are necessary tools, but they are not sufficient. German philosopher Jurgen Habermas, French sociologist Pierre Bourdieu, and French historian Michel Foucault have each proposed different approaches to what we might call "fieldwork" studies, openly ontological studies that attend to the body and to the history of society and of individuals, with a critical and emancipatory interest. Together their work will help us define a program of inquiry to study constitution.

11

Qualitative Research as Critical Inquiry

> Critical theory is a metaphor for a certain kind of theoretical orientation which owes its origin to Kant, Hegel and Marx, its systematization to Horkheimer and his associates at the Institute for Social Research in Frankfurt, and its development to successors, particularly to the group led by Jürgen Habermas, who have sustained it under various redefinitions to the present day.
>
> Rasmussen, 1996, p. 11

The notion of critique takes us back once again to the philosophy of Immanuel Kant and to Georg Hegel's response. We saw in Chapter 7 that Kant considered the primary task of philosophy – perhaps its only task – as critiquing knowledge rather than justifying it. He proposed a "critical philosophy" that would explore the conditions for the possibility of true knowledge. We have seen that for Kant these conditions were transcendental, universal, a priori concepts. Kant maintained that a universal capacity for reason provides the basis for each individual's "Enlightenment" (German: *Aufklärung*) because it can "free our concepts from the fetters of experience and from the limits of the mere contemplation of nature" (Kant, 1784/2000, p. 402):

> Enlightenment is man's emergence from his self-incurred immaturity. Immaturity is the inability to use one's own intellect without the guidance of another. This immaturity is self-incurred if its cause is not lack of understanding, but lack of resolution and courage to use it without the guidance of another. The motto of enlightenment is therefore: *Sapere aude!* Have the courage to use your *own* understanding! (Kant, 1784/2000, p. 401)

For Kant, enlightenment is something one achieves individually, and maturity amounts to independence and the ability to decide for oneself what is right and what is true.

Hegel took a different position, arguing that the conditions for knowledge are historical, so that enlightenment is a *social* process that depends on

273

relationships and participation in community. In this chapter, I will explore what Marx did to the concepts of critique and enlightenment, and how the group of scholars known as the Frankfurt School built on his work to expose the irrationality lurking at the heart of Enlightenment rationality. But first it is important to mention one portion of Hegel's reconstruction of the development of consciousness that has received considerable attention: his account of the "master–slave" dialectic, the struggle between what he called "lord" and "bondsman." This analysis has been influential in psychoanalysis (accounts of the superego), Marxism and critical theory (the class-consciousness of workers), feminist theory (standpoint epistemology), and race studies.

The Dialectic of Master and Slave

Hegel described self-consciousness not simply as becoming aware of one's self through reflection as we become conscious of an object through perception. Reflection in that sense would be simply like looking in a mirror at a self that is just another kind of object. For Hegel, self-consciousness is not this kind of bending backward or self-examination but a completely new way of relating to objects in which there now *is* a self, whereas before there was not. "Hegel undertakes nothing less than to develop a whole new concept of being" (Heidegger, 1988, p. 137). Proof of this lies in the fact that self-consciousness is constructed by the mediation of being-for-another. We cannot exist for ourselves, Hegel suggested, until we first exist for another person, another self-consciousness. And conflict provides this recognition.

In brief, Hegel described self-consciousness emerging in a struggle for recognition. Master and slave are mutually dependent, and each struggles for independence in the form of the other's recognition. The outcome of this paradoxical struggle is that the master comes to recognize that he is dependent on the slave, who, working with his hands to satisfy the master's needs, overcomes his fear and becomes aware of his own powers. Hegel wrote, "Without the formative activity shaping the thing, fear remains inward and mute, and consciousness does not become objective for itself" (Hegel, 1807/ 1977, p. 239).

Of course, this relationship of inequality cannot provide full self-consciousness. The struggle between master and slave is merely a step along the way, and social institutions of real mutual recognition are needed for further development: "In relations of inequality typical of modern society, most of us are no more than partially self-aware, hence no more than partially capable of anything approaching objective knowledge. Full self-consciousness, or mutual recognition, presupposes social equality that is qualitatively different from mere legal equality" (Rockmore, 1997, pp. 201–202).

But in this passage Hegel introduced the important proposal that a human being is *constituted* as a self-aware subject in their interactions with other people. Self-consciousness, the thinking it makes possible, and even the self itself come not from self-reflection but from these interactions, and especially from one's recognition by others. Hegel emphasized that there are "three equally significant, interrelated but irreducible dimensions of self-formation: language, labor, and moral relations" (Forbath, 1998, p. 973). As the subject is constituted in these dimensions, the objectivity of the world is defined, though ultimately the thinking subject will come to recognize that world and self are both part of a single unfolding process.

For Hegel, enlightenment works itself out over the course of human history and critique is inherent in practice. But it was Karl Marx who would transform critical inquiry into an exploration of the conditions of knowledge *and* the unmasking of inequality and exploitation.

THE CRITIQUE OF CAPITALISM: KARL MARX

Karl Marx (1818–1883) is famous as the figure whose work inspired revolutionaries who created communist societies in the Soviet Union, China, and elsewhere. Marx was educated in philosophy, though he considered its task to be to *change* the world, not just to *know* it – "philosophers have only *interpreted* the world; the point is to *change* it" (Marx, 1888/1983). But Marx understood very well that to change the world one needs to know it first. Knowledge and political action were, for Marx, intimately connected. The important and difficult thing is to obtain the *right* kind of knowledge about the world, for Marx maintained that we are constantly misled by ideological illusions. One of these is the illusion that knowledge can be disinterested, politically neutral, and value-free.

Marx's writings provide what Ricoeur (1979) called "a great phenomenology of economic life." When one thinks of phenomenologists Marx does not typically come to mind, but he was strongly influenced by Hegel, especially the latter's *Phenomenology of Spirit*. Marx's method of analysis is phenomenological in the way Hegel defines this. Marx was interested in understanding the new forms of capitalist economies that were developing in the 19th century, especially in Britain as it became the first country to industrialize. Marx took from Kant the notion that a critique uncovers the conditions for the possibility of a phenomenon and from Hegel the notion that human history has a teleology and is developing toward a society that is just and equitable. Whereas Hegel's aim was to offer a sweeping reconstruction of the history and development of mind, and its progression through science, philosophy, and religion to ultimate knowledge, Marx's focus was equally on process, but an importantly different one. Rejecting "German idealism," for him "[t]he object before us, to begin with, [is] *material production*" (Marx, 1857–1858/1973, p. 83, emphasis original). Marx claimed that he had "turned Hegel on his head" by seeing the force behind historical change not as *Geist* but as human agency in the form of labor. This was the activity that Marx saw shaping the human world, shaping human history, and shaping humans, too. An understanding of society required understanding this practical and productive activity, both through history and as presently organized. In his own time material production was the labor of the working class, the proletariat, whose ranks were swelling in Britain as people moved from rural areas into the cities to work in the new "manufactories" that were rationalizing the production of goods.

It was these "commodities" that Marx took to be the key to his critical analysis in _Grundrisse_ and _Capital_. He focused on the "commodity form" and asked where its value came from. For example, mass production was turning the manufacture of pins into quick, unskilled work. Pins could be sold, relatively cheaply but still for a profit, on the free market. Did the value of a pin come, then, from its exchange for money or some equivalent commodity? This was the dominant economic theory of the time, but Marx argued that it was incorrect and in fact was dangerously misleading and politically reactionary. He proposed instead that the value of a pin is created in the process of its production. The value comes from the labor that goes into its manufacture; this is known as his "labor theory of value."

A profit can be obtained from selling something like a pin only if its value is greater than the value of the labor that went into its production. In Marx's time (and ours, too), that labor was "wage labor"; that is to say, workers were hired for factory mass production for an hourly or weekly wage. It was clearly in the interest of the owners of a factory to keep the cost of labor as low as possible, and that meant keeping the workers' wages as low as possible. In this way they would maximize the profit they could obtain by selling their product, the pins.

Marx proposed that capitalism had a particular organization: it was unique and was, at that time, a new "mode of production." (Marx distinguished capitalism from three other previous modes, those of slavery, feudalism, and mercantilism, and from another mode that he predicted would follow, socialism.) Capitalism is a mode of production in which a division has developed between, to put it simply, those who do the work and those they work for. Humankind has achieved great efficiency of production through the division of labor: people perform specialized tasks and then exchange the fruits of their labor. No single person, or single family, has to produce _all_ the necessities of life. But in capitalism one group of people – one _economic class_ – has come to own the raw materials, the factory, and the equipment, and they also own the commodities that are produced in the factory. The other group owns none of these things. All they possess that has any real value is their capacity to labor, their ability to work, and this they must _sell_ to the factory owner for a wage. It is their labor, not the factory owner's, that creates the value of the pin that they manufacture. However, the pin belongs not to them but to the factory owner who hired them, so it is the owner who pockets the profit that comes from selling the pin.

It might be argued that it is the factory owner who must also absorb any *loss* that comes from selling the pins, if for example a competitor forces a cut in prices. But such losses are usually limited by the laws of bankruptcy and other considerations. If workers lose their jobs, and hence their wages, this loss is generally unlimited. Indeed, workers in such circumstances might lose their lives, especially at the time when Marx was writing. Marx argued for better protection of workers' rights – and provisions such as unemployment benefits were achieved by workers who organized to bargain collectively for the recognition of their rights.

What Marx had found, by tracing the commodity form through cycles of production, exchange, and consumption, was a fundamental *condition for the possibility* of a capitalist economy. That condition is the *exploitation* of wage labor. A factory worker *must* be paid less than their labor is worth if a profit is to be made. And although a capitalist economy can absorb individual losses, overall it must be profitable. The less workers are paid, the greater the profit. If investors are to receive the best return, the cost of labor must be minimized. A *surplus* of value must be created.

There are other strategies that a manufacturer can employ to increase profits – upgrading equipment, forcing competitors into bankruptcy or taking them over, or even training workers in new skills. But squeezing labor is the strategy that is central to the capitalist mode of production in the sense that it defines the relationship between the two great classes of people, the working class and the capitalist class, that have been formed in and by this economic system. Working class and capitalist class cannot be defined independently: they are defined in and by their relationship, which is one of antagonism, opposing interests, and exploitation. In Marx's analysis, there is a crucial and unavoidable tension in a capitalist society.

Alienation, Fetishism, and Ideology

The tension inherent in capitalism generally goes unrecognized. Hegel frequently wrote of *alienation* (German: *Entfremdung*) in his *Phenomenology*, using the word (Schacht, 1970) first as a separation between two related terms, such as an individual and society, and second as a surrender or sacrifice that is necessary to overcome such separation. For Hegel, consciousness externalizes its powers and then is confronted by these externalizations as independent forces and agencies. Marx's early writing, "Economico-philosophical Manuscripts" (Marx, 1844/1983), reinterpreted alienation as an economic process, especially in capitalism, a system in which people are divided from one another, from the products of their own labor, and from the activity of work itself. In this system, "this realization of labor appears as the *loss of reality* of the worker; objectification appears as the *loss of the object* and *bondage to* it; appropriation appears as *alienation*, as *externalization*" (Marx, 1844/1983, p. 133, emphasis original). In "alienated labour" the product becomes an "alien being" (p. 133). Alienation is a pervasive and damaging aspect of such an economy and can be overcome only through a radical transformation of the organization of labor (Blauner, 1964): "The theory of alienation is the intellectual construct in which Marx displays the devastating effect of capitalist production on human beings, on their physical and mental states and on the social processes of which they are a part" (Ollman, 1976, p. 131).

One aspect of alienation is "the fetishism of the commodity." In the first chapter of *Capital* Marx wrote that "a commodity appears at first sight an extremely obvious, trivial thing. But its analysis brings out that it is a very strange thing, abounding in metaphysical subtleties and theological niceties" (Marx, 1867/1977, p. 163). Because the commodity is the starting point for Marx's analysis of capitalism, this mystery is of central importance. When wood is made into a table, for example, the table continues to be "an ordinary, sensuous thing." But once the table enters the marketplace as a commodity, "it changes into a thing which transcends sensuousness. It not only stands with its feet on the ground, but, in relation to all other commodities, it stands on its head, and evolves out of its wooden brain grotesque ideas, far more wonderful than if it were to begin dancing of its own free will" (pp. 163–164). Marx explains that the commodity is mysterious because it "reflects the social characteristics of men's own labour as objective characteristics of the products of labour themselves, as the socio-natural properties of these things" (pp. 164–165). Within a capitalist economy, the social relationships among people have been transformed into alien objects that seem to have their own relations. Private individuals laboring independently come into contact only indirectly, when the commodities they have produced are exchanged. The products of human labor become "sensuous things which are at the same time suprasensible or social" (p. 165). The things people make seem to have a life of their own, determining the fate of those who made them. This fetishism can be seen in economic analyses today; for example, when experts explain a drop in stock prices by saying that "[t]he market is anxious."

Alienation is one of several notions that Marx used to explore the ways in which the people who participate in the complex organization of a capitalist society inevitably *misunderstand* how it operates. The working class fails to see its exploitation, and the capitalist class, with its "bourgeois" theories, interprets the arrangements from which they benefit as natural, although in Marx's analysis they are highly artificial and contingent; things could well be otherwise. An *ideology* is a theory or set of concepts that obscures the social interests that created it. Ideology serves the reproduction of an unfair arrangement (Althusser, 1970/1971; Eagleton, 1991; Zizek, 1993).

Marx argued that consciousness is inseparable from humans' "historical life-process" (Marx & Engels, 1845/1988, p. 47) but that sometimes things "appear upside-down" in consciousness. Both philosophy and scientific knowledge reflect what Georg Lukàcs (1923/1988) called "false consciousness." In "the German ideology" of Hegel and his followers, for example, what people come to know "descends from heaven to earth" (Marx & Engels, 1845/1988, p. 47). Marx and Engels insisted that "[m]orality, religion, metaphysics, all the rest of ideology and their corresponding forms of consciousness" (p. 47) arise from real existence, from participation in social practices, even though they take distorted forms.

Marx proposed that it is workers who are in the better position to become conscious of the realities and contradictions of capitalism. He believed that a worker can come to comprehend the kind of society they are living in and grasp the magnitude of their unfair treatment: "[T]he worker stands on a higher plane than the capitalist from the outset, since the latter has his roots in the process of alienation and finds absolute satisfaction in it whereas right from the start the worker is a victim who confronts it as a rebel and experiences it as a process of enslavement" (Marx, 1867/1977, p. 990).

Workers do not have this consciousness automatically. It is achieved collectively as part of their "*class-consciousness*": their slowly growing awareness that they do indeed form a distinct class that occupies a particular position in the capitalist system (Lukàcs, 1923/1988). In particular, workers will recognize that capitalism is not natural or inevitable and appreciate their own power to change it. They will form a political party whose platform – that the means of production would "be worked in common by all and for the account and benefit of all" (Engels, 1845/2008) – will be shared with workers in all countries. Then will come a workers' revolution and the transition to a new kind of economy, a new mode of production, and a just society, a socialist society.

Hegel's influence here was strong, particularly his account of the struggle between master and slave. In this dialectical relationship, it is the *slave* who is in the position to achieve self-consciousness, largely because he must do the work that the master requires (see Box 11.1). This was one small but important part of Hegel's view, largely accepted by Marx, that history has a logic and will one day come to an end in a society in which all contradictions have been resolved.

People as Both Products and Agents of History

For Marx, one of the most important misunderstandings that is generated by the capitalist system concerns the form and genesis of *the individual*. Modern economics is largely based on the concept of the individual consumer making rational personal choices. But Marx was continually critical of what he called

BOX 11.1. *Standpoint Epistemology*

The basic notion that knowledge reflects the particular social position and social perspective of the knower has come to be called "standpoint epistemology." Standpoint theories claim that a specific socially situated position can provide privileged knowledge of the world. In some way, this is straightforward: a professional cook is likely to be in a better position to know about cakes than the average computer programmer. But standpoint theories more often claim that a member of a disadvantaged group knows the nature and causes of social inequalities better than other groups. Those who are privileged will tend to view inequities as natural, even necessary. But the standpoint of those who are disadvantaged enables them, at least in principle, to see that inequalities are contingent and can be transformed. For example, feminists have claimed that the standpoint of women provides a privileged understanding of gender relations and the capability to identify and repudiate sexist attitudes and patriarchy. A feminist standpoint is needed to oppose patriarchy, the oppression of women, and the devaluation of women's knowledge (see Harding, 1991; Hartsock, 1987; MacKinnon, 1999).

"bourgeois economics," which tries to build economic theory on the basis of independent individuals who come together only through contractual agreement; for example, the famous "social contract" of Romantic philosopher Jean-Jacques Rousseau (1712–1778). Marx argued that the idea that people are *naturally* individuals, originally existing independently like Robinson Crusoe on his desert island, is part of an ideology. We have seen that the social sciences also build theories about the "natural" individual. In Marx's analysis, the individual, just like a commodity, is a product of the advanced capitalist economic system. Far from being natural, the individual is actually a product of society and history.

Although in 18th-century civil society people *appeared* to be independent individuals, defined no longer by their family or their position in a feudal hierarchy, in reality social connections still made them who they were. In fact, Marx argued, this kind of society is actually one in which social relations are *most* developed. Humans can never exist outside of social relationships. We may develop in such a way that we view the relationships that define us as external to us, but this is a misunderstanding whose origin and character the social scientist must note and explain, not one that should be accepted and adopted as the basis for investigation:

> The more deeply we go back into history, the more does the individual, and hence also the producing individual, appear as dependent, as belonging to a greater whole: in a still quite natural way in the family

and in the family expanded into the clan; then later in the various forms of communal society arising out of the antithesis and fusions of the clans. Only in the eighteenth century, in "civil society," do the various forms of social connectedness confront the individual as a mere means towards his private purposes, as external necessity. But the epoch which produces this standpoint, that of the isolated individual, is also precisely that of the hitherto most developed social (from this standpoint, general) relations. The human being is in the most literal sense a *zoon politigon* [*political animal*], not merely a gregarious animal, but an animal which can individuate itself only in the midst of society. Production by an isolated individual outside society – a rare exception which may well occur when a civilized person in whom the social forces are already dynamically present is cast by accident into the wilderness – is as much an absurdity as is the development of language without individuals living *together* and talking to each other. (Marx, 1857–1858/1973, p. 84, emphasis original)

In the capitalist system of economic exchange, social relations are primarily *economic* relations, and "the ties of personal dependence, of distinctions of blood, education, etc. are in fact exploded, ripped up (at least, personal ties all appear as *personal* relations); and individuals *seem* independent ..., free to collide with one another and engage in exchange within this freedom" (pp. 163–164, emphasis original). But the independence that this seems to permit "is at bottom merely an illusion, and it is more correctly called indifference" (p. 163). People seem independent individuals "only for someone who abstracts from the *conditions*, the *conditions of existence* within which these individuals enter into contact" (p. 164, emphasis original). At the same time, "the conditions [appear] independent of the individuals and, although created by society, appear as if they were *natural conditions*, not controllable by individuals" (p. 164, emphasis original). Once again, it is crucial to grasp the conditions of existence if we are to understand what people do and how they live.

Followers of Marx differed in the extent to which they considered him to have viewed individuals as active agents or as acted on by forces that operate "behind our backs." His later works, such as *Capital*, are easier to interpret in the latter terms. His earlier writing, which was published later, has a more "humanistic" tone. Some, like Georg Lukàcs, were able to find an acknowledgment of human agency even in *Capital*: Lukàcs's *History and Class Consciousness* (1923/1988) anticipated notions that were later found in "Economico-philosophical Manuscripts." Others, such as Louis Althusser (1918–1990) (with whom Michel Foucault studied), read *Capital* in structuralist terms, as offering an account of social processes and historical transformation in which people are "interpolated" within "ideological practices" in which they come to believe, falsely, that they are free agents (Althusser, 1970/1971).

Economics today is one discipline among many, and the economist is a professional with specialized interests. To see Marx as interested in economics in this sense would be to misunderstand him. Marx saw that the economy is indeed distinct from other aspects of modern society, but he considered this to be a historical development, itself of interest and importance for understanding how we live and organize our lives. The relative independence of the economy is not natural; it is a historical achievement, sustained by ongoing social activity. Marx's analysis of economic phenomena – price fluctuations, devaluations, capital accumulation – was always part of his larger interest in society as a whole, plus his belief that labor, material production, is the *basic* motor of society and history. The modern, seemingly independent economic sphere is a specific way of organizing human labor that has had consequences for every other form of human life, which it would be hard to overestimate. To use Marx's terminology, the "forces of production" are key to the "relations of production." Social and cultural phenomena turn out, in Marx's analysis, to rest on economic factors. One way of putting this is to say that modern society has an economic "base" and built on this, but always dependent on it, a "superstructure" that includes politics, culture, law, and religion. The precise relationship that Marx saw between base and superstructure has been a topic of debate: is the superstructure entirely *determined* by the base, or is the relationship one of *mutual* influence, something more dialectical?

Marx's Method

Much has been written about the method of analysis that Marx employed, especially in *Capital* (e.g., Bologh, 1979; Ollman, 1990, 2003; Sayer, 1987). Marx's analyses began not with concrete things but with relationships. Even the commodity form turns out to be a process of circulation. Although it might seem obvious to begin with "activities, things and people as they appear to the ordinary observer" (Marx, 1867/1977, p. 35), these are really just empty abstractions. They appear to be "just there," but in fact they are grounded in relations and presuppositions. These objects, events, and individuals are embedded in, and are products of, a complex whole, a totality of relationships. The proper aim of analysis is to articulate this totality.

Marx's method was to take a single but central unit of the society of his time, the commodity form, and trace its movements through that society, through the "cycles" of economic production, exchange, and consumption. Every entity – money, a commodity, a laborer – is what it is only because it participates in the system of organization of material production. Marx undertook a *holistic analysis* in which he traced entities as they had developed historically (barter becomes cash exchanges, the gold standard, etc.) and circulated through the contemporary economy. Looked at this way, one small part can tell us about the whole.

Analysis is, of course, an activity of thought, an activity that leads to the formation of theoretical abstractions. It is only by thinking that we can figure out which relations exist. Concepts (or "categories") are the way these relations are represented theoretically. When Marx criticized traditional ("bourgeois") economics, his criticism took the form of an examination of its concepts and a diagnosis of their inadequacies. Traditional concepts are usually one-sided. But Marx didn't simply show that traditional thinking is wrong; he aimed to show how this thinking is possible, itself the product of a specific form of life, and so always has some validity, albeit limited.

The totality of relations that makes up the concrete whole is a form of life – for Marx, more specifically a form of production. This form of life is historically specific; that is to say, it has changed over history, and a current form should not be read back into the past or forward into the future. Thus the concepts – both traditional and dialectical – with which we think about these relations are also historically specific. The concept of "wealth" in modern capitalist society differs from the concept of "well-being" in medieval times. "Money" now is not "money" of 500 years ago. Modern terms such as mortgage and credit show only traces of their original usage as "dead pledge" and "trust." Marx explored how economic concepts are grounded in concrete social relations that the concepts either obscure (traditional economic theory) or illuminate (dialectical analysis). Even abstract concepts – such as "abstract labor" (the common quality of labor in general that can be abstracted from weaving, tailoring, etc.) – have their basis in specific concrete forms of life. Abstract labor is grounded in a form of life "in which labor in reality has become the means of creating wealth in general, and has ceased to be organically linked with particular individuals in any specific form" (Marx, 1857–1858/1973, p. 104). Modern capitalism is a highly developed economic system that can create wealth from *any* kind of productive labor rather than from a few specific kinds, such as agriculture. The concept of abstract labor does have some validity for thinking about other periods in history, precisely because it is abstract. But it possesses *full* validity only for the form of life of which it is a product.

An historical analysis needs to avoid forcing abstract concepts onto past forms of life. Imposing such concepts amounts to treating them as natural instead of historical and obliterates both their history and the specific social relations that produced them. The result is "eternal natural laws independent of history" (Marx, 1857–1858/1973, p. 87), economic laws that present as natural, and thus unchangeable, arrangements that are in fact arbitrary and problematic. Traditional economic theory does this and serves to bolster the status quo. Theory in the traditional social sciences does the same.

Marx suggested that we must understand the present before we can explore the past, as "[h]uman anatomy contains a key to the anatomy of the ape" (Marx, 1857–1858/1973, p. 105). The historical conditions in which a form of life originally became possible were generally different from the current conditions in which that form of life reproduces itself. Marx calls the first the "history of formation," the second "contemporary history" (Bologh, 1979, p. 44; Marx, 1857–1858/1973, p. 459). The analysis of contemporary history also offers "foreshadowings of the future" (Bologh, 1979, p. 46; Marx, 1857–1858/1973, p. 461). Historical analysis doesn't consist in merely tracing the origins of current arrangements but in viewing the contemporary form of life as itself an ongoing historical process, a process in which it must continually renew and reproduce itself but in which it is also engaged in "suspending" itself and preparing the conditions for a new form of life. All these aspects of Marx's method of inquiry have been influential, and their traces will be evident in the chapters that follow.

THE FRANKFURT SCHOOL

Kant foretold what Hollywood consciously put into practice. (Horkheimer & Adorno, 1944/1989, p. 84)

Marx argued that capitalism would end with an uprising of the proletariat, the working class. Exploited to the point where their very survival was at stake, the workers of the world would unite, Marx proposed, and throw off their chains. Dramatic events took place in Europe during the first decades of the 20th century: the First World War (1914–1918) and the end of the Austro-Hungarian Empire (1918); the revolution and civil war in Russia (1917) and formation of a Soviet state (1922); and the growth of the international workers' movement, to name a few. Many people considered these to be signs of a workers' revolution. But at the end of the war capitalism was still thriving. The members of the Frankfurt School suggested that Marxists had failed to appreciate the importance of cultural factors in historical change.

The "Frankfurt School" was the name informally given to a group of social scientists who worked in the Institute for Social Research (Institut fur Sozialforschung), founded in Germany in 1922 in affiliation with the University of Frankfurt am Main (Jay, 1973; Rasmussen, 1996). The major figures were Max Horkheimer (1895–1973), Theodor Adorno (1903–1969), Herbert Marcuse (1898–1979), Walter Benjamin (1892–1940), and Erich Fromm (1900–1980). The rise of the Nazi Party forced members of the institute to leave Germany, first to Switzerland in 1933, then New York in 1935 (creating The New School for Social Research), and California in 1941. This diaspora enabled the group to have a large impact.

We have seen that in Marx's earliest writings, the "Economico-philosophical Manuscripts" – his "humanist" work – he had explored factors such as alienation and ideology, but this work was not published until the 1930s. In his better-known writing, such as *Capital*, Marx moved toward increasingly determinist-sounding descriptions of historical change. Engels, Marx's colleague, had written even more as though change were the direct product of the forces of production, no matter how people responded to them (e.g., Engels, 1926/1960).

The Frankfurt School built on Marx's materialist analysis but worked to restore the experiential dimension they felt it left out, in part by returning to Hegel's dialectical account of the relation of subjectivity and objectivity, in part by drawing on sociologist Max Weber and psychoanalyst Sigmund Freud. Marxism, they felt, had ceased to be a truly critical theory, one that differed from traditional theory in its exploration of the conditions for knowledge claims and fostering of moral autonomy. Like Marx, they felt that a "[c]ritical theory insists that one needs a theory of society grounded in a theory of capitalism to make sense of sociohistorical processes and developments because the dynamics of capitalism play such a constitutive role in social life" (Kellner, 1989, p. 22). But they insisted that such a theory must attend to the way these dynamics are taken up – and generally misunderstood – by the people who live in them. There is, the members of the Frankfurt School insisted, no inevitable progress in history, no final and perfect goal, no end to partiality. The path of history depends on people's ongoing efforts. They argued that the workers' revolution had been prevented by the power of capitalist ideology, and that orthodox Marxism needed to be complemented by a better understanding of how people can be deceived about their situation and their own interests. Furthermore, human reason can never achieve perfection; it will always include unreasonable elements. There will always be a tension between how the world is and what we know of it: "there is no complete picture of reality" (Horkheimer, cited in Held, 1980, p. 179). Reason is a human creation, so it will always contain irrationalities. This means both that critical inquiry should avoid claiming to have a totally objective analysis and that scientific rationality should never be completely trusted.

The Frankfurt School was especially skeptical about the kind of reason Kant had valued. In 1944, Horkheimer and Adorno published *Dialectic of Enlightenment*, in which they argued that Kantian rationality has a dark side (Horkheimer & Adorno, 1944/1989). It is a narrow technical rationality that has fostered the view that our planet is merely a stockpile of raw material to be utilized and that people are a source of labor power to be exploited. They took direct aim at Kant, who they accused of trying to justify an empty formal logic that "permits peace or war, tolerance or repression." They pointed out that Kant's concept of universal reason fails as soon as people contradict one another. His philosophy was full of "unresolved contradictions." Its "secret utopia" (Horkheimer & Adorno, 1944/1989, p. 84) was a society in which people freely come together in complete agreement and harmony, but Kant's concept of reason equally suits self-serving individual calculation. Kantian reason is a system of "calculation, of planning; it is neutral in regard to ends" (p. 88). The result is that "emotion and finally all human expression, even culture as a whole, are withdrawn from thought" (p. 91). With the

Enlightenment, "reason ... became a purposeless purposiveness which might thus be attached to all ends" (p. 89). It treated individuals only as exemplars, with an attitude as heartless as that of an insurance company, to whom what is important is not who dies but the ratio of mortalities to liabilities (p. 84). It led to an "organization of life as a whole which is deprived of any substantial goal" (p. 88), and pretty soon "[p]ure reason became unreason, a faultless and insubstantial mode of procedure" (p. 90).

Kant's effort "to replace enfeebled religion with some reason for persisting in society" (Horkheimer & Adorno, 1944/1989, p. 85) was equally hopeless; just "propaganda" for bourgeois values. The Enlightenment, with its "liberation of forces, universal freedom, self-determination" (p. 93), was "the instrument by means of which the bourgeoisie came to power." But then it turned against them, becoming "a system of domination." Rejecting all forms of authority, all beliefs, Enlightenment turned into "its very antithesis – into opposition to reason," a tool that any authority could use. "With the formalization of reason ... thought appears meaningful only when meaning has been discarded" (p. 93). Kant's "cold Law," which "knows neither love nor the stake" (p. 114), can support fascism, sadism, and the Nietzschian will to power just as easily as liberal democracy.

This was not an attack on reason in general but criticism of a conception of rationality that presumes that the reasoner is autonomous and disengaged. The "autonomous" thinkers of the Enlightenment were a powerful elite of white males, and their philosophical accomplishments, viewed closely, can scarcely be called autonomous. They required the dependent servitude of a large number of people – women, people of color – who were generally quite unfree to determine the course of their own actions. The freedom of the elite was an illusion, for in reality they depended on the services of these others. Their work can scarcely be called individual either because it depended on the organization of a whole society. Kant's neutral, anonymous, and decontextualized "transcendental ego" was a convenient fiction. The Kantian ego "is in fact the product of, as well as the condition for, material existence. Individuals, who have to look after themselves, develop the ego. ... [I]t is extended and contracted as the prospects of economic self-sufficiency and productive ownership extend and contract" (Horkheimer & Adorno, 1944/1989, p. 87). For Horkheimer and Adorno, the Marquis de Sade (1740–1814) and Friedrich Nietzsche (1844–1900) were "the black writers of the bourgeoisie" who "mercilessly declared the shocking truth" about Enlightenment reason – "the indissoluble unity of reason and crime, civil society and domination" (p. 118). Indeed, abstraction and the distinction between subject and object are results of the domination of the slave by the master. "The distance between subject and object, a presupposition of abstraction, is grounded in the distance from the thing itself which the master achieved through the mastered" (p. 13). This was "the history of thought as an organ of domination" (p. 117).

The Frankfurt School struggled to build a different kind of reasoning, a critical theory that would integrate economic, political, cultural, and psychological aspects of life in advanced industrialized societies. Horkheimer described the specific object of critical theory as the overall process of society. Central to this work was the goal of human emancipation, the aim "to liberate human beings from the circumstances that enslave them" (Horkheimer, 1982, p. 244; see also Horkheimer, 1972; Menke, 1996). A number of empirical studies were conducted: investigations of personality, the family, and structures of authority, including *Studies of Authority and the Family* (Horkheimer, Fromm, & Marcuse, 1936) and *The Authoritarian Personality* (Adorno, 1950). The tone of this work became increasingly pessimistic, though the opening lines of *The Dialectic of Enlightenment* had not exactly been cheerful: "In the most general sense of progressive thought, the Enlightenment has always aimed at liberating men from fear and establishing their sovereignty. Yet the fully enlightened earth radiates disaster triumphant" (Horkheimer & Adorno, 1944/1989, p. 3). In particular, the members of the Frankfurt School viewed with suspicion and distaste the totalizing institutions of modern society, especially the culture industries. They saw these eroding subjectivity, encouraging conformity and passivity, threatening "the end of the individual," and undermining any revolutionary consciousness on the part of the working class. In his view that people see the world as they make it, "Kant foretold what Hollywood consciously put into practice" (Holkheimer & Adorno, 1944/1989, p. 84) by offering images to the public that are tailored to suit their predilection. For the Frankfurt School, preserving agency in the face of these threats was crucial for achieving the goal of increasing human freedom and liberty.

CONCLUSIONS

Marx described capitalism as an economic system based on the systematic, structural exploitation of one class of people by another class of people. With Marx, critique became a matter both of showing the conditions for the possibility of a phenomenon and exposing injustice and inequity. Marx's study of capitalism provided a powerful illustration of critical inquiry as an approach to investigation that recognizes and works to overcome the misunderstanding that members of a form of life are subject to.

The aim of this kind of inquiry is to enlighten and emancipate, to diagnose the source of inequities and injustices. Analysis is needed to articulate what is going on and become able to change it. To achieve this aim, Marx traced in detail the processes of capitalist economics, the cycles of production, exchange, and consumption through which move not just commodities but also people. He traced the historical development of these processes because he considered capitalism to be dynamic, both contradictory and changing.

Marx recognized the two-way relationship of constitution, citing Aristotle's definition of the *zoon politikon*. The Frankfurt School held on to Marx's insight that constitution goes both ways. People are not *merely* products of a form of life; their actions sustain that form of life and can transform it. They believed that insisting on this was crucial in any effort to enlighten and emancipate.

Marx's analysis worked toward a sense of the whole in order to identify the positions people occupy within this whole. But Marx's claim to have grasped the "totality" of society, and the truth of this totality, can be seen as the Achilles' heel of his method. It presumes that there is a point from which, and an objective gaze with which, the whole can be seen, all at once. This "totalizing" aspect of his analysis has attracted much criticism. It raises the question of what exactly is the position from which a researcher can see a form of life free from the illusions and false consciousness that handicap its members. What must be added to the practical understanding of a member?

It might be said that Marx assumed that what needs to be added is the "big picture" – a grasp of the whole and its transformation over time. In the following three chapters, I will explore three more conceptions of critical inquiry, each of which tries to grasp the "objective framework" in a different way, focusing in turn on the work of three important figures: philosopher Jurgen Habermas, sociologist Pierre Bourdieu, and historian Michel Foucault. Habermas has tried to bring up to date Kant's proposal that humans have a universal capacity for reason, although he locates this potential not in transcendental categories but in "transhistorical" structures of communication. Bourdieu also insisted that transcendental structures do not exist and that scientific rationality is the product of particular social and historical conditions. But he added that science is a special form of social struggle that can produce knowledge that "escapes" these conditions and has universal application. Foucault argued in contrast that any kind of universal knowledge or disinterested theory is impossible. Social scientific truth claims are always limited in scope, and knowledge is always conditioned, local, and particular. Researchers must acknowledge their involvement in history and culture and take a stand on this involvement. The "constitution of self" and an intervention in the present are linked aspects of critical inquiry: the social scientist can, and in fact must, "dare to know and constitute him- or herself in political opposition to present structures of domination" (Poster, 1989, p. 77).

Each of these three figures has proposed a form of inquiry with emancipatory potential. Each had an interest in the relations among knowledge, ethics, and identity; among work, language, and power. But each wove these together in a different way. These are complex and subtle analyses. By comparing them we can learn about the important issues in a kind of qualitative research that seeks to answer questions about the relation of constitution between human beings and our forms of life.

12

Emancipatory Inquiry as Rational Reconstruction

Against the intellectual grain of the time, Jurgen Habermas's grand theorizing defends the claims of reason, universality and normative validity. Among internationally prominent intellectuals on the left today, Habermas is a rarity: he insists the Enlightenment got it mostly right. For decades, he has been at work reconstructing what he sees as the damaged but vital legacy of Enlightenment rationality, rescuing its practical and emancipatory dimensions from the encroachments of technocratic and instrumental reason.

Forbath, 1998, p. 969

Jurgen Habermas is one of the most important philosophers today. He was trained in the Frankfurt School, and he continued its reappraisal of Marxism, which by the 1960s he felt seemed to be better at *legitimating* authoritarian states than critiquing them. Orthodox Marxism – with its determinist reading of Marx's writings – seemed to be serving an ideological function similar to the one the Frankfurt School had found in positivist science – that of "portraying particular interests and goals as technical necessities" (Mendelson, 1979, p. 46). Four aspects of Habermas's work are important to us here. The first is his elaboration of Marx's proposal that knowledge is never neutral and disinterested but is made possible by "knowledge-constitutive interests." This proposal transformed the image of critical inquiry, showing that it is distinct from both empirical-analytic science and interpretive inquiry but is no less a science for its interest in emancipation. The second is his debate with Gadamer over the requirements for critical inquiry, in which Habermas stated a position that he has never changed: that a researcher needs more than the resources of a member. A frame of reference *outside* the form of life being investigated is necessary. The third aspect of Habermas's work is his exploration of psychoanalysis as a model for critical and emancipatory inquiry, a process in which the therapist carries out a depth hermeneutics aided by a theoretical reconstruction of the consequences of childhood trauma. The fourth and final aspect is Habermas's turn to a collaboration between philosophy and social science to carry out the "rational reconstruction" of human communication and social organization in order to identify their universal underlying norms.

These four aspects center around the crucial question of what kind of special knowledge or stance a researcher needs in order to conduct critical inquiry. Habermas has argued consistently that participation in the practices of a form of life provides insufficient resources to enable a researcher to identify its conditions and inequities. But, at the same time, he doesn't want to return to Kant's transcendentalism. His efforts to find a "third path" between absolutism and relativism have made important contributions to our understanding of critical inquiry. His attempts to establish the normative basis for critique – first in cognitive interests, then in a universal theory of communication, and finally in an ethics of discourse – have been enormously influential, even if many consider them not to have been fully successful.

THE CRITIQUE OF DISINTERESTED KNOWLEDGE

In his book *Knowledge and Human Interests* (1968/1971), Habermas questioned "the concept of theory that has defined the tradition of great philosophy since its beginnings" (Habermas, 1968/1971, p. 301), namely that "the *only* knowledge that can truly orient action is knowledge that frees itself from mere human interests and is based on Ideas – in other words, knowledge that has taken a theoretical attitude" (p. 301). Since the time of the ancient Greeks, genuine knowledge has been seen as *theoretical*, the product of detached contemplation, a "looking on," first at sacred events, then at the cosmos. We have seen that Kant shared this view.

The Greeks' view of knowledge presupposed an ontology that has lasted ever since, in which the human world is a temporal realm that is "mutable and perishable," full of "inconstancy and uncertainty" – of correlation rather than real causation (Habermas 1968/1971, p. 301). Only the heavenly cosmos is perfect and immutable. For the Greeks, the imperfect world could be the source only of *doxa*, of mere opinion, not of true knowledge. *Doxa* (δόξα) was the Greek word for opinion or popular belief, in contrast with *episteme* (επιστήμη), the term for justified knowledge. It followed that in order to obtain genuine knowledge and to live an ethical life, a philosopher must contemplate the immortal order of the cosmos in a detached way and "manifest [its] proportions" (p. 302).

Habermas argued that modern science still adopts this ideology of interest-free knowledge. The empirical-analytic sciences understand themselves in the terms of logical positivism, which we explored in Chapter 1, and their "basic ontological assumption" is that there is "a structure of the world independent of the knower" (Habermas, 1968/1971, p. 304). They are, as Habermas puts it, "committed to a theoretical attitude that frees those who take it from dogmatic association with the natural interests of life and their irritating influence" (p. 303). Their aim is "describing the universe theoretically in its lawlike order, just as it is" (p. 303). The social sciences, too, have often adopted this positivistic self-understanding, with "an unconditional commitment to theory" and "the severance of knowledge from interest" (p. 303) – calling for value-neutrality, for example, and insisting on the distinction between *is* and *ought*, between description and evaluation.

A second type of scientific inquiry is the "*historical-hermeneutic* sciences." These "are concerned with the sphere of transitory things and mere opinion" (p. 303), and this seems on the face of it a very different aim. For example, these sciences seek to describe the symbolic meanings of tradition. "But they, too, comprise a *scientistic consciousness*" (p. 303). They, too, have the goal of "describing a structured reality within the horizon of the theoretical attitude" (p. 303).

Both these kinds of science operate under what Habermas called an "objectivist illusion." They adopt "an attitude that naively correlates theoretical propositions with matters of fact" (Habermas, 1968/1971, p. 307). They fail to recognize that their objects of study are constituted in everyday cultural practices, and they don't understand the "interlocking of knowledge with interests from the life-world" (p. 306). These sciences do not in fact create "pure theory." Indeed, Habermas argues, "pure theory" has *never* been free from interests. On the contrary, even for the Greeks, "[t]heory had educational and cultural implications . . . because it derived *pseudonormative power from the concealment of its actual interest*" (p. 306, emphasis original).

What does this mean? Habermas speculates that in the Greek city-states of the 5th century BC the "illusion of pure theory" (Habermas, 1968/1971, p. 307) served an important social function. It facilitated the "individuation" that was starting at that time by offering an "anchor" for the "identity of the individual ego as a stable entity" (p. 307). Habermas suggests that only by believing that the knowledge they were constructing was objective truth about a stable, ordered, and pure cosmos could Greek thinkers gradually free their sense of their identity from myth and legend. There was, Habermas proposes, an essentially *emancipatory* interest to ancient Greek philosophy, an interest that paradoxically needed to remain hidden to be effective. The claim that true knowledge is disinterested and objective, and that interests and concerns can never be the basis for true knowledge, only for opinion, is itself a myth that in ancient times had important purposes.

But today we are surely mature enough to throw this myth away. Why then does modern science hold on to the illusion of disinterested knowledge? This is especially puzzling because modern empirical-analytic sciences don't claim to be the basis for enlightened or ethical conduct. Why continue to "suppress the transcendental framework that is the precondition of the meaning of the validity of [theoretical] propositions" when modern empirical-analytic and hermeneutic sciences no longer claim to have an emancipatory potential?

TABLE 12.1. *Knowledge-Constitutive Interests*

	Empirical-Analytical Inquiry	Interpretive Inquiry	Critical Inquiry
Underlying interest	Technical	Practical	Emancipatory
Form of possible knowledge	"Information that expands our power of technical control"	"Information that makes possible the orientation of action within common traditions"	"Analyses that free consciousness from its dependence on hypostatized powers"
Illusion	Empirical statements simply describe the facts.	Mental facts are given in direct evidence.	
Hidden motivation	Technical control	Mutual understanding	Emancipation from seemingly natural constraints
Underlying kind of social organization	Work	Language	Power

Habermas's answer to these questions took the form of identifying the "knowledge-constitutive interests" (see Table 12.1) that operate, invisible, in contemporary science. In a critical analysis of scientific inquiry, he explored the interests that ground and lie behind knowledge-claims as conditions for their operation. There are three such interests: "The approach of the empirical-analytic sciences incorporates a *technical* cognitive interest; that of the historical-hermeneutic sciences incorporates a *practical* one; and the approach of critically oriented sciences incorporates the *emancipatory* cognitive interest that, as we saw, was at the root of traditional theories" (Habermas, 1968/1971, p. 308, emphasis original).

The "technical or instrumental interest" works to constitute the "facts" studied by the empirical-analytic sciences. Habermas agreed with the logical positivists that these sciences construct theories in the form of hypothetico-deductive connections among propositions "which permit the deduction of lawlike hypotheses with empirical content" (Habermas, 1968/1971, p. 308). These hypotheses are predictions about observable events, and they are tested by means of operations conducted under controlled conditions. But at this point Habermas parted from the positivist story. The basic empirical statements in empirical-analytic science, he insisted, "are not simple representations of facts in themselves, but express the success or failure of our operations" (p. 308). Our *instrumental actions*, our technical ability to repro-duce phenomena under the controlled conditions of the laboratory with modern technology, provide the regularities about which the empirical-analytic sciences construct their theories. These sciences are organized and made possible by a hidden interest in instrumental action, a "cognitive interest in technical control over objectified processes." This "constitutive interest in the possible securing and expansion, through information, of feedback-monitored action" (p. 309) operates invisibly within these natural and social sciences.

The second knowledge-constitutive interest operates in the sphere of intersubjective communication and the interpretive social sciences. Habermas called it a "practical, communicative interest." The interpretive sciences, as we have seen, aim at the objective and disinterested study of subjective experience. There is an objectivist illusion here, too, this time that of "historicism" (Habermas, 1968/1971, p. 309). By this Habermas means the way "[i]t appears as though the interpreter transposes himself into the horizon of the world or language from which a text derives its meaning" (p. 309). But in fact the interpreter's preunderstanding, arising from their own situation, always mediates the knowledge they obtain. Habermas agreed with Gadamer (Chapter 4) that the interpretation of a text is always *applied* to the interpreter's situation: "The world of traditional meaning discloses itself to the interpreter only to the extent that his own world becomes clarified at the same time. The subject of understanding establishes communication between both worlds. He comprehends the substantive content of tradition by *applying* tradition to himself and his situation" (Habermas, 1968/1971, pp. 309–310).

This suggests that interpretive inquiry actually "discloses reality subject to a constitutive interest in the preservation and expansion of the intersubjec-tivity of possible action-oriented mutual understanding" (Habermas, 1968/1971, p. 310). Interpretive inquiry seeks knowledge not for its own sake but for the sake of fostering "the attainment of possible consensus among actors in the framework of a self-understanding derived from tradition" (p. 310). The hermeneutic sciences are motivated by a fundamental human interest in understanding one another in order to get things done together.

The third knowledge-constitutive interest is an "emancipatory interest," and it operates in the critical sciences. Critical inquiry differs from empirical-analytic and interpretive inquiry because it is concerned "to determine when theoretical statements grasp invariant regularities of social action as such and when they express ideologically frozen relations of dependence that can in principle be transformed" (Habermas, 1968/1971, p. 310).

For example, by using the methodological procedures of the empirical-analytic sciences, we might determine that girls regularly obtain lower scores than boys in high school math classes. We might take another step by using the techniques of the interpretive sciences and discover that girls view math as difficult and irrelevant to their lives. But the question still remains whether this systematic difference represents a simple "fact of life" or is the result of relations of dominance and dependence among the sexes that young men and women are struggling to make sense of but are not merely "natural" and could in principle be changed.

A *critical* social science tries to answer this kind of question. In doing so, it can "take into account that information about lawlike connections sets off a process of reflection in the consciousness of those whom the laws are about. Thus the level of unreflected consciousness, which is one of the initial conditions of such laws, can be transformed" (Habermas, 1968/1971, p. 310). Once we *know* that there is a systematic difference between the math scores of boys and girls, and more important once *they* know this, a transformation has already begun. Habermas insists that humans have a capacity for reflection and self-consciousness so that knowledge of our behavior can in itself transform that behavior. Self-reflection, in Habermas's words, "releases the subject from dependence on hypostatized powers" (p. 310); that is to say, from powers that are viewed as objective and natural when in fact they are contingent and constructed.

Where do these knowledge-constitutive interests come from? Habermas described them as "quasi-transcendental limits" (Habermas, 1968/1971, p. 311) that we cannot get beyond, even when we become aware of them. They provide a "metalogical necessity" to our knowledge of the world and ourselves. Habermas proposed that they function to "establish the specific viewpoints from which we can apprehend reality as such in any way whatsoever" (p. 311). We can neither represent them nor prescribe them but only "*come to terms*" with them. They operate at a level deeper than either logic or empirical observation, and so "they cannot be either logically deduced or empirically demonstrated" (p. 312).

The three interests are rooted in "the natural history of the human species" (Habermas, 1968/1971, p. 312, emphasis removed). This is not a sociobiological argument; Habermas is not suggesting that science simply has survival value, that it serves purely evolutionary functions. Each of the interests is linked to a crucial aspect of our existence as a species and as individuals. As a species, our existence is based on forms of social organization: namely, *work, language,* and *power*. These define, respectively, the productive forces of a society, its cultural traditions, and the forms of legitimation "through which a society interprets itself" (p. 313). (We will see in Chapter 14 that Michel Foucault highlighted these same three aspects of human existence.) Our individual existence depends on our adaptation to environmental conditions, our effective functioning in the "communication system of a social life-world" (p. 313), and our capacity to create an identity in relation to the norms of our social group. The knowledge-constitutive interests are the product of who we are – both our animal nature, says Habermas, *and* the "cultural break" humans have made with this nature. As creatures of culture, we have natural drives but also the means to release ourselves from their constraints. Our search for knowledge doesn't simply serve the reproduction of our collective life; it also serves to "determine the definitions of this life" (p. 313). It can *change* who we are, both as individuals and as a species.

As we have seen, Habermas began by pointing out that Greek philosophy had an emancipatory interest: theory was knowledge that could guide and inform conduct. He concluded by suggesting that this connection between knowledge and "the good and true life" (Habermas, 1968/1971, p. 317) could still hold today, but this would be possible "only on the ruins of [traditional] ontology" (p. 317). To see knowledge differently, we need to see the world differently, and vice versa. We need to see that emancipation doesn't come from *disinterested* inquiry but from inquiry that is appropriately motivated.

THE GADAMER-HABERMAS DEBATE

In showing that all inquiry is organized by knowledge-constitutive interests that have their roots in our productive activity, our communication and culture, and our search for identity, Habermas expanded critical inquiry, rejected the notion that theoretical knowledge needs to be disinterested, and conceived of a kind of inquiry that has an *emancipatory* interest. Habermas was continuing the work of the Frankfurt School in developing Marx's critique of modern capitalist society and his view that emancipation is necessary to overcome the false understanding that arises from participation in the practices of such a society. In the early 1970s, a debate between Gadamer and Habermas served to clarify how Habermas understood this emancipatory inquiry, the part that interpretation played in it, and what a researcher needs to engage in a critical social science (see McCarthy, 1978; Mendelson, 1979; Misgeld, 1976; Ricoeur, 1973). Habermas agreed with Gadamer that critical inquiry needed to be hermeneutic, but he argued that this interpretation needs to be guided by a theoretical reference frame.

In suggesting that critical inquiry needed a hermeneutic component, Habermas was parting company with Horkheimer and Adorno, who had not had much sympathy with hermeneutics, which they considered deeply conservative. Adorno's *The Jargon of Authenticity* (1964/1973) was very critical of Heidegger's existential hermeneutics. But there are similarities between hermeneutics and critical theory. Both are strongly critical of instrumental reason and the reduction of rationality to the technical means–ends planning that aims to control nature for profit. Both Gadamer and Heidegger criticized "scientism," which Habermas defines as "science's belief in itself: that is the conviction that we can no longer understand science as one form of possible knowledge, but rather must identify knowledge with science" (Habermas, 1968/1971, p. 4). Both argue that scientific knowledge is possible only on the basis of the practical know-how of everyday life. And both have criticized the idealist view of the human subject as a transcendental ego rather than an empirical, materially embodied person.

Habermas's turn to hermeneutics was also a rejection of the orthodox Marxist view that politics and culture are a mere "superstructure" that is completely determined by the economic "base." Such a reduction leads quickly to the view that political freedom and a truly democratic society will *automatically* follow from economic change and that centrally organized and planned production can help. Marx's critique of capitalism had become read as an objective statement of universal social laws. Habermas was keen to explore ways that phenomena such as language, communication, and culture could be seen as more than mere epiphenomena. He hoped this would make it possible again for critical inquiry to identify possibilities for collective political action, even in the consumeristic, bureaucratic, and often authoritarian societies of the late 20th century.

Attending to these phenomena meant that critical inquiry could not avoid hermeneutics. In his book *On the Logic of the Social Sciences* (1967/1988), Habermas considered phenomenological (Schutz, Garfinkel), linguistic (Winch), and hermeneutic (Gadamer) approaches to inquiry. As we saw in Chapter 4, Gadamer (following Heidegger's emphasis on "historicity") had viewed *verstehen* not as a method special to the human sciences (Dilthey's position) but as a fundamental aspect of human existence. We have also seen how Gadamer argued that the effort to step out of one's historical and cultural situation (as Dithey had sought to do) is not the way to "objectivity." Gadamer insisted that any historical knowledge *requires* preconceptions and that these cannot simply be dissolved through reflection. Habermas, too, believed that social scientific inquiry must recognize the historical positions of the researcher and the object of inquiry. Social scientists who adopt the model of natural science assume that they can step outside history, but in fact the basic categories of social science – such as "role," "status," and "culture" – reflect the societies within which these scientists operate. Habermas insisted that critical inquiry must reflect on the conditions of its own knowledge.

The task of interpretation, as Gadamer conceived it, is not just to understand and know an object, a text, or an event but also to come to know the tradition in which both the interpreter and the object of interpretation are located. We have seen that Gadamer's view was that we cannot remove ourselves from our particular situation and interests, but we can work to become *conscious* of them. This means becoming aware of one's place in a tradition of history and interpretation. This "nexus of tradition" is the "one great horizon." Interpretation is both oriented by application to the present and guided by the practical task of preserving and transforming tradition. "All acts of interpretation are part of the movement of history in which tradition is preserved and transformed and the horizon of the present constituted" (Mendelson, 1979, p. 56).

We saw in Chapter 9 some of the problems with this unitary conception of tradition. To Habermas, Gadamer's version of hermeneutics eliminated all possibility of critique (Habermas, 1977). Gadamer gave tradition a weight and authority that Habermas considered unwarranted. Prejudgments are not always legitimate; preconception can be prejudice. Reason can never escape from tradition completely, but we do need to draw a distinction between what is traditional and what is rational. Authority and knowledge, Habermas argued, are not the same thing; indeed they are often at odds. It is important to recognize that the authority of a tradition can be imposed, through force or threat, and this means that traditions must *legitimate* themselves in rational discourse. When we reflect, Habermas argued, we are able to *alter* our relationship to our situation and our tradition. In reflection, we can question the claims and imperatives of tradition and decide whether or not to accept them. Whereas Gadamer had proposed that understanding and interpretation contribute to continuing and passing along our tradition, Habermas insisted that interpretation could also shake up a dogmatic tradition (see Table 12.2). We shouldn't reflect on our preconceptions simply to become aware of them; we should also question them and perhaps reject them. Gadamer was correct to say that reflection and reasoning can never enable us to entirely escape from our form of life, but they can help us realize that this form of life is not simply "natural" and thus change our relationship to it as its members.

TABLE 12.2. *The Habermas–Gadamer Debate*

	Hans-Georg Gadamer	Jurgen Habermas
The model for inquiry	The human sciences: concerned with contemporary reinterpretation of a cultural tradition. The humanities.	Critical social sciences: directly aimed against institutional reifications.
What is not acknowledged	Prejudice: borrowed from Romanticism, reinterpreted as preunderstanding.	Interest: Marx interpreted by Lukàcs and the Frankfurt School.
The obstacle to understanding	Misunderstanding: the (inner) obstacle to understanding.	Ideology: the systematic distortion of communication by the hidden exercise of force. Allegedly disinterested knowledge.
The basis for understanding	Consensus: rooted in tradition and language/discourse.	The "regulative ideal" of unrestricted and unconstrained communication. Consensus is rare. We must aim for it.

In Habermas's view, when Gadamer argued that "experiences of truth" and "method" are mutually exclusive, he gave up too much to the scientists who were, in his view, obsessed with method, and he missed the opportunity to work out a hermeneutic methodology for the human sciences. Gadamer's version of interpretation was able to articulate a current form of life and its way of being, but was not able to correct its distortions (including the distortions in its sciences). To Habermas, interpretation must become the *critique of ideology.* Interpretive inquiry "comes up against walls of the traditional framework from the inside, as it were" (Habermas, 1978, p. 44).

Habermas proposed that to accomplish the critique of ideology, "hermeneutic understanding be mediated through critical theory" (Mendelson, 1979, p. 64). Interpretation needs the help of a frame of reference that is outside tradition. With his talk of tradition, Gadamer had failed to appreciate the material conditions of labor and domination that are always aspects of our historicity. Habermas argued that to comprehend these material conditions we need a "theoretical reference-system" in the form of a "reconstruction" of how traditions develop and change. Language and meaning operate within objective systems of human labor and political domination, and these must be considered from a historical point of view, from a viewpoint that is theoretical but at the same time practical. As we reconstruct the past in theory, at the same time we project a possible future in practice.

Gadamer replied that the critique of tradition has no need for such a theoretical reconstruction. The Enlightenment itself, he argued, is a *tradition of critique* in which authority is accepted only when it is judged *legitimate;* that is to say, as rational, noncontradictory, and ethical. Reason is not a matter of getting *outside* any tradition but the ability to judge critically from within. "This principle of reason can be unfolded as the principle of voluntary evaluation on the basis of critical thought and can be counterposed to traditions whose binding power rests on coercion" (Mendelson, 1979, p. 61). Gadamer argued that Habermas was idealizing reflective reason, overestimating its power, and failing to see that it, too, is always embedded in a form of life. He followed Heidegger again in insisting that human *being* is always at a deeper and broader level than *knowing.* We will never have a fully transparent "prejudice-structure," no matter what kind of reasoning we employ. Habermas was failing to distinguish between what we know and who we are.

Gadamer added that human labor and political domination themselves require language, so they are not prior to or outside human interaction and communication. They can be dealt with inside hermeneutics and don't require the additional apparatus of a theory. Work and politics are meaningful to us and no more "objective" or "real" than culture, tradition, or any other part of the "linguisticality" of human existence. Furthermore, any critique of ideology has to be conducted through open dialogue and debate. In none of these respects, then, do we need to go beyond hermeneutics, as Habermas proposed.

Habermas's response was that everyday competence in ordinary language is not sufficient for a critical inquiry. One needs to combine a hermeneutic mode of analysis with a historical mode of analysis. The researcher needs to interpret on the basis of a preunderstanding that is theoretically grounded in "systematically generalized empirical knowledge" about historical events. Only in this way could the radically situational character of understanding, which Gadamer himself insisted on, be reduced. And only when the victims of domination *recognize* themselves in the account offered by critical inquiry can that account be considered valid.

The topic of this debate was an important one, and it is one we will pursue through the remaining chapters. We saw in Part II that ethnomethodology (and conversation analysis) and traditional ethnography highlight the perspective of the members of a form of life. But we have also seen that membership can lead to *mis*understanding, which suggests that to critique a form of life one needs to grasp it in a way that is *different* from a member's point of view. For Marx this meant seeing the big picture: the society as a whole, including its transformation over time. Habermas is suggesting that this big picture should have the form of a *theoretical reconstruction*, one that abstracts what is universal from what is specific to a form of life.

PSYCHOANALYSIS AS A MODEL FOR EMANCIPATORY INQUIRY

Habermas offered an illustration of his proposal that critical emancipatory inquiry requires both interpretation and a theoretical reference frame, in the form of a fresh reading of Freudian psychoanalysis. He suggested that psychoanalysis could offer "guidelines for the construction of a critical social theory" (McCarthy, 1978, p. 195).

Here Habermas was following a path that the original members of the Frankfurt School had mapped, turning to Freud to escape from determinist readings of Marx. He was insisting that history does not follow universal, inevitable laws but is the outcome of human agency. But in viewing psychoanalysis as a metahermeneutics, and in suggesting that it could serve as a model for critical inquiry, he was taking a step that would prove controversial.

Sigmund Freud (1856–1939) never doubted that his creation, psychoanalysis, was a natural science (Freud, 1963). He even suggested that the goals of psychoanalysis might one day be achieved more directly with drugs and other medical techniques. But Habermas proposed that Freud's own writings actually make it clear that psychoanalysis is a form of emancipatory inquiry. Psychoanalysis is the "interpretation of muted and distorted texts by means of which their authors deceive themselves" (Habermas, 1968/1971, p. 252). Psychoanalysis is a *hermeneutic* process, specifically a *depth* hermeneutics.

Freud's own model for psychoanalysis, his "meta-psychology," was formulated in terms borrowed from the natural sciences of his time – terms such as energy, tension, and discharge. It went through various versions – conscious and unconscious; id, ego, superego – but each was inadequate to the task of explaining how psychoanalysis actually works in practice. Habermas argued that psychoanalysis, far from being a "steered natural process" (Habermas, 1968/1971, p. 251), is "*a movement of self-reflection*" that takes place "on the level of intersubjectivity in ordinary language between doctor and patient" (p. 251, emphasis original). The metapsychology described the operation of psychic structures, agencies, and processes that Freud believed he had discovered in the analytic situation. But what it did not include was *the situation itself*, the room and the couch, the "specifically sheltered communication" between client and analyst. And "[t]he conditions of this communication are thus the conditions of the possibility of analytic knowledge for both partners, doctor and patient, likewise" (p. 252). If we want to understand psychoanalysis, we need to understand these conditions of communication. Habermas proposed that what we should call the "metapsychology" of psychoanalysis is not Freud's scientific model but "the fundamental assumptions about the pathological connection between ordinary language and interaction" (p. 254) that can be found in the work of psychoanalysis as Freud practiced it and described that practice. Viewed this way, the metapsychology is a theory of "systematically distorted communication," a "metahermeneutics" that tells us about the conditions of the possibility of the knowledge obtained in psychoanalysis and the power this knowledge has to transform.

Systematically Distorted Communication

Psychoanalysis is an example of a methodology that is, in Habermas's view, equal to those of the natural and cultural sciences but deals with processes of *self-knowledge*. Habermas's reading of Freud "unfolds the logic of interpretation" (Habermas, 1968/1971, p. 254) of this methodology to show how it is a form of emancipatory inquiry.

In psychotherapy, what the client says and does is a text whose meaning is transparent to neither client nor analyst. Freud assumed that it was a result of censorship, with gaps, omissions, deletions, and substitutions. Communication had become systematically distorted. His concept of the unconscious referred to what had been suppressed or removed from public communication. Interpreting such texts requires more than the normal competencies of a language user: it is a practice of decoding and translating. The analyst is aided in this practice by a theory about the origins and purposes of censoring that provides "theoretical perspectives and technical rules" (McCarthy, 1978, p. 197) that go beyond everyday membership. The analyst interprets what is said and done with a "preunderstanding" that directs their attention to a relatively restricted set of possible meanings, namely the client's earliest relations with caretakers, which are assumed to have been disturbed by conflict. On the basis of such assumptions, the analyst listens to what the client says with an ear to double meanings.

The result is that psychoanalytic interpretation "steps back from language as a means of communication and penetrates the symbolic level in which subjects *deceive themselves* about themselves through language and simultaneously give themselves away" (Habermas, 1968/1971, p. 256, emphasis original). In the psychoanalytic situation the client's unconscious motives become evident, and the analyst and patient reflect on them together, reconstruct their origins, and work to undo the distortions. A suppressed symbol becomes "objectively understandable through rules *resulting* from contingent circumstances of the individual's life history." Repressed motives gradually come under the conscious control of the client.

Habermas suggested that psychoanalysis employs a theoretical preunderstanding with three main components: "basic metahermeneutical assumptions about communicative action, language deformation, and behavioral pathology" (Habermas, 1968/1971, p. 255).

1. Communicative action: psychoanalytic practice operates on the basis of preconceptions about the structure of nondistorted communication.
2. Language deformation: distortions in "the text of everyday language games" (p. 255) are traced back to a confusion of prelinguistic and linguistic developmental stages of symbol formation.
3. Behavioral pathology: a theory of deviant socialization processes, especially "flight from a superior partner" (p. 257), which connects early childhood interaction patterns with adult personality organization.

The third of these is the most explicit in Freud's metapsychology; the two others are largely implicit (so Habermas needed to articulate them). Freud proposed basic concepts and assumptions about the connections among communication, distortions in language, and behavioral pathology. He offered a theory of deviant socialization processes that connected interaction patterns in early childhood with adult personality. For example, conflict was interpreted in terms of the concept of "defense." Freud's theory of censorship was that the ego denies its identity with the part of the psyche containing instinctual impulses that it would be dangerous to act on. Many "need interpretations" become "repressed and privatized . . . compelling substitute-gratifications and symbolizations" (p. 255). Impulses slip past the censor only in disguised symbolic form. Language is developmentally important because these psychic symbols become intersubjectively valid in linguistically mediated interactions. Socialization is an initiation into language games, with the result that "motivations are not impulses that operate from behind subjectivity, but subjectively guiding, symbolically mediated, and reciprocally interrelated intentions" (p. 255). When trauma leads to repression, the hidden needs "distort the text of everyday language games and make themselves noticeable as disturbances of habitual interactions: as compulsion, lies, and the inability to correspond to expectations that have been made socially obligatory" (p. 255).

General Interpretations: Narrative Forms

Habermas proposed that the metahermeneutics of psychoanalysis provides the analyst with a theory that takes the form of "general interpretations" of early childhood development, including concepts such as learning mechanisms, defense mechanisms, and the character and consequences of interactions between the child and primary reference persons. These general interpretations play a role similar to that of theories in the natural sciences. In psychoanalysis, "[t]heory can take the form of a narrative that depicts the psychodynamic development of the child as a course of action: with typical role assignments, successively appearing basic conflicts, recurrent patterns of interaction, dangers, crises, solutions, triumphs, and defects" (Habermas, 1968/1971, p. 259). This is a "narrative background" against which the client, with the analyst's help, can fill in gaps to tell a more complete history of their own particular personal development. The "narrative forms" enable the psychoanalyst to offer an "interpretive scheme for an individual's life history" (p. 258).

In psychoanalysis, individual development is viewed as a "self-formative process that goes through various stages of self-objectification and that has its telos in the life-consciousness of a reflectively appropriated life history" (p. 259). The development of the child is viewed as the successive states of a self-forming system. This, Habermas argues, is a "dramatic" model in which "we are at once both actor and critic." The final state of development is achieved "only if the subject *remembers* its identifications and alienations, the objectifications forced upon it and the reflections it arrived at, as the path upon which it constituted itself" (p. 260, emphasis original).

Narratives in general (as we saw in Chapter 5), and psychoanalytic narratives in particular, don't just describe. They explain by showing "how a subject is involved in a history," and Habermas proposed that a psychoanalytic explanation must make use of the way the subject *understands* that involvement. "That is why narrative representation is tied to ordinary language" and its "reflexivity."

Clearly, a general interpretation offers a "schematic" narrative, not a specific person's story. It "contains no names of individuals but only anonymous roles. It contains no contingent circumstances, but recurring configurations and patterns of action. It contains no idiomatic use of language, but only a standardized vocabulary" (Habermas, 1968/1971, p. 264). How has a general interpretation escaped from what seems to be the inevitable particularity of historical narrative? Habermas argued (following Ricoeur) that there is a general relevance to every particular story. When we comprehend the narrative of an individual's history, we abstract its implications, its lessons, for our own situation. We "abstract the comparable from the differences, and concretize the derived model under the specific life circumstances of our own case" (p. 263).

The general interpretations in psychoanalysis were the product of many hours of interpretive work by Freud and his colleagues, "the result of numerous and repeated clinical experiences." They were derived, Habermas suggested, "according to the elastic procedure of hermeneutic anticipations (*Vorgriff*), with their circular corroboration" (Habermas, 1968/1971, p. 259). (*Vorgriff* was Heidegger's term for one of the elements of the fore-structure of interpretation, translated as fore-grasp.) That is to say, they embody in narrative form the professional, technical expertise of the practice of psychoanalysis. They make possible a "translation" of the patient's words: when the patient offers "symbolic expressions of a fragmented life and history" (p. 265), the psychoanalyst translates these with the aid of the schematic narrative into a *particular* story, the patient's autobiographical history. The analyst fills in the gaps, reconstructs the conflicts, and anticipates what the client's reflection will bring to light.

The Validity of Psychoanalytic Interpretation

How does a psychoanalyst know whether their translation is valid? Psychoanalysis has been criticized for not being a genuine science because its assumptions cannot be tested and its predictions are vague (e.g., Grunbaum, 1984). Habermas didn't agree with this criticism, but he did acknowledge that the validity of a psychoanalytic interpretation differs in kind from both the natural sciences and the interpretive sciences.

Verification is not a matter of a psychoanalyst asking their client whether their interpretation makes sense. After all, the analyst "makes interpretive suggestions for a story that the patient cannot tell." These suggestions "can be verified . . . only if the patient adopts them and tells his own story with their aid. The interpretation of the case is corroborated only by the successful continuation of an interrupted self-formative process" (Habermas, 1968/1971, p. 260). That is to say, Habermas insisted that the "narrative presentation of an individual history" (p. 266) finds verification not in the client's simple *agreement* with it but in their *application* of it. It is in this "*realization* of the interpretation" (p. 266, emphasis original) that both its effectiveness and its validation are found. Only when the patient can "know and recognize themselves" (p. 262) in the analyst's narrative does development move forward and the analyst's interpretation gain validity. The analyst as "subject" is trying to gain knowledge of the client as "object," but "[t]he subject cannot

obtain knowledge of the object unless it becomes knowledge for the object –
and unless the latter thereby emancipates itself by becoming a subject"
(p. 262).

So the "double hermeneutic" that Giddens identified (Chapter 6) is
central to psychotherapy, and by implication to all emancipatory inquiry.
Psychotherapy is the study of a *changing* object, an object who becomes a
subject. Its *aim* is to change the person who is the focus of inquiry, and it can
only do this to the extent that its interpretations foster a *self-formative* process.
For Habermas, psychoanalysis offered a model for emancipatory inquiry
because it is a kind of inquiry that fosters enlightenment in the form of self-
knowledge and self-formation. Psychoanalysis seeks "the *conscious* appropriation of a suppressed fragment of life history" (p. 251) with "the intention of
enlightenment, according to which ego should develop out of id" (p. 254).
Where the client had experienced compulsion, they now have a degree of
freedom. When psychoanalysis goes well, "the subject frees itself from a state
in which it had become an object for itself," and this is a process of *reflection* by
the client that can never be replaced by technological intervention (i.e., by
drugs and medicine). This is a model for emancipatory inquiry understood as
a process of dialogue, diagnosis, and self-recovery.

RECONSTRUCTIVE SCIENCE

In his subsequent work Habermas continued to develop the tools he considered necessary for a critical science of contemporary society, one that could
diagnose its problems and help "steer" it toward a more enlightened state. He
insists that he has been searching for a third way, between universalism and
relativism, to establish the normative basis for emancipatory science (Menke,
1996, p. 68). That is to say, when someone makes a claim about a form of life,
on what basis can we judge its truth, its ethical validity, and the speaker's
authenticity? Habermas denied that there are transcendental standards but
also that such judgments are no more than relative to the specific
circumstances.

Psychoanalysis answered these questions by drawing on the reconstruction of the genesis of systematically distorted communication that its "general
interpretations" provided. But the idea that psychoanalysis could serve as a
model for emancipatory inquiry ran into problems. Before considering the
alternative that Habermas turned to, let us quickly review these problems.

Problems with the Psychoanalytic Model

Psychoanalysis is emancipatory inquiry directed toward the individual, not to
a form of life or a society. If one tries to apply it directly as a model for social
critique, this difference quickly becomes clear. Habermas's proposal that

psychoanalysis offered a model for critical inquiry seemed to imply that the *political* process of interpreting and challenging the false consciousness of a social group could be modeled on the interaction between analyst and client in psychoanalysis. This proved to be highly controversial (McCarthy, 1978, pp. 205ff). Gadamer was one of the critics. He pointed out that psychoanalysis operates within an institutional framework that establishes criteria and limits for professional practice, whereas Habermas was trying to define a critical inquiry that could analyze society as a whole. There was a danger that people taking the role of "doctor" could operate without control or legitimacy. To Gadamer, the basic assumption that critique and social change require emancipation involved a dangerously patronizing attitude. Habermas was assuming that victims of oppression and exploitation need expert help to discover what their own actions mean and where their objective interests lie. The view that they are deluded and have a "false consciousness" that must be deciphered with a "depth" interpretation suggests that experts can decide where the truth really lies. (We will see in Chapter 14 that Foucault, too, regarded depth hermeneutics as a form of oppression.) Gadamer proposed instead that truths can be reached only through a dialogue in which everyone remains open to others' points of view.

Another line of criticism was that it is not the oppressed and exploited whose consciousness is false but those who oppress them. Hegel's analysis of master and servant (Chapter 11) suggested that oppressors are more blind to their position and interests than the oppressed. Statements by those in power are often ideological, mask the truth, and need to be interpreted with suspicion and an eye toward interests that the speakers themselves may not recognize. People with power and influence are unlikely to enter conversation willingly, or they will use dialogue as an opportunity to expand their influence.

To be fair, Habermas has never said that *only* oppressed groups have false consciousness. He responded to these and other criticisms (Habermas, 1973) by distinguishing three tasks or "functions" of critique, each of which has a distinct criterion of evaluation and involves a distinct kind of interaction. First, we want to develop true scientific theories through open and level dialogue. Second, we want to apply these theories to stimulate reflection in particular groups of people and validate our theories through their application in this process of "enlightenment." Third, we want to pick prudent strategies and tactics for political struggle. The therapeutic model is appropriate only in the second function, and dialogue here is asymmetrical; we aim to *achieve* a conversation between equals but we don't *start* there. Critical inquiry aims to enlighten people about their position and their objective interests in an unfair society, but the "patient" remains the final authority on the appropriateness of the critic's interpretation. As far as the third function is concerned, Habermas suggested that political action should follow only from debate among people who have become conscious of their own interests and their circumstances. Only such people can willingly take the necessary risks, whether they are undertaking reform or revolutionary struggle. Political groups need democratic organization and democratic debate, and this presupposes that enlightenment has already been achieved.

But even if we restrict Habermas's use of the therapeutic model to the second of these three tasks, there are additional problems. Psychoanalysis works with resistance, but could a critical researcher work toward enlightenment with a group that actively resisted their efforts? Psychoanalysis builds on transference, but what would be the equivalent political process? Therapy takes place in a setting detached from the pressures of everyday life, but what would be the analogous setting for an oppressed group? And successful psychotherapy involves recollection of previously repressed childhood memories, but does a social group have to rediscover its forgotten past?

Universal Pragmatics

If psychoanalysis is not such a good model for critical inquiry with a solid normative basis, what might be the alternative? Increasingly Habermas emphasized everyday communication as the source of reason and morality. In his earliest work he had described how the new "public sphere" that emerged in the 18th century – in newspapers, clubs, and coffeehouses – had provided new opportunities for debate with a critical edge. He considered that today this public sphere is under attack by the mass media. In the 1980s, Habermas introduced his theory of "universal pragmatics," a general theory of communication and the rationality implicit in it, which could "identify and reconstruct universal conditions of possible understanding" (Habermas, 1979, p. 1; see also Habermas, 1981/1984, 1981/1989). The result was an account of the values that Habermas proposed are implicit in every situation of communication. Like Kant, Habermas was maintaining that there is a universal kind of reason, but rather than it being a transcendental property of individual consciousness, he proposed that it is a "communicative rationality" that is inherent in human practices of communication.

An important component of universal pragmatics was what Habermas called the "ideal speech situation." When we communicate, he argued, we appeal to implicit norms of truthfulness, validity, and authenticity. These are universal because the need to communicate is universal. They amount to an image of an ideal kind of communication, a situation in which:

1. All potential participants have equal rights to speak.
2. All who participate have equal opportunity to make arguments for and against.
3. Participants can express equally their attitudes, feelings and wishes.

4. Participants have equal opportunity to order and resist orders, to promise and refuse, to be accountable for their conduct and to demand accountability from others.

(Habermas, 1973)

Such ideal communication does not in fact exist, but when we evaluate what is actually said and done we tacitly refer to it. It is a fiction, but one that actually operates in our interactions.

Reconstructive Science

Rather than propose a distinctive kind of emancipatory science, Habermas began at this time to recommend a division of labor between philosophy and social science. Critique, he argued, could come from uniting mainstream empirical investigation with philosophy's capacity for normative reflection. In collaboration the two would "reconstruct" general human competencies and identify the "deep structures" of both society and the person, as Noam Chomsky had identified the deep structures underlying competence in grammar. Philosophy would propose a rational reconstruction of the preconceptions implicit in important human capabilities, and the sciences would then test it empirically. The universal pragmatics was an example of this kind of reconstruction.

What makes such a reconstruction "rational"? Habermas continued to insist that research in the social sciences cannot avoid interpretation. To understand what he or she studies, the researcher must take the perspective of a participant, with all of the preconceptions that this involves. The researcher judges what happens because that is what it means to be rational; a disinterested and value-neutral stance is just not possible. At the same time, Habermas insisted that reflection on these tacit evaluations is necessary, and it is here that philosophy contributes. Traditional social science tries to be objective and neutral and fails to be reflective. Philosophical reflection on tacit ideals and rules provides investigation with its emancipatory basis, in the form of "standards for a self-correcting learning process." Interpretation is rational when it appeals to these standards of rationality, which in Habermas's view are located in the "deep structure" of human practices.

Whereas Marx had argued that we establish our autonomy in labor, Habermas locates the potential for human emancipation in human communication – in "talking, reason giving, and reasoning itself, conceived as a public and deliberative, consensus-seeking practice that makes government into self-rule" (Forbath, 1998). He argues that the very structure of human language contains a potential for autonomy and responsibility and it is language that "raises us out of nature" (Habermas, 1968/1971, p. 314). "Our first sentence expresses unequivocally the intention of universal and unconstrained consensus" (p. 314). This universal consensus does not yet exist because "only in an emancipated society, whose members' autonomy and responsibility had been realized, would communication have developed into the non-authoritarian and universally practiced dialogue from which both our model of reciprocally constituted ego identity and our idea of true consensus are always implicitly derived" (p. 314). Today violence still deforms our communication. But the ideal of free and unconstrained communication gives us a standard against which to judge what is said and done.

Reconstructive science, then, is a combination of reflection on and empirical study of the conditions presupposed in everyday communication. Just as Chomsky's reconstruction of grammatical competence doesn't tell people what to say but shows what is presupposed when we do speak, Habermas's reconstruction of communication aims to uncover its universal conditions. "Rationalization," in this context, means eliminating force and overcoming systematic distortions in communication.

An important characteristic of this kind of reconstructive science, one that distinguishes it from Husserlian phenomenology or Kantian transcendental analysis (Habermas, 1979, pp. 21ff), is that it is based on empirical data, not on researchers' reflections on their personal experiences and intuitions. Critical studies of social phenomena are theoretically guided, but the theory is applied to concrete forms of sociocultural life using hermeneutic procedures. One might say that critical inquiry is theory applied interpretively. The researcher articulates what the participants presuppose unquestioningly, questions what they recognize unthinkingly, and thus "deepens and radicalizes" the context of communication that is being investigated. The researcher draws on participants' pretheoretical foreknowledge and systematically reconstructs it with the aid of a theory. "A reconstructive science turns know-how into know-that, precisely in so far as it explicates our intuitive abilities to follow rules" (Edgar, 2006, p. 130). Reconstruction is a particular kind of reflection, in which a general account of the constitution of our capabilities is arrived at.

Discourse Ethics

Habermas's work on communicative rationality has led most recently to a "discourse ethics" (Habermas, 1990) in which he has tried to identify formal procedures that people everywhere can use to decide what is moral. A just procedure would be one that people could follow to make a genuinely moral decision, one that all people would agree to. "Discourse," as Habermas uses the term, refers to a specific kind of communication that has become reflective in the sense defined previously, where it involves processes of argumentation that explicitly test the factual and normative assumptions that are tacit in everyday communication. He has written on the implications of his ethics for law and government, both national and international.

The aim of discourse ethics is not to argue for a particular form of life, identified as *the* good society, but instead to judge the validity of different forms of ethical life. Discourse ethics is concerned not with specific views of the good life but universal issues of justice – "universal" in the sense that everyone must agree. Habermas does not want to make claims about how people should live but rather offer procedures that can enable people to decide this for themselves together. But, in general, a form of life is ethical when it enables its members to achieve freedom:

> To be free for Habermas does not signify a mere sophomorish absence of external obstacles or constraints in the satisfaction, fulfillment or enjoyment of just *any* desire. Rather, it signifies a material, model and social state of internal independence to be a certain kind of person, i.e., one capable of participating in those projects that are congruent with the physical, psychological, social and moral *well-being* of individual persons and society as a whole. (Badillo, 1991, p. 93, emphasis original)

Habermas's view of freedom is based on the recognition of the phenomenon of constitution. Freedom *is* a dynamic relation of autonomy and responsibility between individual and society, one that preserves the dignity of the person and the integrity of the community, for "what is at the core of this view of freedom is a dynamic interplay between the individual that *qua* moral agent is capable of affecting the social unit to which he belongs and the social unit that *qua* collective moral agent is capable of affecting its individual members. Society is thereby both the product of its members and the collective agent that fashions them" (p. 93).

Social Evolution

Based on his general theory of communication, Habermas has proposed a "systematically generalized history" of societal change that distinguishes history from "social evolution." The former tells the story of particular events, whereas the latter articulates the rules and structures that Habermas believes underlie all social change. Once again, Habermas's strategy has been to search for universal, underlying processes. The result has been an ambitious "reconstruction of the logic of development" of culture, society, and personality. Habermas views each of these as a process of learning and has gone so far as to draw parallels between the stages of individual development and those of societal change (see Box 12.1).

Marx viewed society as evolving toward socialism. But, for Habermas, socialism itself has been cause for great alarm. The socialist elimination of private property and its regulation of the economy led to social nightmares and authoritarian states. This is no basis for critical inquiry or critical praxis. In Habermas's view, Marxism itself called for "taking a theory apart and putting it back together in a new form in order to attain more fully the goal it has set itself" (Habermas, 1979, p. 95), so he reconstructed Marxism to give a fresh account of social evolution that rejected the values of both capitalist and socialist economies and sought to promote freedom and democracy (Forbath, 1998). Habermas suggests that society as a whole is learning and, as a consequence, evolving toward an egalitarian community of emancipated individuals. Critical inquiry needs to "reconstruct" this process in order to help it along its way (Habermas, 1979).

In Habermas's diagnosis, modern developed societies "do not make full use of the learning potential culturally available to them, but deliver themselves over to an uncontrolled growth of complexity" that consumes non-renewable resources, threatens traditional cultures, and undermines the "infrastructure" of social relations (Habermas, 1981/1989, p. 375). This "rationalization" of everyday life imposes "paradoxical conditions of life" (p. 378) that lead to "real abstractions" that are ignored by both the empirical-analytic and the interpretive social sciences because they fail to be sufficiently reflexive and historically aware.

To Habermas, the most important of these distortions is the "magnificent 'one-sidednesses'" of modern life (Habermas, 1981/1989, p. 397). We have divided our tradition into disconnected kinds of rationality. "Even without the guidance of the critiques of pure and practical reason, the sons and daughters of modernity learned how to divide up and develop further the cultural tradition under these different aspects of rationality – as questions of truth, justice, and taste" (p. 397). Truth is equated with positive science. The good has been reduced to the lawful. Art plays only to a decentered and removed subjectivity. Habermas insists that we must critique these differentiated aspects of reason if we are to overcome their division. Critical social theory must "bring viewpoints of moral and aesthetic critique to bear – without threatening the primacy of questions of truth" (p. 398). The ambitious aim of Habermas's project at this point is to reunite the three domains of work, language, and power that Kant helped divide.

CONCLUSIONS

In the last chapter we started to explore the proposal that qualitative research should be critical inquiry by examining the work of Marx and the Frankfurt School. In this chapter we have focused on the efforts by Jurgen Habermas, one of the Frankfurt School's graduates, to establish a firm normative basis for emancipatory inquiry. Habermas has argued consistently that being a member and participant – in a language, a tradition, or a form of life – is not sufficient. He grants that critical inquiry is a hermeneutic process, but in his debate with Gadamer he explained why he agrees with Marx that interpretation needs to

BOX 12.1. *Critical Research and Ontogenesis*

It has become clear that one of the elements of a critical approach to research is a conception of ontogenesis: the history of the individual person. People act in the present, but they are products of the past. The work of each of the three people considered in detail in Part III includes an ontogenetic component. Each of them felt a need for an account of why a person acts as they do not only because of *present* circumstances but also because of their *past*: their personal history, their development. Habermas drew first on Freud's account of the origins of communicative distortions in early childhood trauma and then on the structuralist models of Piaget (the ontogenesis of instrumental rationality), Lawrence Kohlberg (moral reasoning), and Robert Selman (social reasoning) (see Kohlberg, 1981; Piaget, 1970/1972; Selman, 1980). This has been part of Habermas's general movement away from a theory of distortions in communication that result from childhood trauma, what might be called a negative theory of communication, toward a theory of ideal communication that can be grounded in an account of "normal" ontogenesis. The development of an individual, Habermas argues, "can be analyzed in terms of the capacity for cognition, speech, and action" (Habermas, 1979, p. 100). For Habermas, a rational reconstruction of individual ontogenesis is needed to guide the researcher's interpretation of current situations by providing a theory of the genesis of distorted communication.

In Chapter 13 we will see that Bourdieu also tried to understand the way history is "incorporated" by the individual – how time literally shapes a person's body. Bourdieu described ontogenesis in terms of the "dispositions" a person acquires; the formation of *habitus*. The properties of habitus – that it is inculcated, enduring, and generative – explain why a person's dispositions to act depend not only on their current position but also on their earlier positions in other fields. In other words, habitus accounts for an identifiable "hysteresis" in human conduct. Hysteresis is the phenomenon in which a physical object's response depends on its history. For example, the way a piece of iron reacts in a magnetic field depends on how it has been magnetized in the past. Humans, too, react to circumstances in the present in ways that depend on past experiences. Bourdieu tried to avoid an overly intellectual account of this phenomenon, the view that we accumulate a stock of knowledge from past experiences and simply draw on this stock when facing new situations.

Are these accounts of ontogenesis adequate? Habermas claims that socialization involves two particular kinds of intuitive knowledge. The first is learning the everyday views and standards of rationality of a specific lifeworld. The second is the acquisition of what Habermas considers universal

(continued)

BOX 12.1 *(continued)*

structures of competence in cognition, language, and interaction. For Habermas, these structures get filled with content by each particular culture, but in their general form they are universal. He draws a line between the contingent and the universal, focusing his attention on the latter. The structuralist reconstructions of ontogenesis by Piaget, Kohlberg, and Selman have also claimed to identify *universal stages* of individual development. Their research has been criticized for its lack of attention to the role of culture in development and to differences resulting from class, ethnicity, and gender.

As we will see in Chapter 14, Foucault also explored how a human being becomes a knowing, acting, and judging subject. He began to explore this in terms of techniques and technology for the formation of the self, but his work was cut short. Foucault sketched a way to conceptualize the constitution of multiple kinds of subjects and subjectivities on a background of practical activity that provides each newborn child with the fruits of generations of human history:

> It is always against a background of the already begun that man is able to reflect on what may serve for him as origin. For man, then, origin is by no means the beginning – a sort of dawn of history from which his ulterior acquisitions would have accumulated. Origin, for man, is much more the way in which man in general, any man, articulates himself upon the already-begun of labour, life, and language; it must be sought for in that fold where man in all simplicity applies his labour to a world that has been worked for thousands of years, lives in the freshness of his unique, recent, and precarious existence a life that has its roots in the first organic formations, and composes into sentences which have never before been spoken (even though generation after generation has repeated them) words that are older than all memory. (Foucault, 1966/1973, p. 330)

This articulation of "the already-begun of labour, life, and language" was Foucault's response to the Kantian division among knowledge, practice, and beauty.

Critical analysis and critical inquiry evidently require an account of human development and do not yet seem to have one. It seems very likely that the work of the Russian psychologist Lev Vygotsky could offer much here. Vygotsky, working from a Marxist-Hegelian perspective, also rejected Kant's division, his dualism of appearance/reality, and his representational model of human functioning. Vygotsky's program for a general psychology of human psychological functions, a psychology that would necessarily focus on the ontogenesis of the individual, has aims similar to those I am exploring in this book (Packer, 2008; Vygotsky, 1987).

be guided by a "big picture" of the social whole and its transformation over time. As Thomas McCarthy has put it, we cannot *understand* the *what* without *explaining* the *why* (McCarthy, 1978, p. xiii). Critical inquiry needs to search for *both* understanding *and* causes because social systems and practices have an impact that operates outside members' awareness.

Habermas has made a series of suggestions about the frame of reference that is necessary. His first proposal was that *knowledge-constitutive interests* guide inquiry. Habermas considered psychoanalysis to be a possible model for critical inquiry because it is a hermeneutics that is guided by a "theoretically generalized history" of distorted communication. In psychoanalysis, the analyst speaks the language of the client but also uses a theoretical reconstruction of child development and its typical problems in order to decode distortions in the way the client comprehends and talks about their current situation. This theoretical reconstruction enables the therapist to help the client become self-reflective and self-transformative. In the same way, the critical investigation of societal ills will need more tools than are provided by the researcher's membership in a historical tradition.

But the model of psychoanalysis was not without problems, so Habermas turned to a division of labor between philosophy and empirical inquiry to create a *reconstructive science*. He placed great emphasis on the importance of grasping the "objective framework" within which discourse moves – how our ways of communicating extend beyond our everyday ordinary understanding. Habermas suggested that grasping this objective framework requires a step of analysis that will not simply accept at face value "the intuitive knowledge of competent subjects" but will "systematically reconstruct" this knowledge (Habermas, 1979, p. 9). Hermeneutics is part of critical inquiry, but traditional hermeneutics has a situation-bound character that must be overcome. Critical inquiry must be interpretation guided by a "rational reconstruction" of human society and its history.

There is much of value in Habermas's perspective on critical inquiry. His notion of knowledge-constitutive interests defined a place for critical investigation, research with an emancipatory interest, an interest in empowering oppressed people and transforming society. He has clarified the relationship between critique and interpretation. Critique is rooted in a uniquely human capacity for reflection and self-transcendence, which Habermas sees as the root of human rationality.

But at the same time he has insisted on a distinction between two kinds of critical reflection that is not entirely clear. The first kind articulates the conditions for a form of knowledge or mode of action. The second kind of reflection counters systematic distortions of understanding that are constraining the human self-formative process. We can see that the first has its origins in Kant's critique of knowledge, whereas the origins of the second are in Marx's critique of ideology. The former deals with undistorted communication or "reasonable discourse," whereas the latter deals with distorted communication. Both bring to consciousness factors that were previously operating unconsciously.

These two kinds of critical reflection remain in tension in Habermas's work. On the one hand, Habermas believes that people must talk and work together to resolve issues and determine the direction of social change. He calls this "practical verification." On the other hand, he values philosophy as a reflexive enterprise that can identify general, universal norms and values, and so reconstruct an objective account of historical change and individual development. His work has explored the connection between these two, but it also shows the tension between them. Certainly there is no guarantee that reflection based on membership and participation in everyday practices can overcome the misunderstandings, the ideology, embedded in these practices. On the other hand, if critical inquiry rejects the traditions that operate in particular forms of life and searches for an external frame of reference, how does it make contact with people? How is Habermas's idealized model of consensus through discourse relevant to real occasions of decision making?

Habermas has said that the goal is not to be "outside" every form of life but to achieve a critical "distance." He continues to insist that researchers should not simply be members, but adds that they should not become objective or impartial outsiders. Rather, they must try to achieve what Habermas calls a "negotiated impartiality." The "inevitable involvement in the process of reaching understanding does indeed deprive him [the researcher] of the privileged status of the objective observer or the third person but *for the same reason* also provides him with the means to maintain a position of negotiated impartiality from within" (Habermas, 1983/1990, p. 29, emphasis original).

Such a researcher articulates participants' presuppositions, calls into question what they take for granted, and thus "deepens and radicalizes" the situation. Drawing on a member's know-how, the researcher systematically reconstructs it with the aid of knowledge of the big picture. Working to reconstruct the *reasons* for an action or statement and then *evaluating* these reasons, the researcher builds a rational reconstruction that provides *theoretical* knowledge when it succeeds in reconstructing "very general conditions of validity" (Habermas, 1983/1990, p. 32). The *critical* character of this analysis arises from its attempt to explain deviant cases. Researchers do not claim to have the final word; they must attempt to establish the validity of their reconstruction through dialogue with the participants. Habermas insists that, in the end, "in a process of enlightenment there can be only participants" (cited in McCarthy, 1978, p. 357).

Habermas continues to emphasize four key characteristics of critical inquiry that he first pointed out in his study of psychoanalysis. First, critical inquiry involves dialogue between researcher and researched. Second, it requires interpretation that goes beyond everyday understanding and seeks to fill in the gaps and grasp the parapraxes, the distortions, of communication. Third, this interpretation is informed by a reconstruction of history. Fourth, the criterion of the validity of interpretation is its adoption by the participants, its power to emancipate and increase freedom (what Lather [1986] has called "catalytic validity").

Where Habermas has worked hard, even struggled, is in clarifying the form that the reconstruction of history should take. Is it a narrative, a metahermeneutic, or an abstract theory of universal processes? His emphasis on theory has led some to judge that to Habermas each of us is defined by our intellectual "know-that"; that "Habermasian man has, however, no body, no feelings; the 'structure of personality' is identified with cognition, language and interaction" (Heller, 1982, p. 22). In this chapter, I have emphasized Habermas's treatment of psychoanalysis because it represents a time in his work when he focused on reconstruction that did *not* claim to be universal but was a "theoretically informed immanent critique" (Mendelson, 1979) operating within a form of life. The psychoanalyst is not outside a form of life, not a detached theorist, but a concerned member who has learned *practices* of emancipation. These practices can be employed whether or not they are thematized and reconstructed by a philosopher.

13

Social Science as Participant Objectification

> Bourdieu's theory of practice is a systematic attempt to move beyond a series of oppositions and antinomies which have plagued the social sciences since their inception. For anyone involved in the social sciences today, these oppositions have a familiar ring: the individual versus society, action versus structure, freedom versus necessity, etc. Bourdieu's theoretical approach is intended to bypass or dissolve a plethora of such oppositions.
>
> Thompson, 1991, p. 11

A second perspective on the kind of investigation that attends to the "objective framework" in which people live can be found in the work of French sociologist Pierre Bourdieu (1930–2002). For Bourdieu, we are always "playing a game" but are necessarily unaware of its arbitrary character. Everyday activity is the product of an interaction between what he called "habitus" and "social field," a situated encounter between agents who are endowed with socially structured resources and competencies, and it is frequently the occasion for "symbolic violence" that critical inquiry must expose.

Bourdieu emphasized that the presuppositions Habermas insists we must examine are not matters of individual tacit knowledge. Like Kuhn, Bourdieu argues that the presuppositions are embedded in social practices and embodied in bodily habitus. Reflection will not expose them. We need to study material practices, and because scientific techniques themselves are practices of "objectification," we need to objectify these techniques, turning them on themselves as instruments of reflexivity. Critical inquiry needs to be reflexive rather than reflective, and it achieves this by "objectifying objectification." Reflexivity is also a matter of turning our instruments of objectification on ourselves. This means studying one's own habitus together with the field in which one acquired it (where one grew up) and the field in which one applies it (the academy).

By being reflexive, critical inquiry can avoid objectivism, the error of claiming that one's system of categorization and classification is neutral and uniquely appropriate. It also can avoid subjectivism, the error of merely cataloging the diverse perspectives among the players of a game. The result is a viewpoint that "transcends" the "partial and partisan" point of view of a player but is not the gaze of a "divine spectator." It becomes possible to describe the field of play so as to show how the players have different perspectives because they occupy different positions. The researcher avoids taking sides and, not individually but together with others who practice the forms of communication of reflexive science, is able to produce knowledge that surpasses the local circumstances of its production. The point is not to judge how well people are playing the game, or who is winning and losing, but to judge the game itself – whether it is fair or biased, just or unjust.

HABITUS AND FIELD: THE SOCIAL GAME

As a young man Bourdieu was a star rugby player, and the metaphor of the game runs through all of his work (Calhoun, 2003). For Bourdieu, everyday life resembles a game: governed by conventions that are taken so seriously by the players that they view it as natural, normal, and inevitable. He emphasized that a game is always a struggle on a field of play in which the practical bodily expertise of the player is primary. The game metaphor defined the two central concepts in Bourdieu's work: "habitus" and "field." With the aid of these concepts (deliberately defined as "thin concepts," which could be applied in a wide variety of settings), and additional notions such as "cultural capital" and "symbolic violence," Bourdieu directed and conducted a program of research concerned with the creation and in particular the *reproduction* of social order, especially in circumstances where inequalities would seem to make such repro-duction problematic. For example, he studied how the French educational system helps to reproduce economic inequalities (Bourdieu & Passeron, 1970/1990).

Throughout his work, Bourdieu sought to avoid the dualisms that have dogged social science, such as individual/society, action/structure, and free-dom/necessity. His strategy was to reject and steer between both "objectivist" and "subjectivist" analyses (a strategy that he may have learned from Merleau-Ponty). Subjectivism – for example, Alfred Schutz's phenomenology – fails to see the social space in which people interact. Objectivism – for example, Levi Strauss's structuralism – fails to grasp the role of ongoing practical activities in the formation of this space. Bourdieu's use of the concepts of habitus and field was intended to avoid falling into the traps of either of these two polar positions. The two concepts are defined *relationally* rather than in opposition. We see here "the two central metasociological concerns that have driven [Bourdieu's] research program for forty years, namely the substitution of a relational for a substantialist conception of social reality and the transcen-dence of the fundamental antinomy of subjectivist and objectivist approaches to the study of social life" (Vandenberghe, 1999, p. 61).

HABITUS

In Bourdieu's view, practices – the things people do – are outcomes of an *interplay* between their objective characteristics and those of the social framework they inhabit. The former is "habitus," which Bourdieu defined as embodied expertise. It is a set of *dispositions* that incline a person to act and react in particular ways. Habitus is a way of standing, talking, walking – and at the same time a way of feeling and thinking. Habitus is "inculcated," "structured," "durable," "generative," and "transposable" (Table 13.1). It is *inculcated* in that one's habitus develops in childhood as the body is molded and as particular ways of acting, talking, and so on become second nature. Habitus is "embodied history" (Collins, 1998). It is consequently *structured*: the habitus of a person reflects the social conditions of its acquisition. For example, a person's middle-class origins will be apparent in their tastes, manner of speaking, accent, and so on. Futhermore, habitus is *durable*: because it is ingrained in the body, it endures over time. It cannot be changed simply through a conscious decision; habitus is preconscious and thus escapes deliberate control. Habitus is also *generative*: it is not simply habit but a creative source of new actions. Habitus is not simply a matter of repeating customary actions and activities but a way of doing new things, albeit in an acquired style and manner. And this means that habitus is *transposable*: it operates in new settings, not just those in which it was acquired.

Habitus is what provides each of us with "a feel for the game" that we are playing in everyday life. It is *le sens pratique*, the practical ability to negotiate everyday life, the feeling for what is appropriate and what is not. Our habitus grows out of the game each of us played as children. But, in a complex modern society, there is not only one game being played, and we may find ourselves in situations where the game is different from the one we grew up playing. Then our habitus hinders rather than helps us. We feel out of place. For example, Bourdieu proposes, upper-class individuals have a habitus that fits formal, official occasions. This correspondence, this congruence, between habitus and setting provides these people with a sense of confidence and fluency, with the competence to speak in the manner that is appropriate to such formal

TABLE 13.1. *Characteristics of Habitus*

Inculcated	Habitus is acquired in childhood.
Structured	It reflects one's social origins.
Durable	It endures over time and is not easily altered.
Generative	It is the source of new actions, although these will tend to have the customary style.
Transposable	It can operate in new situations.

occasions and the knowledge that they have this competence. This in turn provides them with a symbolic benefit, which increases their power and prestige, thus reinforcing (reproducing) their higher social status.

It should now be clearer how habitus is a concept designed to avoid the traditional dichotomies of mind and body, reason and emotion, thought and action. Bourdieu emphasizes that habitus is not a state of mind but a state of the *body*, a state of *being*. For Bourdieu, the body *is* a source of intelligence. Our body is the site of our incorporated history. Like Garfinkel, Bourdieu refers to *doxa*: the fundamental, deep-founded, unthought beliefs, taken as self-evident universals, that inform an actor's participation in a particular field. For Bourdieu, *doxa* is located in the body as "a corporeal 'hexis,'" as "practical schemes." Habitus is "political mythology embodied"; it is "a class culture turned into nature" (Bourdieu, 1980/1991, p. 190); it is a system of classificatory schemes that defines one's "tastes."

The concept of habitus places the body at the center of Bourdieu's investigations of social phenomena. The human body of course has the capacity for biological reproduction. For Bourdieu, the body also embodies strategies for its *cultural* reproduction through historical time.

Bourdieu writes of the dispositions of habitus as "openings to the world" that are acquired because the body, being in the world, is constantly at risk of harm and even death. This emphasis on the finitude and mortality of human existence contrasts with the Kantian focus on a transcendental ego that exists outside history and free from harm:

> With a Heideggerian play on words, one might say that we are *disposed* because we are *exposed*. It is because the body is (to unequal degrees) exposed and endangered in the world, faced with the risk of emotion, lesion, suffering, sometimes death, and therefore obliged to take the world seriously (and nothing is more serious than emotion, which touches depths of our organic being) that it is able to acquire dispositions that are themselves an openness to the world, that is, to the very structures of the social world of which they are the incorporated form. (Bourdieu, 1997/2000, pp. 140–141)

THE SOCIAL FIELD

The second central concept in Bourdieu's work is that of the "social field." The French word *champ* can be translated as "market" or "game" as well as "field." This is the objective framework, a *structured space of positions* within which resources are distributed, usually unequally. A field is a *site of struggles*, the space of the competitive game that is being played. Bourdieu insists that a field is defined relationally by the positions of the various players. What he seems to have in mind is the configuration that a player experiences in the midst of a game; we should think not so much of a playing field as a field of play: "A field exists where people are struggling over something they share, where something specific is at stake (for example in the literary field how to write and judge literature), where specific investments and entrance-fees are expected from new pretenders, where there are specific rules of the game, specific stakes, rewards and signs of authority" (Broady, 2002, p. 383).

Again Merleau-Ponty seems to have been an influence. In *The Structure of Behaviour*, Merleau-Ponty wrote of the game of football in similar terms to emphasize how in practical action a person "becomes one" with the milieu and their consciousness *is* the relation between setting and action:

> For the player in action the football field is not an "'object," that is, the ideal term which can give rise to an indefinite multiplicity of perspectival views and remain equivalent under its apparent transformations. It is pervaded with lines of force (the "yard lines"; those which demarcate the "penalty area") and articulated in sectors (for example, the "openings" between the adversaries) which call for a certain mode of action and which initiate and guide the action as if the player were unaware of it. The field itself is not given to him, but present as the immanent term of his practical intentions; the player becomes one with it and feels the direction of the "goal," for example, just as immediately as the vertical and horizontal planes of his own body. It would not be sufficient to say that consciousness inhabits this milieu. At this moment consciousness is nothing other than the dialectic of milieu and action. (Merleau-Ponty, 1942/1963, pp. 168–169)

In Bourdieu's view, a society is composed of multiple fields, each one relatively autonomous of the others, though homologies – structural similarities – exist among them. These fields are the distinct social spaces that result from the differentiation of social activities. No single field (of economic activity, for example) dominates. In each field social actors struggle to occupy the dominant positions and compete for the stakes.

Habitus and field define each other, and the researcher can only study the two-way relationship between them. There is a mutual constitution of habitus and field, so they cannot be identified with subjectivity and objectivity. Habitus enacts the organization of the field, and the field sustains habitus. Habitus is material *and* discerning (taste; see, e.g., Bourdieu, 1979/1984). Field is independent of any *individual* consciousness, but at the same time it exists only as a tacit agreement *among* agents engaged in struggle. Bourdieu noted that, paradoxically, a struggle always presupposes a fundamental accord between opposing sides. The two teams in a game of rugby have to agree that they are playing the same game, even though they have opposing interests.

Bourdieu made the seemingly paradoxical point that objectification is needed to gain adequate access to an agent's habitus. A researcher can only hope to gain some access to someone's "creative project" by first reconstructing, by means of work of objectification, the social field in which the agent is participating in order to lay out the range of possible positions and thereby identify the particular one the actor occupies (or more likely the actor's trajectory through a series of such positions). Bourdieu recommended that research be composed of two "minutes." The first is a stage in which one looks at the relations that define the structures of the social field. The second is an analysis of social agents' dispositions to act and the categories of perception and understanding that result from their inhabiting the field. When studying the work of an author, for example, "we can only be sure of some chance of participating in the author's subjective intention (or, if you like, in what I have called elsewhere his 'creative project') provided we complete the long work of objectification necessary to reconstruct the universe of positions within which he was situated and where what he wanted to do was defined" (Bourdieu, 1992/1996, p. 88).

The study of the field often involves fieldwork. For example, a student and colleague of Bourdieu, Loic Wacquant, explored firsthand the world of prize-fighting (Wacquant, 2004b). He joined a gym, trained as a boxer, and fought in several matches. Wacquant's question was why boxers choose such a demanding career, and his method was to explore how they are constituted. He described the "silent pedagogy" "that transforms the totality of the being of the fighter by extracting him from the profane realm and thrusting him into a distinctive sensual, moral, and practical cosmos that entices him to remake himself and achieve (masculine) honor by submitting himself to the ascetic rules of his craft" (Wacquant, 2004a, p. 183). But before "probing the visceral quality of this 'ontological complicity' that binds agent and world" (Wacquant, 2005, p. 443), Wacquant first articulated the "cosmos," the field of boxing, principally by tracing the work of the "matchmaker," a key figure who, using connections with trainers, managers, and fighters, is able to arrange the contests in which the boxers compete (Wacquant, 1998). Wacquant's interest was in how the field is "assembled" through the matchmaker's practical skills as broker, booking agent, and negotiator.

Bourdieu also often used the statistical technique known as "correspondence analysis" to objectify a social field and describe its positions (Benzecri, 1992). Correspondence analysis is a nonparametric form of factor analysis – that is to say, a numerical technique that identifies axes of covariation among data that are ordinal in character. It is "nothing else but the operational materialization of the relational mode of thinking which characterizes [Bourdieu's] generative structuralism" (Vandenberghe, 1999, p. 46). Bourdieu used it to analyze survey and archival data, generally identifying a pair of axes that define a two-dimensional plane. He equated this plane with the field in which survey respondents were located and acting. Correspondence analysis illustrates how a field can be identified from data collected from individuals that are presumed to reflect their positions and positioning (Figure 13.1). It also illustrates the use of what

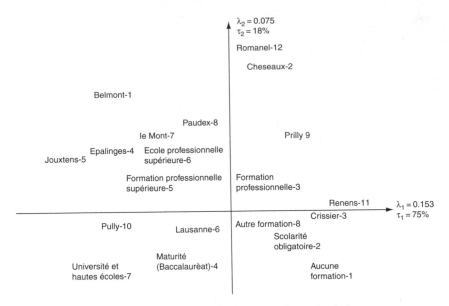

FIGURE 13.1. Results of a Correspondence Analysis

Bourdieu calls "techniques of objectification" in social science. Bourdieu says that "[t]he diagram is a model of 'reality' as we know it or, more accurately, as it is revealed to us, in ordinary experience, in the (veiled) guise" (Bourdieu, 1984/1990, p. 24). It is an "objectified, codified form of the practical patterns of perception and action which orient the practices of the agents best adjusted to the immanent necessity of their domain" (p. 24).

CULTURAL CAPITAL

Every game is played for stakes of some kind, and Bourdieu referred to these stakes as "capital." Each field has its own specific species of capital. Although he was borrowing from Marx, Bourdieu saw capital as taking many forms, not simply a financial one. He saw capital as something people seek to accumulate rather than a form of value they produce. (This was the basis of Willis's [1981] critique that Bourdieu lacks a conception of "cultural production.") And Bourdieu departed considerably from Marx in not theorizing capitalism as a distinct social formation (Calhoun, 1993).

Fields permit the conversion of capital from one form to another. Economic capital can be cashed in to obtain what Bourdieu called "cultural capital." Cultural capital can be embodied in its owner as personal skills and tastes (for particular styles of music, writing, etc.), or it can exist in certified form (as degrees or credentials), or objectified in possessions such as works of art, books, and so on.

Illusio

I have mentioned that Bourdieu emphasized that although the competing parties in a field are opposed to each other ("dissensus") they share common fundamental presuppositions: there is "consensus in the dissensus." He also proposed that any game requires its own "*illusio*." The people participating in the game of everyday life are subject to certain "vital errors and illusions" that are *necessary* if they are to keep playing the game.

This "practical faith" is an "undisputed, pre-reflexive, naïve, native compliance with the fundamental presuppositions of the field which is the very definition of doxa" (Bourdieu, 1980/1991, p. 68). A principal consequence of *illusio* is that inequities are allowed to continue unchecked. Consequently, the kind of social science Bourdieu was practicing had as one of its central aims the effort "to expose misrecognitions and false oppositions":

> His typical goal was to demystify the ways in which seemingly neutral institutions in fact make it harder for ordinary people to learn the truth about the state or public affairs. He has called for an "internationale" of intellectuals (to replace the old Internationale of the working class movement). . . . More generally, Bourdieu's mode of intervention has been to use the methods of good social scientific research to expose misrecognitions that support injustice. A prime example is the enormous collective study of "the suffering of the world" produced under his direction. . . . This aimed not simply to expose poverty or hardship, but to challenge the dominant points of view that made it difficult for those living in comfort, and especially those running the state, to understand the lives of those who had to struggle most simply to exist. (Calhoun, 2003, pp. 284–285)

SYMBOLIC VIOLENCE

We all know that often games are not played fairly, and play may become violent. Bourdieu referred to forms of coercion that operate without resorting to physical force as "symbolic violence." When a person who holds symbolic capital uses the power it confers against a person who holds less, and seeks thereby to alter their actions, they exercise symbolic violence. This is "gentle, invisible violence, unrecognised as such, chosen as much as undergone, that of trust, obligation, personal loyalty, hospitality, gifts, debts, piety" (Bourdieu, 1980/1991, p. 127). For example, social scientists have often studied gift-giving, viewing it as the social glue that holds society together (e.g., Mauss, 1960). But in Bourdieu's view gift-giving, although disguised as hospitality, is actually a matter of debt and obligation. It illustrates how power is seldom exercised in overt, physical form but is instead transmuted into a symbolic form and as a consequence becomes invisible and misrecognized. The result is that its victims accept its legitimacy and consider the hierarchy that power sustains to be something natural. The dominated come to accept as normal the arbitrary relations of domination to which they are subjected.

This phenomenon, in Bourdieu's view, is central to the reproduction of social order. One of the ways the institution of schooling, for example, serves the maintenance of a hierarchy of power is through the symbolic violence it practices, in which evaluation and recognition serve to reproduce the inequities of the larger social order and ensure that the arbitrariness of this order is ignored or misrecognized as natural. Those children born into families with more than their fair share of economic capital acquire a habitus that the school recognizes and, equally unfairly, rewards with scholarly credentials. These can later be traded for further economic advantage. This is an illustration of how the social reproduction of social institutions and the personal positions within them occurs *tacitly*, through the inculcation of habitus, and *inequitably*, through symbolic violence. Bourdieu and Passeron (1970/1990) described in detail how academic credentials sustain social inequalities and provide a justification (all the more powerful because it is practical rather than theoretical) of the established social order. Those who are victims of this cycle of reproduction are willing participants because they have acquired a habitus that predisposes them to choose courses of action that perpetuate their situation. Freedom and necessity are not opposed but relationally related: "Necessity can only be fulfilled, most of the time, because the agents are inclined to fulfill it, because they have a taste for what they are anyway condemned to" (Bourdieu & Passeron, 1970/1990, p. 178).

Kant's Symbolic Violence

One example of symbolic violence, from Bourdieu's point of view, was Kant's philosophical analysis of knowledge, ethics, and aesthetics. In each case, Kant gave priority to detached contemplation over involved participation. Bourdieu addressed Kant directly in the appendix to *Distinction* (1979/1984), a book that dealt with taste and aesthetic judgment. The thrust of his argument was that what lies at the root of the distinctions people make – between good and bad food, entertainment, clothing, decoration, and so on – is the degree of distance from necessity that is permitted by their social position. The poor, living from hand to mouth, make distinctions in which bodily appetite is an important factor. The wealthy, lacking for nothing, have the luxury of adopting a detached attitude of aesthetic appreciation. For them, "taste" has nothing to do with sensual flavor on the tongue, for example. Kant, who described aesthetic judgment as a detached appreciation that doesn't seek to possess or consume, was in fact universalizing and naturalizing the taste of the dominant class.

The Violence of Analysis

Bourdieu argued that much of social science itself amounts to not much more than symbolic violence. He was critical of formal linguistics and "semiotic" or "semiological" analyses from Ferdinand de Saussure (1915/1959) to Roland Barthes (1973/1975) because these ignore the social-historical conditions of the production and reception of linguistic forms. They also take for granted the position of the analyst; that is to say, they fail to be *reflexive*. As a result, they merely express the position of the analyst in the intellectual division of labor. Analytical classifications reflect the distinctions, the tastes, that are customary to the classifier as a result of their history and their position in a social space of classifications: "Taste classifies, and it classifies the classifier. Social subjects, classified by their classifications, distinguish themselves by the distinctions they make, between the beautiful and the ugly, the distinguished and the vulgar, in which their position in the objective classifications is expressed or betrayed" (Bourdieu, 1979/1984, p. 6).

The Violence of Coding

Bourdieu made the important point that every social field, every field of human activity, involves competing and conflicting classifications of entities and their values. There will be "rival, sometimes hostile representations, which all claim the status of truth and thereby the right to exist" (Bourdieu, 1984/1990, p. 13). This calls into question the use of any coding system, for when a researcher adopts one classification system – and only one – they have taken sides in the competition. Once a coding system is objectified, "the code changes from an instrument of analysis to a subject of analysis. . . . The objectified product becomes ... the immediately readable trace of the operation of the construction of the object" (p. 7). The researcher can avoid this, Bourdieu insists, only by being clear about their principle of classification and how it relates to those used by agents. We must avoid "universalizing an individual [i.e., particular] viewpoint" (p. 32).

The Violence of Linguistics

Bourdieu also rejected the famous distinction, originally made by Saussure, between "language" (*langue*) and "speech" (*parole*) (see Box 5.1). His criticism focused on the fact that "the language" is an *idealization* of linguistic practices that have specific social conditions of existence and have changed over time. For example, France is one of many countries in which there have been battles over the definition of a national language. Any notion of "the" French language is a product of concrete political agendas, located at particular points in history. The structuralist concept of *langue* feeds an "illusion of linguistic communism," the notion that "we all speak the same language." It is a disguised *normative* model. Why should one particular way of speaking be the dominant form, the only one considered legitimate? Only because it is "the victorious language," the winner in a political struggle. Access to the dominant language is not granted equally to all. In the United States, for example, Black English is an example of speaking by a dominated group that has frequently been judged linguistically "impoverished" by "objective" researchers (Dillard, 1973). Bourdieu emphasized that "the language" is a *pre*constituted object. When researchers set out to study "language," they are accepting, without question, the outcome of a political struggle.

It follows that the division between linguistics and sociology is artificial. "What characterizes 'pure' linguistics is the primacy it accords to the synchronic, internal, structural perspective over the historical, social, economic, or external, determinations of language" (Wacquant, 1989, p. 45). Linguistics views language as an object to be studied from an impartial perspective "rather than using it to think and to speak with" (p. 45). It treats language as a *logos*, stripped of practical and political functions, instead of a *praxis*. (*Logos* was a Greek word meaning both speaking and thought or hypothesis. *Praxis* means practical activity.) Language is assumed to be an autonomous formal code, and speaking a language is reduced to learning this code. Once the distinction has been drawn between language and speech, between competence and performance, it becomes impossible to grasp what people actually *do* with language, and at the same time the social scientist has become a judge of the objective meaning of talk and practice.

On the other hand, most analyses of *speech* (parole) also ignore its social conditions. Bourdieu argued that speech should not be considered simply the output of a preexisting underlying linguistic faculty. Grammatical competence is insufficient for the production of appropriate speech; a *practical* competence is also needed. Bourdieu pointed out that a "practical sense" of how to talk is necessary to make oneself heard, believed, obeyed, and so forth. Learning a language is a matter of coming to have the right thing to say at the right time.

It might seem that the speech act theory of Austin (1975) and Searle (1969) would be the kind of nonobjectifying analysis of speech that Bourdieu was advocating (see Box 10.1). After all, it is an analysis of language in action, of the pragmatics of talk. But Bourdieu was strongly critical of speech act theory, pointing out that both Austin and Searle ignored the fact that what they called the "fulfillment conditions" of performative utterances are defined by *institutions*; that is to say, in Bourdieu's view, by durable sets of social relations that endow certain people with power, status, and resources. Symbolic devices other than speech are involved, too, such as wigs, uniforms, and badges. Although Austin and Searle noted that "conventions" are crucial to language use, they didn't examine them as social phenomena, embroiled in conflict and struggle.

So in speech act theory, too, an important normative aspect of the phenomenon of language is overlooked. And when it is overlooked, it goes unquestioned, its reproduction unchallenged.

We must recognize that "[e]ven the simplest linguistic exchange brings into play a complex and ramifying web of historical power relations between the speaker, endowed with a specific social authority, and an audience, which recognizes this authority to varying degrees, as well as between the groups to which they respectively belong" (Wacquant, 1989, p. 46). To grasp what goes on in an interaction, the researcher needs to pay attention to "the totality of the structure of the power relations." This is the case with "even the content of the message itself" (p. 46). For example, when settlers and natives talk in a colonial or even post-colonial context, the choice of language is an important issue. If the settler speaks the native language, this is a "strategy of condescension" in which the dominant denies his dominant position while profiting from it by drawing attention to it. If, on the other hand, the dominant language is spoken, "the dominated speaks a *broken language* ... and his linguistic capital is more or less completely devalued" (p. 46): "In short, if a French person talks with an Algerian, or a black American to a WASP, it is not two persons who speak to each other but, through them, the colonial history in its entirety, or the whole history of the economic, political, and cultural subjugation of blacks (or women, or workers, etc.) in the United States" (Wacquant, 1989, p. 47).

Using Marx's phrase, Bourdieu says that this structure of power relations "is unconscious and works almost wholly 'behind the backs' of locators" (Wacquant, 1989, p. 47). It is also ignored by most researchers. As we learned in Chapters 2 and 3, we have to stop thinking of talk as the expression of an inner state and think of it instead as an aspect of activity. The social scientist must study "the laws defining the social conditions of acceptability" (Bourdieu, 1981/1991, p. 76) of language in the marketplace of talk. Talk is not language so much as discourse, and Bourdieu maintained that what is said must be analyzed as a matter of position, of habitus, and of the state of the game that is being played.

The Violence of Academic Discourse

Bourdieu illustrated these points with a critical analysis of scholarly language – the way academics talk and write. In his view, a successful academic must be a skilled speaker and writer who accurately assesses the prevailing conditions in the academic game in order to produce appropriate utterances – that is to say, in Bourdieu's view, utterances that have been suitably "euphemized." For example, Bourdieu analyzed the writing of Martin Heidegger (Bourdieu, 1988/1991). He pointed out that Heidegger's language was highly euphemized: it borrowed words from ordinary language – words such as "care" and "solicitude" (see Chapter 8) – and introduced these words into a philosophical field that previously had excluded them. In doing so, it transformed the words, adapting them to prevailing philosophical conventions. The result is that although Heidegger's book *Being and Time* (1927/1962) had the appearance of professional autonomy, it was alluding to and hence dependent on ordinary language and at the same time concealed this dependence. Its academic success was, in Bourdieu's evaluation, a consequence of the fact that it managed to achieve both loftiness and simplicity, to be both ennobled and ordinary, to speak two languages at the same time. Bourdieu's analysis highlights the sociopolitical conditions of the production of this philosophical text and offers a rich reading of the text itself as grounded in "a social aristocratism" and a "contempt for the social sciences" (Wacquant, 1989, p. 49).

The Violence of Ethnography

Bourdieu was equally critical of most anthropological fieldwork. The problem with ethnography, in Bourdieu's view, is that the fieldworker is neither participant nor reflective observer. As we saw in Chapter 9, an ethnographer *cannot* participate in cultural practices the way a member does. In general, however, they don't reflect on the practices of objectification that define their observing. The result is a "theoretical distortion" in which the fieldworker attributes to the members the kind of "decoding" that the fieldworker has to do:

> The anthropologist's particular relation to the object of his study contains the makings of a theoretical distortion inasmuch as his situation as an observer, excluded from the real play of social activities by the fact that he has no place (except by choice or by way of a game) in the system observed and has no need to make a place for himself there, inclines him to a hermeneutic representation of practices, leading him to reduce all social relations to communicative relations and, more precisely, to decoding operations. (Bourdieu, 1972/1977, p. 1)

Properly understood, the task of fieldwork is not to come to know the ways that participants in a form of life understand and describe their everyday activity but to *break* with this everyday understanding. Ethnography needs to explore the *limitations* of an insider's perspective: "The task, as [Bourdieu] began to argue didactically and to exemplify in all his work, was to combine intimate knowledge of practical activity with more abstract knowledge of objective patterns, and, using the dialectical relation between the two, to break with the familiar ways in which people understand their own everyday actions" (Calhoun, 2003, p. 281). The similarity with Habermas's project is clear.

The everyday understanding that participants display, together with the everyday accounts they give each other and the fieldworker, are inescapably strategic. They operate within a social field that, as we have seen, participants do not and cannot grasp as a whole, and they function unwittingly to reproduce this social field and its inequities, even when the person may be struggling to overcome these inequities. The task of the fieldworker is not merely to *obtain* members' first-person accounts but to *explain* them:

> These everyday accounts always contain distortions and misrecognitions that do various sorts of ideological work. . . . Bourdieu's project was to grasp the practical strategies people employed, their relationship to the explanations they gave (to themselves as well as to others), and the ways in which people's pursuit of their own ends nonetheless tended to reproduce objective patterns which they did not choose and of which they might even be unaware. (Calhoun, 2003, p. 281)

Bourdieu proposed that ethnographic fieldwork requires not participant observation but what he called "participant objectification." (His term *objectivacion* has been translated both as *objectification* and *objectivation*. I have chosen the former and modified quotations accordingly.) This is neither immersion as a member in the form of life being studied nor detached observation. Participant objectification is the exploration of the social conditions of existence in a form of life, coupled with the study of the act of scientific objectification. Scientific ethnography involves not some romantic participation but the ethnographer's objectification of the objectifying distance they inevitably maintain as a scientific researcher:

> One does not have to choose between participant observation, a necessarily fictitious immersion in a foreign milieu, and the objectivism of the "gaze from afar" of an observer who remains as remote from himself as from his object. Participant objectification undertakes to explore not the "lived experience" of the knowing subject but the social conditions of possibility – and therefore the effects and limits – of that experience and, more precisely, of the act of objectification itself. It aims at objectifying the subjective relation to the object which, far from leading to a relativistic and more or less anti-scientific subjectivism, is one of the conditions of genuine scientific objectivity. (Bourdieu, 2003, p. 282)

Research puts practical know-how and objectified knowing in dialectical relationship. The researcher needs some grasp of the practical sense of the game being studied (enough to *recognize* what's done, if not *produce* it themselves, as we saw in Chapter 10), but also the ability to "break" with familiar practical activity and "objectificate" it as something foreign and peculiar. Researchers have to find ways to question not only the practical, preconscious *doxa* of the people studied but also their own. Statistical techniques offer one way to do this. But such "objective" knowledge by itself is not sufficient; the researcher must also grasp the practical understanding, the *doxa*, that is necessary to playing the game.

Bourdieu emphasized that science itself is not just a matter of theory but a form of practice. The practice of scientific research is a matter of continually facing and resolving difficulties – rather like a rugby player in a long and difficult match: "The logic of research is an intermeshing of major or minor problems which force us to ask ourselves at every moment what we are doing and permit us gradually to understand more fully what we are seeking, by providing the beginnings of an answer, which will suggest new, more fundamental and more explicit questions" (Bourdieu, 1984/1990, p. 7).

Theoretical concepts – in this case, concepts such as habitus, field, and capital – can facilitate this practice by guiding the articulation of detailed empirical analyses of specific locales. They should not be the basis for a grand theoretical edifice; the point was not, Bourdieu insisted, to produce universal theory. At the same time, his interest was not in local studies for their own sake but in order to grasp general aspects of social life and social worlds.

REFLEXIVE RESEARCH

Participant objectification formed the center of what Bourdieu called "reflexive sociology" (Bourdieu & Wacquant, 1992; Wacquant, 1989). To avoid committing symbolic violence in social scientific inquiry, a *reflexive* stance is needed that will not be satisfied with taking at face value either members' accounts of their own activities or the typical practices of objectification used by academic researchers. Bourdieu insisted that social science must be critical, and this means being sensitive to the operations of power and privilege in their varied and subtle forms, both in the phenomena being studied and in the practices of the researcher. The researcher must have respect for the agents who make up the social world while at the same time appreciating that there are limits to what they can say and do. These limits must be understood as socially instituted, not as personal failures.

Reflexivity does not mean personal reflection, let alone celebration of one's subjectivity. Whereas Habermas reflected as an individual philosopher, Bourdieu insisted that researchers must call each other to account for their social positions and the objectifying practices they employ in their investigations. This involves investigation of the social matrices in which habitus is acquired. Reflexive researchers will recognize that they, too, are playing a game, that science is a game in which practices of objectification play a central role.

Although objectification is often "a violence or imperialism," Bourdieu proposed that it could also be liberating. We can turn our techniques of analysis on ourselves. Indeed, *only* if we objectify ourselves can we avoid reproducing in our position *toward* the world the interests that correspond to our social position *in* the world: "Any position adopted towards the social world orders and organizes itself from a certain position in the world, that is to say from the viewpoint of the preservation and augmentation of the power associated with this position" (Bourdieu, 1984/1990, p. 13). We must study the way we study; we must examine the techniques and practices of research in which fluid aspects of the social world get turned into what seem to be objects, and turn those techniques and practices into objects for examination themselves.

Reflexivity can be defined, then, as "the inclusion of a theory of intellectual practice as an integral component and necessary condition of a critical theory of society" (Bourdieu & Wacquant, 1992, p. 36). It is focused not on the individual researcher but on "the *social and intellectual unconscious* embedded in analytic tools and operations." Reflexivity requires collective rather than individual efforts, and its aim is to "*buttress the epistemological security of sociology*," not to undermine it (p. 36, emphasis original). The focus of reflexivity is the way researchers' positions in the intellectual field limit the knowledge they produce. The social origins of the researchers, their positions in the academic field, and the intellectualist bias with which they "construe the world as a *spectacle*" (p. 39) must all be examined. The last of these is central because academic researchers are constantly "collapsing practical logic into theoretical logic" (pp. 39–40). Kuhn (Chapter 1) would have agreed with Bourdieu that the presuppositions of science are practical rather than intellectual or theoretical: they are "built into concepts, instruments of analysis (genealogy, questionnaires, statistical techniques, etc.), and practical operations of research (such as coding routines, 'data cleaning' procedures, or rules of thumb in fieldwork)" (p. 40). Embedded in these practices is the "epistemological unconscious" of a scientific inquiry that "constructs" what are then taken to be the independent objects of research. Reflexivity aims to objectify this objectifying practice in order to examine it, question it, and if necessary change it.

Bourdieu insisted that science is a historical, indeed a political, activity, but he argued that this doesn't mean it cannot produce universally valid truths. The reflexivity of good science sets it apart from other games and helps establish the conditions for a rationality and knowledge that "escape" from history, and Bourdieu complained that reflexivity of this type is not practiced sufficiently. Reflexivity "historicizes reason without dissolving it" (Bourdieu & Wacquant, 1992, p. 47): "In Bourdieu's view, then, reason is a historical product but a highly paradoxical one in that it can 'escape' history (i.e., particularity) under certain conditions, conditions that must be continually (re)produced by working very concretely to safeguard the *institutional bases for rational thought*" (p. 48, emphasis original). Reflexivity is a matter of "uncovering the social at the heart of the individual, the impersonal beneath the intimate, the universal buried within the most particular" (p. 44). For Bourdieu, the practices of public debate, review, and critique in science are its institutionalized reflexivity. The game of science constructs its objects differently than does, say, a religion.

In this spirit, while he conducted fieldwork with the Kabyle in Morocco, Bourdieu also returned to his native village in France to conduct fieldwork there and understand how it had shaped his habitus. Later he undertook a detailed investigation of his own primary social field, the French academic system.

A Reflexive Study of the Academic Game: *Homo Academicus*

Homo Academicus (Bourdieu, 1984/1990) is an important work that illustrates Bourdieu's use of the thin concepts of habitus and field, his image of reflexive inquiry, and his views of science and knowledge more broadly. It also offers an analysis of the institution that many researchers work in, and thereby contextualizes the enterprise of research.

As Bourdieu views it, a university education provides students not merely with knowledge or skills but more importantly with the habitus to position themselves in the social world in a particular way. For professors, the university is an academic game in which one must struggle for tenure and prestige and which one's actions serve to reproduce, whether one intends this or not. Educational institutions function first to reproduce the societal status quo and then to reproduce themselves.

Home Academicus is a reflexive investigation because it is a study of the history of the primary field in which Bourdieu himself was an agent. He used the methods of his science to study his own constitution. This was not, he points out, a transcendental reflection but an examination "of the historical conditions of his own production" with the aim "to trap *Homo Academicus*, supreme classifier among classifiers, in the net of his own classifications" (Bourdieu, 1984/1990, p. xi). He sought to "gain theoretical control" through "detached scrutiny" (p. xi). In the preface, Bourdieu confesses "the special place" this project had for him, "no doubt explained by the peculiar force with which I felt the need to gain rational control over the disappointment felt by an 'oblate' faced with the annihilation of the truths and values to which he was destined and dedicated, rather than take refuge in feelings of self-destructive resentment" (p. xxvi). The fact that Bourdieu would portray himself as an oblate – literally a young recruit to a religious order – who is disillusioned in his values and beliefs raises important questions about his view of science, questions I shall return to shortly.

Bourdieu's book begins with the violent protests by French students, professors, and workers in the 1960s. Bourdieu explains that these events provided just the kind of break with *doxa* that he needed as a researcher in order to be able to study the institution in which he worked: "A crisis affecting an institution which has the function of inculcating and imposing forms of thought must weaken or ruin the social foundations of thought, bringing in its wake a crisis of faith, a veritable, practical *epoche* of doxa, which encourages and facilitates the appearance of a reflexive awareness of these foundations" (Bourdieu, 1984/1990, p. xxv).

Like Heidegger and Garfinkel (Chapter 8), Bourdieu was using a breakdown in practical activity as the opportunity to see the situation with more clarity, as a *practical* version of Husserl's *epoche*. A dramatic increase in the number of young people seeking entry to higher education – more than tripling between 1958 and 1968 – had led to the hiring of many non–tenure-track lecturers to teach large classes. The situation changed the power relations between old and new generations of academics, and a conflict developed between established professors and these young lecturers. The pace and rhythm of academic life were rapidly and radically transformed. Scholars who understood the new situation were able to write a thesis in far less than the normal ten years and catapult themselves into new faculty positions. But these newcomers found that if they were to succeed in their newly won positions, the rules of the game would have to be changed. This provoked a reactionary backlash from the status quo, with the ironic consequence that they became vulnerable as their tacit operations came into the light. The newcomers in turn began "to take up a struggle which we may call revolutionary in so far as it aims to establish alternative goals and more or less completely to redefine the game and the moves which permit one to win it" (Bourdieu, 1984/1990, p. 172).

This was a "critical moment" for the system of higher education, a "qualitative leap." The increased number of students meant that educational qualifications were devalued, and this threatened the career prospects in particular of children of the dominant class. Disenchanted subordinate teachers converged with disappointed students, and synchronized coincidences across the different worlds of distinct social fields spread the crisis from arts and social science faculties into "the whole of the university field," aided by institutions for the mass distribution of cultural goods: the media, cinema, advertising, and others.

The outcome took a radical form because "those consecrated by a bankrupt institution" had inculcated ambitions that, if they were to fulfill them, obliged them "to break with the derisory and henceforth untenable roles which it [the system] assigned to them ... to invent new ways of playing the part of the teacher ... by lending him the strange features of an intellectual master of reflection who reflects on himself and in so doing, destroys himself qua master" (Bourdieu, 1984/1990, p. xxv). The perspective of these "self-destructive masters" harmonized with that of the rebelling students. And because education plays a key role in reproduction of the social order as a whole, the crisis in the universities provoked a crisis in society. The results were student strikes in several universities and high schools, clashes with police and school authorities, and street battles with police after president Charles de Gaulle tried to repress the protests, followed by a general strike by 10 million workers, around two-thirds of the workforce. De Gaulle created a military operations headquarters to respond, dissolved the National Assembly and called for parliamentary elections, and even took brief refuge in Germany.

Axes of Power and Conflict in the University

The conflicts served to highlight various forms of academic power. A university employs multiple axes of evaluation – scientific, administrative, academic, and intellectual – allowing, as Bourdieu puts it, everyone to hide from the truths that anyone knows. Its hierarchies are vague and plural. Scientific prestige vies with administrative power, internal recognition with external renown. Bourdieu identifies three distinct axes of conflict. The first pits academic disciplines against one another in terms of economic or intellectual leverage. The second sets academic generations – old-timer and newcomer – at odds. The third is a conflict between the orthodox and the heretical.

First, the university operates with "two competing principles of legitimation": economic power and cultural power. Law and medicine have the former but lack the latter, while the reverse is true for the arts, humanities, and social sciences. Faculty in departments with high *economic* capital tend to come from privileged backgrounds and are vested in preserving the status quo. Such faculty "have a taste for order" (Bourdieu, 1984/1990, p. 51). Theirs is "knowledge in the service of order and power, aiming at the rationalization, in both senses, of the given order" (pp. 68–69). Faculty in the departments that hold *intellectual* capital tend to have more critical, even "heretical," views. They "occupy very different realities" (p. 54). The different departments understand and use the terms "research," "teaching," and "scholarship" in incommensurate ways. What counts as research in medical faculties, for example, is often a bureaucratic matter of directing a laboratory, providing the university with strong financial returns on social capital.

Different positions on this first axis are associated with different ethical and intellectual dispositions. Departments high in economic capital "have the function of training executive agents able to put into practice without questioning or doubting, within the limits of a given social order, the techniques and recipes of a body of knowledge which they claim neither to produce nor to transform" (Bourdieu, 1984/1990, p. 63). Those departments with intellectual capital, in contrast, "are destined to arrogate to themselves, in their need to establish a rational basis for the knowledge which the other faculties simply inculcate and apply, a freedom which is withheld from executive activities, however respectable these are in the temporal order of practice" (p. 63). In the latter case, what counts as knowledge is an analysis of public order that, through historical contrasts, presents it as merely a particular case. The result is a lack of support for the status quo that "for the guardians of order, is already a critical break, or even evidence of irresponsibility" (p. 69). Bourdieu proposes that the social sciences have a "quite singular position . . . playing the part of a Trojan horse in their struggle to impose a new definition of legitimate culture" (pp. 120–121), doubly subordinate beneath the natural sciences and law/medicine.

The second "principle of division" operates between older, established professors and newcomers. Academic power provides "control of the instruments for reproduction of the professorial body" (Bourdieu, 1984/1990, p. 78) through examinations, appointments, thesis supervision, and so on. Senior faculty have obtained positions that provide them with control over the processes of hiring and firing, recruiting and promoting, and nurturing and advising new faculty, thus enabling them to reproduce the game they play. They have obtained "temporal" capital and use it to foster dispositions similar to their own among the next generation of academics.

This means they will "hardly encourage heretical breaks with the artfully intertwined knowledge and power of academic orthodoxy" (Bourdieu, 1984/1990, p. 105). Academia harbors "a secret resistance to innovation and to intellectual creativity" (p. 95). And this is the third major axis of conflict in the university, between those faculty who hold intellectual viewpoints that are orthodox and institutionally approved and those who hold "heretical" views. Faculty who acquire scholarly prestige (in large part by avoiding committee work, thesis supervision, and of course teaching) lack power over the mechanisms of reproduction, but in Bourdieu's view these "heretics" can establish intellectual authority in the larger marketplace for intellectual products.

CONCLUSIONS

Rather surprisingly, given his portrait of academic life, Bourdieu, like Habermas, said that he would "defend science and even theory" against the "antiscientism" that he felt was the vogue (Bourdieu & Wacquant, 1992, p. 47). He insisted that scientific truth is both desirable and possible, but rather than locating reason in "transhistorical universals of communication" (p. 188), as Habermas wants to do, he insisted that reason must be historicized. Bourdieu emphasized that only *particular* social arrangements of communication can expand our knowledge. His position was one of "historicist rationalism": he believed that scientific rationality "is thoroughly historical without ... being relative or reducible to history" (p. 188). Science, said Bourdieu, is a unique activity: it can produce knowledge that transcends the historical conditions of its production. "I do not think that reason lies in the structure of the mind or of language. It resides, rather, in certain types of historical conditions, in certain social structures of dialogue and nonviolent communication" (p. 189).

In Bourdieu's view, intellectual freedom can be won within the politics of the historically situated space of scientific struggle. Working together in this space, scientists can "push reason forward ... toward a little more universality" (Bourdieu & Wacquant, 1992, pp. 190–191). Bourdieu insisted that knowledge must always be submitted to critique, that systems of categorization are always social, that discourse is conditioned by objective structures. But, in his view, knowledge and power should not be completely conflated. His position has been called a "sensible third path between universalism and particularism, rationalism and relativism, modernists and postmodernists – the whole linked series of problematic dichotomies" (Calhoun, 1993, p. 62), between, some have said, Habermas and Foucault.

We have seen that for Bourdieu the "principal virtue" (Bourdieu, 1984/1990, p. 7) of critical inquiry is that it *objectifies objectification*. Bourdieu claimed that reflexivity enables the researcher to avoid two dangers. The first is claiming that their categories and classifications are the only ones possible. The researcher needs "to renounce his claim to objectivist absolutism and be satisfied with a perspectivist recording of the viewpoints at issue (including his own)" (p. 15). The second danger is using "the technical yet also symbolic force of science to set himself up as a judge of the judges, and impose a judgment which can never be entirely free from the presuppositions and prejudices associated with his position in the field which he claims to objectify."

When the researcher is able to turn "the power" of the scientist's "theoretical and technical methods of objectification" on himself, this provides "freedom in the face of the social determinations which affect him." The scientific viewpoint is neither the "partial and partisan viewpoint of agents engaged in the game" "*nor* the absolute viewpoint of a divine spectator" (Bourdieu, 1984/1990, p. 31, emphasis original). The researcher "transcends ordinary visions" (p. 31) without, however, being able to go so far as to "grasp historical reality as such." Reflexive, critical inquiry provides an analysis of a form of life that is objective without being absolute; that is "the most systematic totalization which can be accomplished." For Bourdieu, being objective is distinct from being objectivist, for in an objective analysis the researcher does not impose a single scale, index, or measure. But it is also distinct from being perspectivist, for this would be merely recording the diverse viewpoints of the players.

Equally, the researcher should not judge the outcome of the game. Bourdieu's proposal is that the researcher should avoid setting himself up as an arbiter, judging which is the right (correct/just) way to act in a social field. The proper task of research is to *describe the logic* of the struggles to define reality and judge the game rather than the players. As I noted earlier, the researcher's job is to evaluate not who is winning and who is losing but the fairness of the game.

Again, this amounts to avoiding – Bourdieu called it "transcending" (Bourdieu, 1984/1990, p. 17) – both objectivism and subjectivism. Objectivism is the mistake of claiming that there is only one classification. Subjectivism is "perspectivism," the approach "that would settle for recording the diversity of hierarchies, treated as so many incommensurable viewpoints" (pp. 17–18). The researcher should instead delineate "regions in the space of positions" and show how these regions, and the objective relations among them, *define* each viewpoint. Again we see that a sense of the whole is needed.

Bourdieu explicitly linked his conception of critical social inquiry to the work of Marx: "Marx suggested that, every now and then, some individuals managed to liberate themselves so completely from the positions assigned to them in social space that they could comprehend that space as a whole, and transmit their vision to those who were still prisoners of the structure" (Bourdieu, 1984/1990, p. 31). But how exactly does reflexivity lead to this kind of liberation? For example, can an academic researcher describe the game of the university – which, as we have seen, is certainly not free from symbolic violence – without continuing to play it? In *Homo Academicus* Bourdieu asks the reader not to attribute any interest to him, or to suspect that he is still playing the academic game. He insists that he wants only to "break with ordinary usage" and offer a "scientific rhetoric." But at the same time he notes that "science gives those who hold it, or who appear to hold it, a monopoly of the legitimate viewpoint" (p. 28). Science has "social effects." It might be argued that *declaring* himself objective and neutral enabled Bourdieu to play the academic game that much more effectively:

> This ... move by which the author of *Homo Academicus* attempts to create a third position for himself is, however, problematic, first, because he himself has clearly stated that there is and can be no independent position on the field within the field itself and, second, because the reflexive application of his own sociology to his own sociology unmasks his epistemological position as an ideological position and, thus, as a move within the field itself. (Vandenberghe, 1999, p. 60)

Bourdieu's view of this kind of tension between participation and objectification seems to have been that science is a *special* kind of game. Bourdieu argued that science is neither an undifferentiated community of scholars nor a direct competition among pure ideas. Far from being disinterested, scientific activity has its own specific interest and *illusio*. Scientific authority among peers and social authority in an institution are its two principal forms of capital, and there is continual conversion between the two. Like Kuhn, Bourdieu believed that changes from one scientific view of the world to another are more like conversion than they are matters of rational calculation. In addition, he suggested that as a scientific field becomes more autonomous it begins to generate "ordinary revolutions," eventually achieving a state of permanent revolution – in which new constructions must both surpass and preserve the old (unlike art, for example, where the old can simply be discarded).

But, for Bourdieu, although science is a game in which there is a struggle for power like any other social field, its players must use strategies that are both social *and* intellectual. Science is not only closed and selective but also open and public. Paradoxically, in Bourdieu's view, we can avoid relativism only by refusing to separate the social from the epistemological aspects of science. (In this sense, sociology of science *is* scientific reason turning on itself.) This is because scientists' social "strategies produce their own transcendence, because they are subjected to the crisscrossing censorship that represents the constitutive reason of the field" (Bourdieu, 1991, p. 20). In other words, communication in science is organized so that scientists can pursue their individual interests *only* by criticizing and correcting one another. In Bourdieu's view the competition among the products of scientific practice becomes so intense that "logic is instituted as the mandatory form of social struggle" (p. 23). Far from requiring elimination of all relations of domination (as Habermas believes), it is the struggle for dominance in science that ensures that criticism becomes so effective that a "real symbolic policing" (p. 22) produces "reason as the best weapon" (p. 23) and leads to a common "interest in truth" (p. 22). In the case of science, historical agents, acting in a historical field, are able to produce universal truths. What underlies these universal scientific truths are neither Kant's "categorical structures" nor Habermas's "knowledge-forming interests" (p. 24), except insofar as the latter are products of specific historical conditions. This was Bourdieu's account of what makes science a different kind of game.

Bourdieu insisted that social science has an inescapable ethical dimension. It cannot and should not tell us what to do, but it can foster moral agency by showing us where there are alternative courses of action. To achieve emancipation and enlightenment, it will not be sufficient to eliminate "systematically distorted exchanges" and remind people "to abide by the universals rediscovered by the philosopher but ignored and violated by the ordinary person" (Bourdieu, 1991, p. 24) if we do not also abolish "the social bases of the abuse of symbolic power" through a "politics of reason" and in this way allow new kinds of cognitive interest to emerge. It is only through coming to recognize the key role played by the social determinations of knowledge that these can be consciously grasped, mastered, and deliberately accepted or rejected.

Wacquant notes the parallel between this "socio-analysis" and psychoanalysis: reflexive sociology "can help us unearth the *social* unconscious embedded into institutions as well as lodged deep inside of us" (Bourdieu & Wacquant, 1992, p. 49, emphasis original). Bourdieu wrote of both "individual systems of defense" and "collective defense mechanisms" (p. 19), products of and reproductions of "an effort to persevere in a social identity." A critical social science debunks the myth that arrangements of power and domination are natural and inevitable by showing how they are unfair, how they have come into being, and how they are actively reproduced. Bourdieu's sociology "attempts to make possible the historical emergence of something like a rational subject via a reflective application of social-scientific knowledge" (p. 49, emphasis removed):

A greater understanding of the mechanisms which govern the intellectual world ... should teach him [the intellectual] to place his responsibilities where his liberties are really situated and resolutely refuse the infinitesimal acts of cowardess and laxness which leave the power of social necessity intact, to fight in himself and in others the opportunist indifference or conformist ennui which allow the social milieu to impose the slippery slope of resigned compliance and submissive complicity. (Bourdieu, 1984/1990, pp. 4–5)

Bourdieu's supporters see an emancipatory potential in his work:

> Most basically, Bourdieu's theory asks for commitment to creating knowledge – and thus to a field shaped by that interest. This commitment launches the very serious game of social science, which in Bourdieu's eyes has the chance to challenge even the state and its operational categories. In this sense, indeed, the theory that explains reproduction and the social closure of fields is a possible weapon in the struggle for more openness in social life. (Calhoun, 2003, p. 726)

His critics, on the other hand, accuse him of placing undue emphasis on reproduction:

> If Bourdieu wants to bring his sociological theory in line with his political intentions, he should open up his system, avoid deterministic descriptions of stable reproduction, and give voluntarism its due. This presupposes that the creativity of the habitus is openly recognized and that culture is not only seen as symbolically sublimated violence, not only as an instrument of domination, but also as an instrument of liberation. After all, a critical theory is not only a theory that uncovers the arbitrary nature of social necessity (domination), but also one that is able to reveal the possibility of the improbable (emancipation). (Vandenberghe, 1999, p. 62)

This criticism seems unfair. As we have seen, Bourdieu insisted that the reproduction of inequitable institutions is the result of freely chosen, albeit unreflexive, courses of action, not of domination or determination. Bourdieu himself insisted that science is "an instrument which enables one truly to constitute oneself – at least a little bit more – as a free subject" (cited in Bourdieu & Wacquant, 1992, p. 49), and that social science necessarily takes sides in political struggles by uncovering the way the social order maintains itself through mechanisms that ensure its misrecognition. His suggestion that a reflexive inquiry can, by objectifying its techniques of objectification, have an emancipatory impact while at the same time producing transcendent truths is an impressive claim. But saying that scientists have a self-interest in keeping each other truthful sounds rather like claiming financiers are able to keep each other honest. It surely cannot *always* be true, and Bourdieu himself noted that reflexive inquiry produces transcendent knowledge only "under certain conditions" (Bourdieu & Wacquant, 1992, p. 48).

Whether a research investigation leads to transcendent truths or whether it imposes symbolic violence is something contingent; there are no guarantees. Knowledge, even scientific knowledge, remains a product of history and culture. The "game" of science (see Box 13.1) is always played in specific places and specific times.

BOX 13.1. *The Game Metaphor*

A game analogy runs through much social scientific investigation of forms of life. We have seen that Ludwig Wittgenstein wrote of "language games." Games play a central role in John Searle's explanation of constitutive rules. Clifford Geertz analyzed the game (or perhaps sport) of cockfighting. Emanuel Schegloff drew our attention to the "turns" and "moves" of everyday conversation. Pierre Bourdieu, himself a rugby player, viewed social reality as a field of play. Michel Foucault wrote of "games of truth." This game metaphor has a long history. Aristotle, suggesting that "man is by nature a political animal," added that "the natural outcast ... may be compared to an isolated piece at draughts" (Aristotle, 1995, p. 5). Outside society, a human being has no game to play and, like an isolated game piece, is without purpose, without identity.

Jean-Francois Lyotard (1979/1997) places the game metaphor at the heart of the philosophical position he calls "post-modernism," seeing, as Fred Jameson has put it,

> language itself as an unstable exchange between its speakers, whose utterances are now seen less as a process of transmission of information or messages, or in terms of some network of signs or even signifying systems, than as (to use one of Lyotard's favorite figures) the "taking of tricks," the trumping of a communicational adversary, an essentially conflictual relationship between tricksters. (Jameson, 1979/1997, p. xi)

What does the notion that social interaction is like playing a game imply for the conduct of research? Like any metaphor, it offers a way of seeing in which certain aspects will stand out. What stands out in the game metaphor? A game is a socially arranged activity in which interaction has a particular recognized and orderly organization. A game has players, who make moves and often take turns. The players have different *avatars*, specific ways they are embodied or manifest in the game: goalkeeper, center, right wing, and so on in soccer; rook, pawn, bishop, and so on in chess. These occupy different positions on a field of play. Players are often both in competition (agonism) and collaboration. There are stakes to be gained along the way, and winners and losers are often defined ("the presence of 'turns' suggests an economy, with turns for something being valued – and with means for allocating them, which affect their relative distribution, as in economies"; Sacks,

Schegloff, & Jefferson, 1974, p. 696). And perhaps most important, although a game is a conventional and contingent way of interacting, it will come to be viewed as natural and inevitable by its players.

Alisdair MacIntyre has drawn several important conclusions about the practice of social science from the game metaphor. He suggests that there is usually not just one game being played but several: "the problem about real life is that moving one's knight to QB3 may always be replied to with a lob across the net" (MacIntyre, 1984, p. 98). While one person is playing chess, the other is playing tennis, or a game of chess can at any moment be transformed into a game of tennis. MacIntyre points out that in a game, as in real life, each player tries to be as unpredictable to opponents as possible, and the most successful players are those who can best deceive other players and observers. Attempts to predict what players will do usually fail, though we can understand their actions in retrospect. MacIntyre also points out that players don't have a clear view of one another or their field of play. This means that "analyses of past determinate situations [cannot be transferred] to the prediction of future indeterminate ones." General Lee wasn't playing the board game *Battle of Gettysburg*, and playing such a game, with clearly positioned players and a map of the setting, is very different from the actual experience of battle. When we reconstruct real life as a game, we need to remain aware that we will fail to capture its true unpredictability.

14

Archaeology, Genealogy, Ethics

[The task] consists of not – of no longer – treating discourses as groups of signs (signifying elements referring to contents or representations) but as practices that systematically form the objects of which they speak. Of course, discourses are composed of signs; but what they do is more than use these signs to designate things. It is this *more* that renders them irreducible to language (*langue*) and to speech. It is this "more" that we must reveal and describe.

Foucault, 1969/1972, p. 49

The third approach to critical inquiry we shall consider is that of Michel Foucault (1926–1984). Foucault developed a form of critical inquiry that explored knowledge, power, and human being itself as products of history and culture. His work throws light on the central aspects that define a form of life, as well as how investigation itself always arises in a form of life. Foucault reversed Kant's critique: whereas Kant had claimed to show how seemingly contingent aspects of human experience, such as causality, are actually necessary and universal, Foucault aimed to show how apparent necessities are actually contingent. Unlike Habermas, he didn't view language as a transcendental or quasi-transcendental domain. Unlike Bourdieu, he didn't view science as a field that produces knowledge that can transcend its circumstances. Indeed, Foucault's historical critique was directed especially at the "universal truths" of the modern biological, psychological, and social sciences. These turn out, in his analysis, to be outcomes of contingent historical events. But we shall see that Foucault did not consider truth to be indistinguishable from opinion, politics to be merely the play of power, or ethical judgments to be culturally relative. Whereas Habermas focuses on individual know-that, on reflection and decision making, and Bourdieu emphasized know-how, the embodied knowledge of habitus, Foucault recognized *both* formal knowledge and informal practical know-how and explored the relationship between them, as well as their relation to the acting knower.

Foucault is sometimes described as a postmodernist and is sometimes accused of being antimodern. It is more accurate to say that all his work was motivated by the aim to understand how modern society operates. He was no more interested in telling a history of the failure of what he calls "our modernity" than in telling a history of its successes. Rather, his interest was in understanding how modernity works as a game of truth, as an exercise of powers, and as the place for specific ways of being human. Modernity "endlessly reformulates" its thematics: it won't inevitably exhaust itself or automatically replace itself. But nor will it resolve its central contradictions because these in fact define it. Modern society will not be transformed through a process of "learning," for it defines what counts as learning. It is a game without a clear end.

Foucault's approach to history is novel. He rejected the common historical focus on individual agents – especially "great men." He was equally critical of the focus on subjectivity in Husserl's phenomenology and Jean-Paul Sartre's existentialism (which Foucault called "transcendental narcissism"). But his work clearly drew from Hegel, Heidegger, and Merleau-Ponty, and in many ways it was a reply to Immanuel Kant. Foucault himself said:

> What I have studied are the three traditional problems: (1) What are the relations we have to truth through scientific knowledge, to those "truth games" which are so important in civilization and in which we are both subject and objects? (2) What are the relationships we have to others through those strange strategies and power relationships? And (3) what are the relationships between truth, power, and self? (Martin, 1982/1988, p. 15)

These "three traditional problems" correspond to the three Kantian critiques: of *pure reason* (our relationship to objects), *practical reason* (our relationship to other people), and *judgment* (our relationship to ourselves). The three problems – truth games, power, and formation of the self – are not independent topics of inquiry but three "axes" (Foucault, 1985/2006, p. 202), which each of Foucault's investigations explored with shifting emphasis.

ARCHAEOLOGY, GENEALOGY, ETHICS

Foucault developed first an "archaeology," then a "genealogy," and finally an "ethics" (Table 14.1). The archaeology was a way of studying the different "games of truth" of a society. The genealogy studied the power relations in these truth games. It did not replace archaeology, but it shifted the emphasis of analysis. Toward the end of his life Foucault began to study the *ethos* of how people form themselves. These three kinds of analyses were interrelated: in a manuscript published posthumously, Foucault defined a program of research that he called "a historical ontology of ourselves," which was "genealogical in

TABLE 14.1. *Archaeology, Genealogy, and Ethics*

Form of Inquiry	Key Concepts
Archaeology (1950s–1970s)	Study of games of truth, of problematization, of how being offers itself to be thought. Human beings as objects of knowledge. Discursive formation: dispersion of statements. Episteme. *Connaissance*: professional knowledge. Constituted objects and their relations. Enunciative modalities of subjectivity. *Madness and Civilization* (1961), *Birth of the Clinic* (1963), *The Order of Things* (1966), *The Archaeology of Knowledge* (1969).
Genealogy (1970s–1980s)	Study of relations of power, of power/knowledge, of *savoir*. Action on the actions of another. Constitution in history. Technologies of subjectivation. Biopower and the body. Governmentalization and resistance. Human beings as objects to themselves. *I, Pierre Riviere . . .* (1973), *Discipline and Punish* (1975), *History of Sexuality 1* (1976)
Ethics (1980s)	Study of case of the self. Techniques and technology of self-constitution; self-cultivation and transformation. Human beings as subjects. *History of Sexuality 2* (1984), *History of Sexuality 3* (1984)

its design and archaeological in its method" (Foucault, 1984b, p. 46) and had an ethical interest or aim.

Foucault insisted that he was not writing histories of ideas but histories of *thought*, and we will need to look more closely at what he considered thinking to be. He proposed that "the proper task of a history of thought, as against a history of behaviors or representations . . . [is] to define the conditions in which human beings 'problematize' what they are, what they do, and the world in which they live" (Foucault, 1984/1986, p. 10). This involves "analyzing, not behaviors or ideas, nor societies and their 'ideologies,' but the *problematizations* through which being offers itself to be, necessarily, thought – and the *practices* on the basis of which these problematizations are formed" (p. 11). Central to this kind of history was a view of language not merely as representing things but as doing something "more": "Of course, discourses are composed of signs; but what they do is more than use these signs to designate things. It is this *more* that renders them irreducible to language (*langue*) and to speech. It is this 'more' that we must reveal and describe" (Foucault, 1969/1972, p. 49, emphasis original).

Just as Garfinkel asked of formal analysis, "'what more?'" can we see, Foucault, too, was interested in what is missed in the typical analyses of scientific knowledge and ethical practice. Just as Wittgenstein insisted that speech is used in many different ways in the language games of everyday life,

Foucault set out to describe the multiple uses of discourse in forms of life. The result is one of the most creative and interesting programs of investigation, one that focused once again on – that *problematized* – the constitution of human beings, first as objects to others, then as objects to themselves, and finally as subjects.

THE ARCHAEOLOGY OF KNOWLEDGE

An "archaeology of knowledge" is an investigation that examines artifacts unearthed in an excavation, but the kind of artifact is not bone, pottery, or metalwork but what people said and wrote in the past: their "statements" (in French, *énoncé*: what has been enunciated or expressed). Foucault was interested not in commonplace things said or written but in "serious speech acts" (Dreyfus & Rabinow, 1982, p. 48), pronouncements made by figures of authority in positions of power. Such statements are noticed, collected and passed around, and preserved so they can later be excavated. The archaeologist's task is to try to reconstruct an understanding of the past by asking what made it possible to make *these* statements but not others.

The archaeologist does this by studying the *space* within which statements circulated. Archaeology is "an enquiry whose aim is to rediscover on what basis knowledge and theory became possible; within what space of order knowledge is constituted" (Foucault, 1966/1973, pp. xxi–xxii). It is a *critical* investigation that aims to discover the social and historical conditions for the possibility of specific forms of knowledge. It is the study of "games of truth." Foucault developed archaeology to study those particular games of truth in which the objects of knowledge are human beings, but he suggested later that every form of life can be considered a truth game. His interest was the human sciences – the "sciences of man" – such as medicine, psychiatry, and penology; his intention was to write a history of the ways that humans in Western culture have developed knowledge about themselves. How is it that a human being can become constituted as an object of knowledge? Foucault proposed that the human sciences constitute both their objects of study and the experts who may speak about these objects. They give status to those who may speak and deny it to those who must remain silent. They exclude and suppress certain people, certain *types* of people: those who are identified as the mad, the sick, or the criminal. It is important that we do not take the knowledge of these sciences at face value but reconstruct the conditions that made it possible.

Discourse and Discursive Formation

Foucault called the variety of statements that characterize a particular human science its "discourse." Each science has its particular mode of discourse, and archaeology studies how statements circulate, gain value, are attributed, and

are appropriated. For example, one finds various k
discourse of 19th-century doctors: "Qualitative de
accounts, the location, interpretation and cross-check
by analogy, deduction, statistical calculation, experime
more (Foucault, 1969/1972, p. 55). These statements, these
amount to a "population of events" whose existence ar
another can be studied. Such statements show a systematic
logical nor linguistic.

One then asks what defines the unity of the discourse
statements one can undoubtedly find objects referred to, co
style of writing, themes and theories. Can we define the discour
these elements? Foucault examined each in turn and rejected it a
forward basis for the identity of a discourse. There is no common
shared concepts or themes, no single style. What we find instead is a
of each of these elements. They are spread over and move in "a comm
a reciprocal functioning" (Foucault, 1969/1972, p. 37) that Foucault
"discursive formation" (p. 38). This space is not a form of knowledge b
"truth game" that unites and dictates what counts as knowledge (Fou
1984c). Following Bourdieu, one might say that it is the field of play.

Many languages draw a distinction between two kinds of knowledge th
is difficult to express in English. Spanish has *saber* and *conocer*; French *savo*
and *connaissance.* The former is a tacit, practical *know-how*, the latter a
theoretical knowledge that one might call *know-that.* Foucault came to see
that the theoretical and scientific discourse about madness, for example, was
dependent on a different kind of knowledge, practical and embodied in
complex institutional systems. Foucault (like Merleau-Ponty) called this practical know-how *savoir* and the official knowledge *connaissance. Savoir,* implicit
knowledge, is the condition for the possibility of the professional theories,
opinions, and judgments – the *connaissance* – of a truth game. As Foucault put
it later, "Thus the theoretical problem that appears is that of an autonomous
social knowledge [*savoir*] which does not take individual conscious learning
[*connaissance*] as a model or function" (Foucault, 1969/2006, p. 8).

A discursive formation itself is neither true nor false because it defines
which statements count as true and false. What is important is not what
statements *say* or *mean* – their images, themes, or concepts – but what they
do. And what a statement does depends on its location among other statements within the discursive formation.

OBJECTS ARE CONSTITUTED

Foucault famously proposed that the discourses of the human sciences do not
simply represent an independent world but *constitute* the objects that define
their domains of inquiry. For example, 19th-century psychopathology

recognized objects such as hallucinations, hypnosis, sexual deviations, lesions of the brain, mental deficiencies, and so on. Each science forms its own objects of study (*empiricites*), how these objects are related to one another (*positivities*), and how the science relates to them: "[T]he object does not await in limbo the order that will free it and enable it to become embodied in a visible and prolix objectivity; it does not preexist itself, held back by some obstacle at the first edges of light. It exists under the positive conditions of a complex group of relations" (Foucault, 1969/1972, p. 45).

For Foucault, knowledge is always a linking of the articulable and the visible (Deleuze, 1986/1988, p. 39). His interest lies not merely in what can be said but in how what we can say organizes what we can see. His book *The Order of Things*, originally titled *Word and Things*, is about neither words nor things but how they are linked: how an ordering of words is used to order what can be seen. This interest in the link between how we can speak and what we can see – and therefore think about – becomes clearer in the genealogies, where Foucault explored spaces like the prison as material organizations of visibility, but it began in the archaeologies.

This means that archaeology is not a history of the *referents* of statements (asking questions such as, "Who was mad?"). Its aim is "to dispense with 'things.' To 'depresentify' them" (Foucault, 1969/1972, p. 47), to abandon the notion that things were there "anterior to discourse." Archaeology is certainly not the phenomenology whose slogan was "to the things themselves!" But equally, it is not the study of the meaning of words (asking questions like, "What did 'madness' mean?"). Archaeology studies neither "words" nor "things," but how things are ordered using words. Foucault explained: "I would like to show that discourse is not a slender surface of contact, or confrontation, between a reality and a language (*langue*), the intrication of a lexicon and an experience" (p. 48).

Constitution

An episteme is a "mode of being of things, and of the order that divided them up before presenting them to the understanding" (Foucault, 1966/1973, p. xxii). Kant proposed that the a priori categories of mind order things before we can reason about them, and thus constitute what we take to be "reality." Foucault – like Hegel, Heidegger, Merleau-Ponty, and Garfinkel – located the power to constitute in the *practices* of discourse, as something ontological. The discursive practices "order" objects and thereby *form* them. Foucault proposed that the space of a discursive formation imposes certain conditions both on the statements that can be made within it and on the "elements" of these statements: their objects, subjects, concepts, and theories.

In his early writing, Foucault proposed that a discursive formation could be viewed as a set of rules: "Archaeology tries to define not the thoughts, representations, images, themes, preoccupations that are concealed or revealed in discourses; but those discourses themselves, those discourses as practices obeying certain rules" (Foucault, 1969/1972, p. 138). These "rules of formation are conditions of existence (but also of coexistence, maintenance, modification, and disappearance) in a given discursive formation" (p. 38). The rules define neither the things nor the words but "the ordering of objects" (p. 49). They are "a priori" but this "does not elude historicity" because, in Foucault's view, they are not fixed and unchanging: "[T]hese rules are not imposed from the outside on the elements that they relate together; they are caught up in the very things that they connect; and if they are not modified with the least of them, they modify them, and are transformed with them" (p. 127).

Let us consider an example in more detail:

> [T]he unity of discourses on madness would not be based upon the existence of the object "madness," or the constitution of a single horizon of objectivity; it would be the interplay of the rules that make possible the appearance of objects during a given period of time: objects that are shaped by measures of discrimination and repression, objects that are differentiated in daily practice, in law, in religious casuistry, in medical diagnosis, objects that are manifested in pathological descriptions, objects that are circumscribed by medical codes, practices, treatment, and care. Moreover, the unity of the discourse on madness would be the interplay of the rules that define the transformations of these different objects, their non-identity through time, the break produced in them, the internal discontinuity that suspends their permanence. (Foucault, 1969/1972, pp. 32–33)

Later Foucault described the task not as finding rules but as of grasping the "problematization" of some particular aspect of "labor, life, and language." What is seen as a problem? What do anxiety, discussion, and reflection focus on? What are the "thematics" of the discourse?

Each science employs its own mode of representation, which it regards as the only valid way of relating words to things. Each science claims to have a neutral language that merely reflects how things are. But for Foucault each is a version of the compulsion to use the order of words to order things. Words themselves are things, and to privilege one kind of representation is inevitably to misunderstand the power of language. Recall Garfinkel's list of the *ways* that language can be used to represent, the "multitude of sign functions": marking, labeling, symbolizing, indicating, miniaturizing, simulating, analogies, anagrams, and more. Recall Wittgenstein's invitation to "[t]hink how many different kinds of thing are called 'description': description of a body's position by means of its co-ordinates; description of a facial expression; description of a sensation of touch; of a mood" (Wittgenstein, 1953/2001, p. 10).

Hayden White suggests that "what Foucault has done is to rediscover the importance of the projective or generational aspect of language, the extent to which it not only 'represents' the world of things but also constitutes the modality of the relationships among things by the very act of assuming a posture before them" (White, 1973b, p. 48). Before language can "represent" things, it first "constitutes" them and how they relate, how they are "ordered." In this sense, there is no "ontological category of reality behind and prior to discourse" (Mohanty, 1993, p. 35). Later, Foucault would make this point very clearly: his approach was an analysis of "the games of truth and error through which being is historically constituted as experience; that is, as something that can be and must be thought" (Foucault, 1984/1986, pp. 6–7). It is important to recognize that this claim that a discursive formation defines "the ordering of objects" is not saying something negative. It is true that within a discursive formation "it is not easy to say something new; it is not enough for us to open our eyes, to pay attention, to be aware, for new objects suddenly to light up and emerge out of the ground" (pp. 44–45). But, at the same time, a discursive formation provides the "positive conditions" for objects to become apparent.

Subjects Are Dispersed

Foucault had two principal goals in these archaeological investigations. The first was to challenge the assumption that scientific *knowledge* is the result of an entirely logical process whose history is the growth of understanding of independent objects. Instead, he emphasized the irrational space within which science operates and the discontinuous way this space has changed. His second goal was to question the *subject* who engages in scientific investigation. When we survey the variety of different statements in a discourse, we should ask about their speaker and their status. We should ask about the sites in which they speak and their position: as a questioning, listening, or observing subject. The answers to these questions will show not *the* unifying function of *a* subject but the *dispersion* of subjectivity in "various enunciative modalities." Foucault explained that "I shall look for a field of regularity for various positions of subjectivity" (Foucault, 1969/1972, p. 55): "I showed earlier that it is neither by 'words' nor by 'things' that the regulation of objects proper to a discursive formation should be defined; similarly, it must now be recognized that it is neither by recourse to a transcendental subject nor by recourse to a psychological subjectivity that the regulation of its enunciation should be defined" (p. 55).

This is not a complete rejection of the human subject but an insistence that we humans are not the origin or center of the practices in which we participate. Subjectivity is not the *source* of meaning. Foucault emphasized that conceptions of human being are, in fact, local constructions that have appeared and disappeared abruptly. Humans act with initiative, but within specific conditions:

> The positivities that I have tried to establish must not be understood as a set of determinations imposed from the outside on the thought of individuals, or inhabiting it from the outside, in advance as it were; they constitute rather the set of conditions in accordance with which a practice is exercised, in accordance with which that practice gives rise to partially or totally new statements, and in accordance with which it can be modified. These positivities are not so much limitations imposed on the initiative of subjects as the field in which that initiative is articulated . . . rules that it puts into operation . . . relations that provide it with a support. . . . I have not denied – far from it – the possibility of changing discourse: I have deprived the sovereignty of the subject of the exclusive and instantaneous right to it. (Foucault, 1969/1972, pp. 208–209)

The Archaeologies

Foucault undertook a series of archaeological studies of the "disciplines" of the sciences of man. *Madness and Civilization* (Foucault, 1961/1988) explored psychiatry, and Foucault argued that what professionals and the public see as an objective scientific fact – that madness is mental illness – is a product of the interventions of a specific disciplinary institution that exercised its power from the 17th to the 19th century. Psychiatry excludes through confinement and imposes a silence on those whom it diagnoses as insane. Foucault described this as a specific problematization of madness through particular social and medical practices:

> I tried to analyze the genesis, during the seventeenth and eighteenth centuries, of a system of thought as a matter of possible experiences; first, the formation of a domain of recognitions [*savoir*] that constitute themselves a specific knowledge [*connaissance*] of "mental illness"; second, the organization of a normative system built on a whole technical, administrative, juridical, and medical apparatus whose purpose was to isolate and take custody of the insane; and finally, the definition of a relation to oneself and to others as possible subjects of madness. (Foucault, 1985/2006, p. 202)

Here, looking back, Foucault was describing the linkages between his early archaeological analyses and later genealogical and ethical studies. *The Birth of the Clinic* (1963/1975) offered a similar analysis of medicine and the "medical gaze." *The Order of Things* (subtitled *An Archaeology of the Human Sciences*) (1966/1973) explored the distinct historical periods of social science, each with its own conditions for knowledge and truth. Whereas his other books of this period dealt with forms of "difference," here Foucault looked at the shifting criteria by which professionals decide what things are "identical."

In *The Archaeology of Knowledge* (1969/1972), Foucault reflected on his methodology and described his earlier books as "a very imperfect sketch" of the newly emerging enterprise of archaeology, in which "questions of the human being, consciousness, origin, and the subject emerge, intersect, mingle, and separate off."

Episteme

In *The Order of Things*, Foucault suggested that the various serious bodies of learning that exist in a society at a particular time share "common, but transformable criteria" (Foucault, 1985/2006, p. 203). Nineteenth-century natural history, economics, and philosophy shared an overall configuration, a common "archaeological system" (Foucault, 1966/1973, p. xi) – an *episteme*. An episteme is a system of possible discourse that dominates a historical era and dictates what counts as knowledge. For example, psychiatry appeared in the 1700s, when changes also occurred in law, literature, philosophy, and politics. A history must pay attention to these broader conditions. Foucault explained, "[W]hat I am attempting to bring to light is the epistemological field, the *episteme* in which knowledge . . . grounds its positivity and thereby manifests a history which is not that of growing perfection, but rather that of its conditions of possibility" (p. xxii). Foucault used the French word *épistémè* rather than *épistémé* (notice the different accent on the last letter; *épistêmê* is the Greek word most often translated as knowledge, in contrast to *technê*, translated as either craft or art). An *épistémè* is not explicit or theoretical knowledge but

> the strategic apparatus which permits of separating out from among all the statements which are possible those that will be acceptable within, I won't say a scientific theory, but a field of scientificity, and which it is possible to say are true or false. The episteme is the "apparatus" which makes possible the separation, not of the true from the false, but of what may from what may not be characterised as scientific. (Foucault, 1980, p. 197)

Foucault identified three major epistemes in the history of Western culture: the medieval, the classical, and the modern. Each marked a distinct way of seeing the world and a manner of using language: as a book to be read; as a timeless order that must be classified with words; as centered around Man in his activities of living, working, and speaking as both knowing subject and object of knowledge. Foucault granted that where one locates the breaks is "relative to the field of one's enquiry"; there is no *single* history, either of continuity or discontinuity – history and time are themselves multiple and of historical origin (Mohanty, 1993, p. 39). The shift from one episteme to another is not a smooth transition. Just as an archaeologist finds distinct layers in the historical record whose boundaries mark sudden reorganizations in the arrangements of life, Foucault found that epistemes were separated by breaks, ruptures, and disjunctions. He proposed that archaeology is a kind of history in which "the problem is no longer one of tradition, of tracing a line, but one of division, of limits; it is no longer one of lasting foundations, but one of transformations that serve as new foundations, the rebuilding of foundations" (Foucault, 1969/1972, p. 5). Traditional history tries to eliminate discontinuity and disclose a hidden continuity to events; Foucault's archaeology looked for rupture and discontinuity. Rather than tracing events back to their origins and "silent beginnings," he looked for displacements and thresholds among the "various fields of constitution and validity" (p. 4). Discontinuity was no longer an obstacle to the work of the historian but "the work itself," both an instrument and an object of research. This was an end to "total history," in which homogenous relations are assumed to underlie seemingly disparate events, and the beginning of a "general history" (p. 9), which distinguishes different "series" – the economy, religion, science, and so on – and tries to determine the relations, the "interplay of correlation and dominance," among them. A total history organizes all phenomena around a single center – "a principle, a meaning, a spirit, a world-view, an overall shape." A general history, in contrast, will "deploy the space of a dispersion" (p. 10). This new history raises new questions about "threshold, rupture, break, mutation, transformation" (p. 5)

Foucault was not claiming that nothing has changed since the 18th century. His point was that we shouldn't *assume* that things have changed simply because time has passed. Perhaps the human sciences have merely moved around the board of their truth games, occupying different positions but not creating anything new. To find out, we need to study them carefully. Nor did Foucault claim that the move from one episteme to another was progress. For example, in the rupture between the classical and modern epistemes, it is "[n]ot that reason made any progress: it was simply that the mode of being of things, and of the order that divided them up before presenting them to the understanding, was profoundly altered" (Foucault, 1966/1973, p. xxii).

Foucault considered Immanuel Kant to have occupied an important position in the modern episteme:

> [T]he Kantian critique ... marks the threshold of our modernity: it questions representation ... on the basis of its rightful limits. Thus it sanctions for the first time that event in European culture which coincides with the end of the eighteenth century: the withdrawal of knowledge and thought outside the space of representations. That space is brought into question in its foundation, its origin, and its limits. (Foucault, 1966/1973, p. 242)

Kant's critique "opens up the possibility" of raising the question of "all that is the source and origin of representation" (p. 243). The problematic of the modern episteme became: What makes representation possible? What makes *valid* representation possible? What makes *knowledge* possible? As we have seen, ever since Kant, "human knowledge seems not to stand in an empirically valid relationship with reality" (Rawls, 1996, p. 431). Foucault proposed that "Kantian doctrine is the first philosophical statement" of the "displacement of being in relation to representation" (Foucault, 1966/1973, p. 245), a displacement that created the possibility for natural sciences of the "hidden depths" of objects and for philosophies of the "hidden depths" of the knower. It was now possible to explore the *basis* for representation and consider that knowledge itself might be historical and cultural. New empirical fields emerged with "a transcendental theme" – fields such as sociology and anthropology. New philosophical analyses – such as those by Heidegger and Merleau-Ponty – explored what makes representation possible and the relationship between subjectivity and objectivity. The danger in taking these steps was that of undermining the validity of genuine knowledge: "[E]ven when modern thought makes knowledge essentially historical, it must retain some functional equivalent of Kant's transcendental realm to guarantee the normative validity of knowledge" (Gutting, 2008).

At the same time, empirical science "questions the conditions of a relation among representations from the point of view of the being itself that is represented" (Foucault, 1966/1973, p. 244), in its explorations of the "enigmatic reality" that constitutes the objects of human knowledge in "the force of labor, the energy of life, the power of speech" (p. 244).

These explorations of subject and object defined the episteme of modernity. What followed was a divide between the formal and the empirical sciences (followed both by repeated efforts to reunite them and by protests that this is impossible) and the appearance within empirical science of reflections on "subjectivity, the human being, and finitude" (Foucault, 1966/1973, p. 248) as it took over this task from philosophy.

Foucault had a jaded view of the human sciences because they have often adopted a view of "Man" that is doubly essentialist. They assume, paradoxically, that humans are at one and the same time both objects and subjects; that people are both determinate and have free will. They study humans merely as the being that forms representation. They investigate humans within the systems of life, labor, and language and explore what humans know (albeit often unconsciously) within these systems. That is to say, they study "subjectivity." But they fail to consider life, labor, and language as "the basis on which man is able to present himself to a possible knowledge" (Foucault, 1966/1973, p. 362). They fail to ask what makes subjectivity possible. They merely *duplicate* the sciences in which the human being is given and studied as an object (as we saw in Part I). They are "sciences of duplication, in a 'meta-epistemological' position" (p. 355).

At this time, Foucault identified two "counter-sciences" that he suggested avoided this kind of study of representations and instead explored what makes representation possible. Psychoanalysis traces the birth of representation in desire, the law of the father, and death. And ethnology explores the origin of representation in social norms, rules, and systems. Foucault envisaged not exactly a "cultural psychology" but a kind of "psychoanalytic anthropology" that would offer "the double articulation of the history of individuals upon the unconsciousness of culture, the historicity of these cultures upon the unconscious of individuals" (Foucault, 1966/1973, p. 379).

The Researcher as Archaeologist

Archaeology requires what some have called a "double bracketing":

> Not only must the investigator bracket the *truth* claims of the serious speech acts he is investigating – Husserl's phenomenological reduction – he must also bracket the *meaning* claims of the speech acts he studies; that is, he not only must remain neutral as to whether what a statement asserts as true is in fact true, he must remain neutral as to whether each specific truth claim even makes sense, and more generally, whether the notion of a context-free truth claim is coherent. (Dreyfus & Rabinow, 1982, p. 49)

In his transcendental phenomenology, Husserl bracketed the *reference* of his experiences, beliefs, and perceptions, which is to say the objects toward which they seemed to be directed. Foucault did the same, refusing to assume that the objects referred to by a statement are real. But Foucault also went a step further: he refused to assume that a statement has a stable *meaning*. He rejected the idea that a statement has a single meaning or fixed content; it has meaning only in relation to other statements within a specific discursive formation. The notion of "content" is, of course, part of the conduit model of language. Foucault was very clear that analysis extracting the contents of a statement was insufficient:

> The analysis of lexical contents defines either the elements of meaning at the disposal of speaking subjects in a given period, or the semantic structure that appears on the surface of a discourse that has already been spoken; it does not concern discursive practice as a place in which a tangled plurality – at once superposed and incomplete – of objects is formed and deformed, appears and disappears. (Foucault, 1969/1972, p. 48)

Foucault focused instead on the "mode of existence" of statements: the basis on which they appear and circulate. The first question to ask of discourse, he insisted, is not about its structure, organization, or coherence but "What *is* it?"

This amounts to asking, "By what right, on the basis of what conditions, does a science manage to *claim* to have 'unity' and 'coherence'?" For Foucault (as for Iser, Barthes, and Gadamer and Garfinkel), meaning and truth are *effects*. Rather than ask what a statement *means*, we should ask what it *does* or *is*, and this will depend on its linkages and affiliations to other statements and to the social institutions in which it is produced and received. Whether a statement is "true" is a matter for debate, contention, and even retribution. The discursive formation is what makes it possible for a statement to appear meaningful, and it is this that the archaeologist must pay attention to. In his archaeological studies, then, Foucault was "simply describing an open logical *space* in which a certain discourse occurs" (Dreyfus & Rabinow, 1982, p. 51). Rather than focus on either the content (meaning) or the referent (object) of statements, Foucault focused on their historical conditions. Dreyfus and Rabinow interpret this as meaning that in investigation understood as archaeology the researcher needs to be doubly detached and neutral.

But Foucault's stance seems less like Husserl's and more like Merleau-Ponty's, slackening the threads that attach us to the world. The researcher's methodological distance is in the service of goals that are more than epistemological. The archaeologist tries to show the way to "escape": "What I am trying to do is grasp the implicit systems which determine our most familiar behavior without our knowing it. I am trying to find their origin, to show their formation, the constraint they impose upon us; I am therefore trying to place myself at a distance from them and to show how one could escape" (quoted in Simon, 1971, p. 201).

Foucault has been accused of suggesting that knowledge and truth are completely relative, that "all possible truth-conditions are equal, depending merely on context or interpretative perspective" (Hook, 2001, p. 525). But the preceding statement shows that such an accusation is inappropriate, that in fact "Foucault views truth-conditions as extremely stable and secure, as situated in a highly specific and idiosyncratic matrix of historical and sociopolitical circumstances, which give rise to, and are part of, the order of discourse" (p. 525). We have seen that an *episteme* is a stable, although contested, way of ordering the world by speaking about it. It is a distinct "regime of truth" with its own specific systems of possible discourse – the discursive formations – that dominate a historical era. Sometimes a resemblance has been noted with Kuhn's notion of a paradigm (Brenner, 1994; Piaget, 1970/1988). Although Foucault emphasized that the grounds on which knowledge becomes stable and statements become meaningful and true are not fixed, that they have changed abruptly in the past and will undoubtedly change again, he insisted that these grounds can seem rock solid. Truth, at least at this point in Foucault's work, was a matter of showing how the truths and even the meaning of the claims of the social sciences (in particular) are conditioned by implicit systems whose operations we take for granted and fail to recognize.

THE GENEALOGY OF POWER/KNOWLEDGE

An archaeology could compare different discursive formations and show that previous ages had written and talked very differently but equally effectively. But it could not explore the transitions from one episteme to another or explain how an episteme becomes firmly established and difficult to change. By 1971, Foucault was developing an approach he called "genealogy" (from the Greek γενεα, genea, meaning "family," and λογοσ, logos, meaning "knowledge"), which traced the pedigree or line of descent of currently accepted systems of thought (those disclosed by archaeology). Traditional history tells a story of progress, development, and fulfillment. Although Charles Darwin wrote of "the origin or descent of man" (Darwin, 1871), most evolutionary accounts see humans as rising above our primate ancestors. But a family tree is a record of chance encounters, irrational attractions, and accidents of fertility and mortality. A genealogy discloses these accidents and the coincidences, surprises, and struggles that produced a descendant. It involves writing "histories of the present" with an emphasis on "the ensemble of historical contingencies, accidents, and illicit relations." This is change not as "an acquisition, a possession that grows and solidifies; rather, it is an unstable assemblage of faults, fissures, and heterogeneous layers that threaten the fragile inheritor from within and from underneath" (Foucault, 1971/1984, p. 82).

Foucault explained how he was influenced by the historical critique of moral and religious prejudices in Friedrich Nietzsche's (1844–1900) *On the Genealogy of Morality* (1877/1998):

> A genealogy of values, morality, asceticism, and knowledge will never confuse itself with a quest for their "origins," will never neglect as inaccessible the vicissitudes of history. On the contrary, it will cultivate the details and accidents that accompany every beginning; it will be scrupulously attentive to their petty malice; it will await their emergence, once unmasked, as the face of the other. (Foucault, 1971/1984, p. 80)

Such histories cast doubt on the legitimacy of contemporary concepts and theories, and they celebrate forgotten forms of knowledge that were spurned or declared illegitimate: "Let us give the term genealogy to the union of erudite knowledge and local memories which allows us to establish a historical knowledge of struggles and to make use of this knowledge tactically today" (Foucault, 1980, p. 83). Foucault's focus broadened from the study of the production of statements to include the relations between "systems of truth and modalities of power" (Davidson, 1986, p. 224). This meant studying the struggles, conflicts, and battles in which forms of knowledge become dominant.

But, like the archaeology, this was an approach in which the historian explored a phenomenon more fundamental than the individual subject. Foucault was directly taking up the "problems of constitution." He wrote:

I wanted to see how these problems of constitution could be resolved within a historical framework, instead of referring them to a constituent object (madness, criminality, or whatever). . . . I don't believe the problem can be solved by historicizing the subject as posited by the phenomenologists, fabricating a subject that evolves through the course of history. One has to dispense with the constituent subject, to get rid of the subject itself, that's to say, to arrive at an analysis which can account for the constitution of the subject within a historical framework. And this is what I would call genealogy, that is, a form of history which can account for the constitution of knowledge, discourses, domains of objects, etc., without having to make reference to a subject which is either transcendental in relation to the field of events or runs in empty sameness through the course of history. (Foucault, 1980, p. 117)

With the view that history always involves struggle and accident, it is no surprise that Foucault began to focus on a new dimension of games of truth, that of power. He began to explore the interconnections between knowledge and power.

THE GENEALOGIES

Discipline and Punish (1975/1977) showed Foucault's shift in emphasis from archaeology to genealogy. This book explored how the "criminal" has been marked as different and managed in various ways at various times. Foucault told the history of the penal system and the radical shift from torture to imprisonment. He highlighted Jeremy Bentham's design for the "Panopticon," an ideal prison in which prisoners are isolated and cannot see one another but are always visible to an unseen guard in a central tower. Because the prisoners never know when they are being observed, they must develop internal control of their activities. Foucault considered this an illustration of the way criminology – like psychology and medicine – had become a "discipline" in a double sense: as a system of knowledge and a system of management. Together such disciplines formed a vast "carceral system" intent on "normalizing" people by molding their conduct and creating "docile bodies." Submitting people to norms through examination and diagnosis, they defined who was "normal" and disciplined those who were not. Their new ontology of "man as machine" was ideal for a society of industrial production.

In Volume 1 of *The History of Sexuality* (1976/1980) Foucault traced the genealogy of sexuality in the practices of confession to priests, doctors, and psychiatrists. He called these practices "*technologies.*" Foucault focused on sexuality because he considered it a historically singular mode of experience that illustrated the formation of subjectivity. Sexuality exemplifies technologies in which subjects become objects for others and for themselves, and in which subjectivities are formed. At the clinic, in the confessional, and on the psychiatrist's couch, the person is exhorted to view him- or herself as an object through specific procedures of "government" and thereby discover the "truth" about their sexual desires.

How did psychoanalysis go from being an exemplary "anti-science" to a "timid" enterprise? The answer seems to lie in Foucault's recognition of the "normalizing functions" (Foucault, 1976/1980, p. 5) of psychoanalytic discourse about sex and repression. The institutionalized practices that claim to search for hidden truths (the clinic, the psychotherapy session, the confessional) must themselves be analyzed. They turn out to establish a relation between an authority and a submissive and to practice a depth hermeneutics that damages both parties. Foucault now rejected the search for "depth" in the form of "underlying" rules or the hidden truth sought by a depth hermeneutics. His genealogy has been described as moving *beyond* both structuralism and hermeneutics (Dreyfus & Rabinow, 1982).

In *I, Pierre Riviere, having slaughtered my mother, my sister, and my brother. . .* (Foucault, 1973/1975), Foucault and his students compiled a dossier of documents from multiple discourses: a memoir that was part autobiography, part confession; witnesses' statements and other court documents; and medical reports. Their focus was on how these diverse and incompatible discourses struggled with one another to define and obtain their principal object: a French peasant tried for parricide in the 1830s. Was he criminal, or insane, or . . . ?

Power and Biopower

Central to genealogy is a focus on power, but Foucault's conception of power challenged the traditional view that it is concentrated in an authority and exercised in a punitive way. He argued that the sovereign's power to take life has been replaced by a power to *foster* life, as society's capacity for control and manipulation has grown to include the life processes of both individuals and populations. Power should be viewed as something positive as well as negative. The "threshold of modernity" was crossed when life processes became subject to political practices. The mechanisms of power now generate and order personal and social forces rather than repressing them, "working to incite, reinforce, control, monitor, optimize, and organize" these forces (Foucault, 1976/1980, p. 136). Within modernity:

> methods of power and knowledge assumed responsibility for the life processes and undertook to control and modify them. Western man was gradually learning what it meant to be a living species in a living world, to have a body, conditions of existence, probabilities of life, an individual and collective welfare, forces that could be modified, and a space in

which they could be distributed in an optimal manner. For the first time in history, no doubt, biological existence was reflected in political existence. (Foucault, 1976/1980, p. 142)

Life is no longer a purely biological process; society can now direct the life of both the individual and the species, and the political process of modern society now "places [human] existence as a living being in question" (p. 143). Power is no longer dealing merely with legal subjects but with living beings whose life processes can be mastered. Life and history have intersected, and both are now topics of explicit calculation and planning.

This "biopower" has grown in two ways. Scientific disciplines have appeared that focus on the body as machine, disciplines whose procedures of power seek to maximize the body's productivity and render it useful and docile. And institutions focus on the life of the species to intervene and regulate populations. Political power now has "the task of administering life" (Foucault, 1976/1980, p. 139); we now have the knowledge to transform our lives or destroy our species. State institutions are the great instruments of society that insert bodies into the economy as producers and consumers, though the techniques of biopower operate "at every level of the social body" (p. 141) from family to school to military. Foucault challenged the view that the state exercises power only in the interests of the totality, or a particular class. He emphasized that "the state's power ... is both an individualizing and a totalizing form of power" (Foucault, 1982, p. 213):

> I don't think that we should consider the "modern state" as an entity which was developed above individuals, ignoring what they are and even their very existence, but on the contrary as a very sophisticated structure, in which individuals can be integrated, under one condition: that this individuality would be shaped in a new form, and submitted to a set of very specific patterns. (Foucault, 1982, p. 214)

This "tricky combination" is perhaps unique in human history, and reflects the way the state has adopted techniques of "pastoral power" that focus on individuals, their inner truth, and their ability to direct their own conscience. This power, once limited to the Christian church, has now "spread out into the whole social body" (p. 215).

The two sides of biopower can be seen in the expansion of institutions to educate, train, reform, and discipline, and in the concern with problems of birthrate, migration, public health, and so on. These two techniques of power were linked by concrete arrangements to form "the great technology of power in the nineteenth century" (Foucault, 1976/1980, p. 140). One of these arrangements is the "deployment of sexuality," an important illustration of the way "individualization" is accomplished through a combination of "subjectivation" and "objectivation." Human beings become objects for medicine and the social sciences, and subjects for the interpretive sciences such as psychiatry. Foucault now viewed the human sciences as instruments of bio-power. The traditional social sciences objectify people, and although the interpretive social sciences seek a new approach, they "subjectify" people in ways that are equally problematic. These interpretive social sciences claim to reveal hidden truths about the "subjects" of their research and at the same time to stand outside the operations of power. Their claims to neutrality, disinterestedness, and objectivity would make it appear that power is irrelevant to their practices, but Foucault was skeptical. To offer this kind of "deep" interpretation, he argued, *is* to exert power. The interpretive social sciences take for granted the particular historical practices they operate within – those of academic psychology, for example. These practices include acts that confer academic status and prestige – graduation, hiring, tenuring, and so on – acts in which power is inevitably involved.

POWER IS EVERYWHERE

Although state institutions certainly employ power, it is dispersed throughout society, not focused at central points. Although it is perfectly legitimate to study power by looking at institutions, there are limits to such an approach. First, the picture is muddied by the mechanisms for their reproduction that institutions need to employ. Second, the temptation is to see the institutions as the source of power when in fact power is the basis for an institution. Third, there is a tendency to assume that power always reflects the regulations and apparatus of an institution. Although power relationships are "embodied and crystallized" in institutions, their origins are elsewhere: "Power is everywhere; not because it embraces everything, but because it comes from everywhere" (Foucault, 1976/1980, p. 93). It is "always local and unstable" (p. 93). Power relations are to be found "rooted deep in the social nexus" (Foucault, 1982, p. 222), "rooted in the system of social networks" (p. 224). What makes power possible is "the moving substrate of force relations" (Foucault, 1976/1980, p. 93).

Power relations, then, are the roots of the order of any society, as much in the details of intimate interaction as in the scale of large state institutions. Foucault was not *reducing* social relationships to relations of power. He acknowledged that there are relationships of communication along with activities of "work, and the transformation of the real" (Foucault, 1982, p. 218). These are three *aspects* of relationships, and they overlap, interact, and are coordinated in changing ways. Foucault viewed power as one aspect of any interaction, just as Habermas saw an emancipatory interest everywhere.

Foucault emphasized "domination" rather than "emancipation": the possibility of a confrontation like that between Hegel's master and slave. Power relations are relations of strategy and even force: "power must be understood in the first instance as the multiplicity of force relations" (Foucault, 1976/1980, p. 92). But power is a kind of force that, unlike violence, grants the other the freedom to be defiant and disobedient: "at the heart of power relations and as a permanent condition of their existence there is an insubordination and a certain essential obstinacy on the part of the principles of freedom" (Foucault, 1982, p. 225). The "strictly relational character of power relationships" (Foucault, 1976/1980, p. 95) can be seen in the fact that power *needs* resistance in some form or another, as "adversary, target, support ..." (p. 95); these are power's "irreducible opposite" (p. 96). In a relationship of power, the other is always "thoroughly *recognized* and maintained to the very end as a person who acts" (Foucault, 1982, p. 220, emphasis added). In a power relationship, one person acts *on the actions* of another: "[W]hat defines a relationship of power is that it is a mode of action which does not act directly and immediately on others. Instead it acts upon their actions: an action upon an action, on existing actions or on those which may arise in the present or in the future" (p. 220). The basis of power is not violence, though this may be its instrument or its result. Power "is always a way of acting upon an acting subject or acting subjects by virtue of their acting or being capable of action" (p. 220). In the final chapter, I will explore how this view fits with ethnomethodology and conversation analysis.

Foucault was careful not to push the emphasis on confrontation too far, and action on another's action should not be viewed as negative. "The term 'power' designates relationships between partners ... an ensemble of actions which induce others and follow from one another" (Foucault, 1982, p. 217). Power "is less a confrontation between two adversaries or the linking of one to the other than a question of government" in the sense of the 16th-century use of this term: "the way in which the conduct of individuals or of groups might be directed: the government of children, of souls, of communities, of families, of the sick" (p. 221). "To govern, in this sense, is to structure the possible field of actions of others.... Power is exercised only over free subjects, and only insofar as they are free" (p. 221). Again, power always grants the possibility for the refusal to submit.

This action on another's action is everywhere, it is continual, and it is essential to participation in a society. "[T]o live in society is to live in such a way that action upon other actions is possible – and in fact ongoing" (Foucault, 1982, p. 222); "the possibility of action upon the action of others" is "coextensive with every social relationship" (p. 224). So although "power relations have been progressively governmentalized, that is to say, elaborated, rationalized, and centralized in the form of, or under the auspices of, state institutions" (p. 224), the origin of power and the best place to study power relations is in the interactions of the "social nexus."

How do we study power, seen this way? We have legal models that address legitimation, economic models that treat power as a commodity, and institutional models of state power. Foucault recommends that instead we begin with resistance, "using this resistance as a chemical catalyst so as to bring to light power relations, locate their position, find out their point of application and the methods used (Foucault, 1982, p. 211). For example, "to find out what our society means by sanity, perhaps we should investigate what is happening in the field of insanity" (p. 211). We will find "mobile and transitory points of resistance" that have profound consequences: "producing cleavages in a society that shift about, fracturing unities and effecting regroupings, furrowing across individuals themselves, cutting them up and remolding them, marking off irreducible regions in them, in their bodies and mind" (Foucault, 1976/1980, p. 96). We should start with those "oppositions" that assert the right to be different, with "struggles against the 'government of individualization'" (Foucault, 1982, p. 212). These oppositions and resistances question and attack a technique or form of power "which makes individuals subjects" in the dual sense of "subject to someone else by control and dependence" and "tied to his own identity by a conscience or self-knowledge" (p. 212). Here again we hear echos of Hegel, both in Foucault's suggestion that it is by understanding those who have been subjected and who resist this subjection that we will come to understand how power operates and in his proposal that "individualization" is a process that makes people both dependent *and* self-conscious.

Power and Knowledge: Power Produces Reality

Let us return to the relationship between power and knowledge. The famous formula "power/knowledge" (French: *pouvoir-savoir*, literally "to be able to know") refers to the way the two axes of knowledge and power define a space within which discourse can exist. The "*régime du savoir*" is "the way in which knowledge circulates and functions, its relations to power" (Foucault, 1982, p. 212). As Foucault explained:

> No knowledge is formed without a system of communication, registration, accumulation, and displacement that is in itself a form of power, linked in its existence and its functioning to other forms of power. No power, on the other hand, is exercised without the extraction, appropriation, distribution, or restraint of a knowledge. At this level there is not knowledge [*connaissance*] on one side and society on the other, or science and the state, but basic forms of "power-knowledge" [*"pouvoir-savoir"*]. (Foucault, 1994/2006, p. 17)

Foucault continued to explore the

> peculiar level among all those which enable one to analyze systems of thought – that of discursive practices. There one finds a type of systematicity which is neither logical nor linguistic. Discursive practices are characterized by the demarcation of a field of objects, by the definition of a legitimate perspective for a subject of knowledge, by the setting of norms for elaborating concepts and theories. (Foucault, 1997/2006, p. 11)

But he came to see that "[d]iscursive practices are not purely and simply modes of manufacture of discourse. They take shape in technical ensembles, in institutions, in behavioral schemes, in types of transmission and dissemination, in pedagogical forms that both impose and maintain them" (p. 12). The genealogist looks at discourse as it is immersed "in the field of multiple and mobile power relations" (Foucault, 1976/1980, p. 98), for "[i]t is in discourse that power and knowledge are joined together" (p. 100). The concept of power/knowledge combines and unifies "the deployment of force and the establishment of truth" (Foucault, 1975/1977, p. 184). Critical theory before Foucault recognized linkages among truth, power, and subjectivity, but it viewed them in limited terms: power leads to ideology, which is a distortion of the truth; power, exercised as domination and exploitation, deforms subjectivity. What Foucault saw was that power is at the heart of the *production* of truth, and furthermore the *production* of human beings who understand themselves to be particular kinds of subjects. Power is productive and enabling: "We must cease once and for all to describe the effects of power in negative terms: it 'excludes,' it 'represses,' it 'censors,' it 'abstracts,' it 'masks,' it 'conceals.' In fact, power produces; it produces reality; it produces domains of objects and rituals of truth. The individual and the knowledge that may be gained of him belong to this production" (p. 194).

Genealogy still attends (as in the archaeology) to what is said and what is not said, to who speaks and from which positions, but it also attends to the power relations that are involved, to the way "discursive elements" are used both strategically and tactically. "Discourses are tactical elements or blocks operating in the field of force relations. There can exist different and even contradictory discourses within the same strategy: they can, on the contrary, circulate without changing their form from one strategy to another, opposing strategy" (Foucault, 1976/1980, p. 101). Foucault's analysis was still a relational one, exploring how knowledge and politics are interwoven. One might study the relationship between psychiatry and political institutions as independent entities or explore the ethical implications of psychiatric treatment of patients:

> But my goal hasn't been to do this; rather I have tried to see how the formation of psychiatry as a science, the limitation of its field, and the definition of its object implicated a political structure and a moral practice: in the twofold sense that they were presupposed by the progressive organization of psychiatry as a science, and that they were also changed by this development. (Foucault, 1984/2006, p. 119)

For example, in the first volume of the *History of Sexuality*, Foucault's "objective" was "to analyze a certain form of knowledge regarding sex, not in terms of repression or law, but in terms of power" (Foucault, 1976/1980, p. 92). His interest was to explore specific discourses on sex in terms of "the most immediate, the most local power relations at work" (p. 97). He concluded that, as a political issue, sex lies "at the pivot of the two axes along which developed the entire political technology of life" (p. 145). At the intersection of the disciplines of the body and the regulation of the population, a problem for both the productivity of the individual body and the well-being of the whole social body, sex gave rise to "an entire micro-power" as well as "comprehensive interventions." "Broadly speaking, at the juncture of the 'body' and the 'population,' sex became a crucial target of a power organized around the management of life rather than the menace of death" (p. 147).

The linkages between knowledge and power are not unidirectional or static. "Discourse can be both an instrument and an effect of power, but also a hindrance, a stumbling-block, a point of resistance and a starting point for an opposing strategy" (Foucault, 1976/1980, p. 101). This means that we cannot "read off" the strategy from discourse: "We must not expect the discourses on sex to tell us . . . what strategy they derive from, or what moral divisions they accompany, or what ideology – dominant or dominated – they represent" (p. 102). Instead, we must "interrogate" the discourse at two levels: "tactical productivity (what reciprocal effects of power and knowledge they ensure)" and "strategical integration (what conjunction and what force relationship make their utilization necessary in a given episode of the various confrontations that occur)" (p. 102).

Tactics and *strategy* are two sides of a single coin: "on the one hand, rational forms, technical procedures, instrumentations through which to operate, and, on the other, strategic games that subject the power relations they are supposed to guarantee to instability and reversal" (Foucault, 1985/2006, p. 203).

A Response to Kant

If Foucault's archaeology was a response to Kant's first critique, with its image of the individual constituting the world by making mental models, his notion of power was a response to Kant's second critique, with its view that moral action is achieved by a universal subject who comes to recognize its abstract duties. Foucault suggested that Kant should be considered an example of pastoral power. Kant (in "What Is Enlightenment?") was the first philosopher to ask the question "Who am I?" in and of "a very precise moment in history" (Foucault, 1982, p. 216). Kant was articulating (and legitimating) precisely the tactics of individualization that were becoming central to the pastoral power of the Enlightenment. Each person now had to ask him- or herself, "Who am I?" They had to reply, "I must *become* a unique but universal historical subject" (p. 216). "I must *become* the individual that I know I am. Within the totalization of the modern state, I must become rational and free."

A double bind is a paradoxical injunction: an order, given by someone who has power over another, to do something freely. The order "be spontaneous!" is a classic double bind (Watzlawick, Beavin, & Jackson, 1967). Double binds have been found at the root of schizophrenia (Bateson, Jackson, Haley, & Weakland, 1956). Foucault countered Kant with the recommendation that we challenge the paradoxical injunctions of individualization and totalization in modern society: "Maybe the target nowadays is not to discover what we are, but to refuse what we are. We have to imagine and to build up what we could be to get rid of this kind of political 'double bind,' which is the simultaneous individualization and totalization of modern power structures" (Foucault, 1982, p. 216).

The Body

The body occupies an important place in genealogy. The genealogies focused on the body (tacit, material, and sentient) rather than the "subject" (cognitive, disembodied, and conscious). To Foucault, the involvement of the body illustrated the fact that the operation of power requires no representation: "Power relations can materially penetrate the body in depth, without depending even on the mediation of the subject's own representations. If power takes hold of the body, this isn't through its having first to be interiorized in people's consciousness" (Foucault, 1980, p. 186). Foucault articulated knowledge and mastery that amounted to a "political technology of the body" (Foucault, 1975/1977, p. 26). The body is the locus of practices of subjectification and objectification, and genealogy maps a historical descent whose primary object is the body:

> Finally, descent attaches itself to the body. It inscribes itself in the nervous system, in temperament, in the digestive apparatus; it appears in faulty respiration, in improper diets, in the debilitated and prostrate bodies of those whose ancestors committed errors. ... [T]he body maintains, in life as in death, through its strength or weakness, the sanction of every truth and error, as it sustains, in an inverse manner, the origin – descent. (Foucault, 1984a, p. 82)

A body is a location with a surface and a volume that become articulated and inscribed. The location becomes the basis for an illusory unitary subject, the surface is marked by the passage of events, and the volume bears the impact of their force:

> The body is the inscribed surface of events (traced by language and dissolved by ideas), the locus of a dissociated self (adopting the illusion of a substantial unity), and a volume in perpetual disintegration. Genealogy, as an analysis of descent, is thus situated within the articulation of the body and history. Its task is to expose a body totally imprinted by history and the process of history's destruction of the body. (Foucault, 1984a, p. 83)

One of the central tasks of genealogy is to "expose" the body as site and surface but also as agency.

The Researcher as Genealogist

With genealogy, it is clear that Foucault did not maintain that the researcher could or should be a detached spectator of excavated discourse. "Foucault introduces genealogy as a method of diagnosing and grasping the significance of social practices from within them" (Dreyfus & Rabinow, 1982, p. 103). He "realizes and thematizes the fact that he himself – like any other investigator – is involved in, and to a large extent produced by, the social practices he is studying" (p. 103). Genealogical investigations have a practical, political aim: a genealogy aims not for knowledge for its own sake but for its critical or political potential, its capacity to contest what is taken to be "truth." As a critical inquiry, genealogy tells "effective history." Genealogists write history aware of their own position and interests; there can be no objective, disinterested account. This requires

> someone who shares the participant's involvement but distances himself from it and does the hard historical work of diagnosing the history and organization of our current cultural practices. The result is a pragmatically guided reading of the effect of present social practices which does not claim to correspond either to the everyday understanding of being in those practices nor to a deeper repressed understanding. (Dreyfus, 1984, p. 80)

The researcher's distance from involved participation, though, is now above rather than below.

The genealogist questions and tries to unsettle the knowledge that has achieved the accepted status of "truth," challenging the power that this knowledge supports. "A certain fragility has been discovered in the very bedrock of existence ... in those aspects ... that are most familiar, most solid and intimately related to our bodies and everyday behaviour" (Foucault, 1980, p. 80). For example, "it wasn't as a matter of course that mad people came to be regarded as mentally ill; it wasn't self-evident that the only thing to be done with a criminal was to lock him up; it wasn't self-evident that the causes of illness were thought to be sought through the individual examination of bodies" (Foucault, cited in Smart, 1983, p. 77). Foucault insisted that "the real political task in a society such as ours is to criticize the working of institutions which appear to be both neutral and independent; to criticize them in such a manner that the political violence which has ... exercised itself obscurely through them will be unmasked, so that one can fight them" (Foucault, 1974, p. 171).

A good example of this kind of questioning is Foucault's criticism of depth hermeneutics, both as a form of subjectification and as a method in the human sciences. We have seen that Foucault examined how the institutions that establish and support interpretation (institutions such as the church, with confession, and the medical profession, with therapy) exercise power. The *alleged* hiddenness of what is sought, Foucault now saw, plays a directly visible role:

> Whereas the interpreter is obliged to go to the depth of things, like an excavator, the moment of interpretation [in genealogy] is like an overview, from higher and higher up, which allows the depth to be laid out in front of him in a more and more profound visibility; depth is resituated as an absolutely superficial secret. (Cited in Dreyfus & Rabinow, 1982, pp. 106–107)

In what can be read as a reply to Schleiermacher and other forms of Romanticist hermeneutics, Foucault wrote: "If interpretation can never be complete, this is quite simply because there is nothing to interpret. There is nothing absolutely primary to interpret, for after all everything is already interpretation" (Foucault, 1967/1999, p. 275). The most powerful form of critique comes, in Foucault's view, not from the depth hermeneutics of suspicion such as Marxism and psychoanalysis but from local and specific settings, from microcontexts. Critique should arise from "subjugated knowledges," not from totalizing theoretical systems. Critique cannot be transcendental; only immanent, local. Yet, as was the case for Marx, critique for Foucault combined an interest in the conditions for the possibility of a phenomenon with an interest in freedom. He wrote that "criticism – understood as analysis of the historical conditions that bear on the creation of links to truth, to rules, and to the self – does not mark out impassible boundaries or describe closed systems; it brings to light transformable singularities" (Foucault, 1985/2006, p. 201). But Foucault's version of critique does not aim to emancipate the oppressed or diagnose their false consciousness. Foucault suggested that to view the exploited and marginalized as analogous to patients in psychoanalysis, as Habermas did, is to perpetuate their problems. The presumption that they need an expert to tell them what their own actions really mean and where their interests truly lie amounts to another form of domination. Instead, an "effective history" is one that opposes a specific oppressor by challenging their authority.

It has been said that Foucault's genealogies tell us "the truth about truth" (Margolis, 1993); that is, "the truth that there is no capitalized 'Truth,' no 'truth of truth'" (Caputo, 1993). Foucault's inquiry may have been directed toward the question of who we are, but he wanted to keep the question open and avoid trying to give a *single* true answer. Any positive theory of the individual would have us "trapped in our own history" (Foucault, 1982, p. 210). The aim instead was "to promote new forms of subjectivity through the refusal of this kind of individuality which has been imposed on us for several centuries" (p. 216). In other words, the aim was *ethical*.

AN ETHICS OF FORMATION OF THE SELF

For some critics, something important was missing from Foucault's histories: human freedom. In both his archaeologies and his genealogies, Foucault had insisted that there is no "subject" – either individual or collective – who moves history. The benefit of this assumption was to avoid seeing history as the work of a few powerful people, as in much historical writing. The danger was that it seemed to preclude the possibility of people acting freely.

Foucault came to focus on how the human subject becomes an object to him or herself. This called for attention to the variety of ways in which someone "is led to observe himself, analyse himself, interpret himself, recognize himself as a domain of possible knowledge" (Foucault, 1984c). Sex and sexuality remained a central case, but Foucault recast subsequent volumes of *The History of Sexuality* to focus on the techniques in which subjectivity has been fostered "if by 'subjectivity' one means the way in which the subject experiences himself in a game of truth where he relates to himself" (Foucault, 1984c).

The basic question was why for Western society sexual practices have been for such a long time a *moral* concern. Why should sexuality, but not say digestion, be a moral issue? Foucault's interest was not so much in sexuality as in morality. He wrote "I am much more interested in problems about techniques of the self and things like that than sex ... sex is boring" (Foucault, 1983/2006, p. 253), and explained that "[t]he general framework of the book about sex is a history of morals" (p. 263).

The focus was on how in their sexual activity a person becomes an "ethical subject," or, as Foucault phrased it, the "reflective practice of freedom" (see Poster, 1993). In various "practices of the self," models are offered for how a person should examine and come to know him- or herself and the transformations they should attempt. *Ethics* here should be understood in the sense of *ethos* (Greek: nature or disposition). This third phase of Foucault's work was connected with Kant's third critique, *The Critique of Judgement*, and indeed he called it "an aesthetics of existence" (Foucault, 1984/1986, p. 12). But it was still a genealogy, and it incorporated the archaeological study of forms of problematization of sexuality. Foucault told the history of "forms of moral subjectivation," of how people "not only set themselves rules of conduct, but also seek to transform themselves, to change themselves in their singular being, and to make their life into an *oeuvre* that carries certain aesthetic values and meets certain stylistic criteria" (pp. 10–11). These forms amounted to a "cultivation of the self" and "a stylistics of existence" (Foucault, 1984/1988, p. 71). Volume 2 of *The History of Sexuality* became a study of carnal pleasure in ancient Greece (the original plan had been for a study of female hysteria in the 19th century). The domain of analysis was those "practical texts" that gave advice to the reader, viewed as "functional devices that would enable individuals to question their own conduct, to watch over and and give shape to it, and to shape themselves as ethical subjects" (Foucault, 1984/1986, p. 13). Once again, the body was of central importance; it is the focus of the work of self-creation, the ethical substance that we work on to become particular kinds of humans.

Self-Constitution

In Volume 3 of *The History of Sexuality*, a study of "care of the self" in later antiquity (Volume 4 was unfinished at the time of his death), Foucault began to elaborate an account of human being as a site of multiple practices. In the various games of truth, humans come to understand themselves, to grasp their "nature" – the "truths" about what it is to be a subject. But being human is also the result, the effect, of a variety of technologies: the practices and apparatuses of home, work, medicine, and others. And, not least, it is the result of projects of self-management, of self-governing, in terms of physical and mental health, education and training, and religious and spiritual well-being. This third dimension is what Foucault referred to as "the care of the self."

Kant argued that moral conduct requires that the individual recognize that they are a universal subject. Foucault proposed that what Kant actually described is a process of self-*constitution*, not just "self-awareness" but "self-formation as an 'ethical subject'" (Foucault, 1984/1986, p. 28). This "mode of subjection" is "the way in which the individual establishes his relation to the rule [of moral conduct] and recognizes himself as obliged to put it into practice" (p. 27).

As we have seen, Kant took on the problem of how valid knowledge is connected to ethical action. As Foucault notes, this connection was greatly debated in the Enlightenment. Kant argued that the universal capacity for reason enabled every individual to identify both true knowledge and moral conduct. "Kant's solution was to find a universal subject that, to the extent it was universal, could be the subject of knowledge, but which demanded, nonetheless, an ethical attitude – precisely the relationship to the self which Kant proposes in *The Critique of Practical Reason*" (Foucault, 1983/2006, p. 279). We have seen that Foucault located Kant at the "threshold of modernity," first for questioning the limits of representation and then in the context of the technologies (and double-binds) of pastoral care that began to spread through society with the advent of biopower. Now he argued that Kant offered new answers to old questions in the form of a new way (within the Western cultural tradition that drew from the Greeks and then Christianity) for people to *constitute* themselves:

Kant says, "I must recognize myself as universal subject, that is, I must constitute myself in each of my actions as a universal subject by conforming to universal rules." The old questions were reinterpreted: How can I constitute myself as a subject of ethics? Recognize myself as such? Are ascetic exercises needed? Or simply this kantian relationship to the universal which makes me ethical by conformity to practical reason? Thus Kant introduces one more way in our tradition whereby the self is not merely given but is constituted in relationship to itself as subject. (Foucault, 1983/2006, pp. 279–280)

Foucault insisted that this constitution of the self – both in power relations and through care of the self, both as an object for others and as an object for oneself – is observable. The self "is constituted in real practices – historically analyzable practices." We can observe "a technology of the constitution of the self" (p. 277). This technology "can be found in all cultures in different forms." It is, however, somewhat *difficult* to study, for two reasons. First, the techniques often don't require material apparatus, so "they are often invisible techniques." Second, they are often tied up with "techniques for the direction of others. For example, if we take educational institutions, we realize that one is managing others and teaching them to manage themselves" (p. 277). But it is nonetheless empirically observable, and this is the task of an ethics (or a "genealogy of ethics"), or what could be called *ethology*.

Foucault proposed that there are four main aspects to this crafting of one's relation to self and to others. First is the part of the self concerned with moral conduct; the "ethical substance" – the material that is going to be worked over by ethics. Second is "the mode of subjectivation," "the way people are invited or incited to recognize their moral obligations." Third is "the means by which we can change ourselves in order to become ethical subjects" (Foucault, 1983/2006, p. 265), "the self-forming activity"; that is to say, asceticism in a broad sense. Finally, there is the kind of being we aspire to be.

THE HISTORICAL ONTOLOGY OF OURSELVES

In a manuscript published after his death, Foucault sketched a program of inquiry that he called "a historical ontology of ourselves" (Foucault, 1984b). This broad and bold program would explore how we human beings have become particular kinds of persons, how we have become subjects. It would ask not (or not only) about the various "concepts" we have of a person but the more fundamental ontological question of how we *become* a person. The different ways of *being* human would be its focus, and these would be examined as historically (and culturally) contingent "products." In particular, Foucault emphasized, the Enlightenment established the conditions for the constitution of human beings as "autonomous subjects":

> We must try to proceed with the analysis of ourselves as beings who are historically determined, to a certain extent, by the Enlightenment. Such

an analysis implies a series of historical inquiries that are as precise as possible; and these inquiries will not be oriented retrospectively toward the "essential kernel of rationality" that can be found in the Enlightenment and that would have to be preserved in any event; they will be oriented toward the "contemporary limits of the necessary," that is, toward what is not or is no longer indispensable for the constitution of ourselves as autonomous subjects. (Foucault, 1984b, p. 43)

Foucault's description of this program shows that his ethics was not separate from his earlier archaeological and genealogical work. His interest in the techniques by which people form themselves had always been intertwined with his interest in truth games and power/knowledge. Indeed, he had written earlier that "the goal of my work during the last twenty years" had been "to create a history of the different modes by which, in our culture, human beings are made subjects" (Foucault, 1982, p. 208). He had focused, he explained, on "three modes of objectification": that of the human sciences; the "dividing practices" that separate mad from sane, sick from healthy; and how people recognize themselves as subjects.

Truth games, power/knowledge, and care of the self are three interrelated aspects, "three axes," of investigation: "the play between types of understanding [*savoir*], forms of normality, and modes of relation to oneself and others" (Foucault, 1985/2006, p. 202) (Figure 14.1). Any specific form of *know-that* is made possible by a kind of *know-how* that is embodied in and sustained by an apparatus of power/knowledge and requires and provides the techniques for a particular kind of relation to oneself. Foucault now described his earlier work in terms of all three aspects, though each study put more weight on one than the others. He now described *Madness and Civilization*, his first major archaeological investigation, in these terms:

I tried to analyze the genesis, during the seventeenth and eighteenth centuries, of a system of thought as a matter of possible experiences; *first*, the formation of a domain of recognitions [*connaissances*] that constitute themselves as specific knowledge [*savoir*] of "mental illness"; *second*, the organization of a normative system built on a whole technical, administrative, juridical, and medical apparatus whose purpose was to isolate and take custody of the insane; and *finally*, the definition of a relation to oneself and to others as possible subjects of madness. (Foucault, 1985/2006, p. 202, emphasis added)

These interrelations are also evident in Foucault's description of the subjects we become:

[T]he historical ontology of ourselves has to answer an open series of questions; it has to make an indefinite number of inquiries which may be multiplied and specified as much as we like, but which will all address the questions systematized as follows: How are we constituted as subjects of

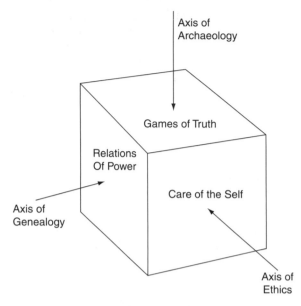

FIGURE 14.1. Three Axes of a Form of Life

our own knowledge? How are we constituted as subjects who exercise or submit to power relations? How are we constituted as moral subjects of our own actions? (Foucault, 1984b, p. 49)

This threefold description of the human subject – in terms of knowledge, power relations, and morality – obviously owes much to Kant's three critiques of knowledge, social practice, and ethics, but Foucault offers a completely different solution to Kant's problem (Allen, 2003). Building on the work of Hegel, Heidegger, Merleau-Ponty, and others, Foucault insists that there is something *more* to our relationship to the world. We don't merely form mental models of our environment; we live *in* the world and we are *of* the world. Like Garfinkel, Foucault returned to *practices* to discover how *order* is achieved: the *ordering* of objects in knowledge, of subjects in power, and of ourselves (as *orderers*) in ethics. This is "a historical investigation into the events that have led us to constitute ourselves and to recognize ourselves as subjects of what we are doing, thinking, saying." It is a program that envisages three interconnected domains:

> First, a historical ontology of ourselves in relation to truth through which we constitute ourselves as subjects of knowledge; second, a historical ontology of ourselves in relation to a field of power through which we constitute ourselves as subjects acting on others; third, a historical ontology of ourselves in relation to ethics through which we constitute ourselves as moral agents. So, three axes are possible for genealogy. (Foucault, 1984d, p. 350)

The methodology for the historical ontology of ourselves *brings together* archaeology, genealogy, and ethics: it will be "genealogical in its design and archaeological in its method" and ethical in its goal "to give new impetus, as far and wide as possible, to the undefined work of freedom." It is:

> Archaeological – and not transcendental – in the sense that it will not seek to identify the universal structures of all knowledge or of all possible moral action, but will seek to treat the instances of discourse that articulate what we think, say, and do as so many historical events. And this critique will be genealogical in the sense that it will not deduce from the form of what we are what it is impossible for us to do and to know; but it will separate out, from the contingency that has made us what we are, the possibility of no longer being, doing, or thinking what we are, do, or think. (Foucault, 1984b, p. 46)

We see here how archaeology and genealogy relate and together form a critique. Transcendental critique is rejected in favor of an archaeological "method" that explores surfaces rather than striving to rise above them.

For example, a school classroom could be studied with attention to the kinds of knowledge that are accepted: both the explicit know-that of the curriculum and the tacit know-how of the "hidden curriculum" (Jackson, 1968). We might then investigate its apparatus of power/knowledge: the normative techniques of authority and the strategies of resistance. And we can also study the kinds of persons that children become in the school class-room, attending to the practices of comportment that shape bodies, form minds (Martin, 1993), and enable a child to be a good or bad student. Foucault's point is that these three are always interconnected. Teachers grasp this point, first on a practical level. Researchers tend to miss the point and try to study the conceptions (know-that) of individual children (Packer & Greco-Brooks, 1999).

The practices in which truth is defined, power/knowledge is generated, and selves are formed are everyday and taken for granted, are complex, and give rise to their own misunderstandings. Even qualitative researchers fre-quently fail to grasp the complexity that Foucault drew to our attention. Foucault's program for a historical ontology of ourselves offers a way to conceive of a methodology for qualitative investigation that asks relevant and important questions, employs a sophisticated approach to analysis, and has important goals.

To Foucault, the principal goal of this kind of investigation was *enlight-enment*, though of a kind completely different from what Kant envisioned. Critical investigation would show that the seemingly natural and inevitable is actually contingent and changeable. It would be a "permanent critique of ourselves" (Foucault, 1984b, p. 43) that would require a "historico-critical attitude" (p. 47). Foucault wrote:

[I]t seems to me that the critical question today has to be turned back into a positive one: in what is given to us as universal, necessary, obligatory what place is occupied by whatever is singular, contingent, and the product of arbitrary constraints? The point in brief is to transform the critique conducted in the form of necessary limitation into a practical critique that takes the form of a possible transgression. (Foucault, 1984b, p. 45)

This practical criticism would be immanent, not transcendental. Its product would be not a corpus of knowledge but an "experiment" on ourselves:

The critical ontology of ourselves has to be considered not, certainly, as a theory, a doctrine, nor even as a permanent body of knowledge that is accumulating; it has to be conceived as an attitude, an ethos, a philosophical life in which the critique of what we are is at one and the same time the historical analysis of the limits that are imposed on us and an experiment with the possibility of going beyond them. (Foucault, 1984b, p. 50)

The historical ontology of ourselves, in short, would be a program of research both practical and ethical.

CONCLUSIONS

Whereas Habermas tried to establish the normative basis for critical inquiry in the reflection of the individual philosopher and Bourdieu tried to do so for sociology by studying his own discipline and origins, Foucault undertook a detailed critical study of the human sciences in which he showed how scientific knowledge is not simply a way of knowing but also a way of acting and a way of being, social before it is individual. Foucault's work can be considered precisely the detailed reflexive examination of the human sciences that Bourdieu called for. Habermas acknowledged that "even when a person seems to retire into himself to live among his own ideas, he is living really with the others who have thought what he is thinking" (Habermas, 1981/1989, pp. 95–96).

It is important to return to the fact that Foucault insisted he was not writing histories of ideas or histories of people. Looking back on his work, he described it as "a critical history of *thought*" (emphasis added). What, then, is *thought*?

By "thought," I mean what establishes, in a variety of possible forms, the play of true and false, and consequently *constitutes* the human being as a knowing subject [*sujet de connaissance*]; in other words, it is the basis for accepting or refusing rules, and constitutes human beings as social and juridical subjects; it is what establishes the relation with oneself and with others, and constitutes the human being as ethical subject. (Foucault, 1985/2006, p. 200, first emphasis added)

Foucault, just like Kant, worked backward from our knowledge of the world to figure out what kind of subject we must be. "The problem is to determine what the subject must be, to what condition he is subject, what status he must have, what position he must occupy in reality or in the imaginary, in order to become a legitimate subject of this or that type of knowledge" (Foucault, 1984c). This required rethinking the subject and also rethinking thinking. We have seen that for Foucault every human being is the product of specific discursive practices, so there are many forms of subjectivity. Thinking, too, is a social practice, and a way of relating to a situation.

Thought assumes "concrete forms" in institutions, practices, and systems of representations (maps, calculi, etc.). "Internal" action on *mental* representations is only *one* form of thought, and it requires the formation of a particular *kind* of self as a knowing subject. Foucault reminds us that, for Kant, becoming a Cartesian subject required a certain kind of ethical stance, a stance in which one viewed oneself as a *universal* subject. Thinking for Kant *should* be individual, detached, and formal.

But Foucault has emphasized – and once again the influence of Merleau-Ponty, Heidegger, and Hegel is evident – that thought is found in *all* action, whether it is intellectual, ethical, or self-conscious:

> "Thought," understood in this way, then, is not to be sought only in theoretical formulations such as those of philosophy or science; it can and must be analyzed in every manner of speaking, doing, or behaving in which the individual appears and acts as knowing subject [*sujet de connaissance*], as ethical or juridical subject, as subject conscious of himself and others. In this sense, thought is understood as the very form of action – as action insofar as it implies the play of true and false, the acceptance or refusal of rules, the relation to oneself and to others. (Foucault, 1985/2006, pp. 200–201)

Rethinking the subject and rethinking thinking go hand in hand with rethinking the nature of knowledge and how something or someone becomes an object of knowledge:

> [I]t is also and at the same time a question of determining under what conditions something can become an object for a possible knowledge [*connaissance*], how it may have been problematized as an object to be known, to what selective procedure it may have been subjected, the part of it that is regarded as pertinent. So it is a matter of determining its mode of objectivation, which is not the same either, depending on the type of knowledge [*savoir*] that is involved. (Foucault, 1984c, p. 942)

Foucault proposed that thinking is a way of relating to a situation that objectifies it and allows us to raise the question of its character, to consider it as a problem, and to choose our actions accordingly. As Rabinow and Rose note, "By definition, the thinker is neither entirely outside of the situation in question nor entirely enmeshed within it without recourse or options" (Rabinow & Rose, 2003, p. 13). Indeed, thought "is what allows one to step back from this way of acting or reacting, to present it to oneself as an object of thought and to question it as to its meaning, its conditions, and its goals. Thought is freedom in relation to what one does, the motion by which one detaches oneself from it, establishes it as an object, and reflects on it as a problem" (Foucault, 1984/2006, p. 117).

In other words, thought is *problematization*. The task of a history of thought, in Foucault's view, was to "define the conditions in which human beings 'problematize' what they are, what they do, and the world in which they live" (Foucault, 1984/1986, p. 10). This is a matter of analyzing "the *problematizations* through which being offers itself to be, necessarily, thought – and the *practices* on the basis of which these problematizations are formed" (p. 11). Problematizations develop when how we live in the world becomes a matter of concern, an element for reflection, a matter of anxiety, discussion, and truth. Problematization is a definition both of a way of thinking and of a domain of objects: a way in which "being offers itself to be thought." We can see now why for Foucault's kind of history we must recognize that language is not merely representation but something "more."

Problematization is the lighting up of objects for thought – for scientific analysis, political assessment, or ethical reflection. To study problematization is to trace the history of thought and the dynamics of human thinking not as a history of ideas or representations or a history of mentalities (of worldviews or attitudes) but as the constitution of subjects and objects. So "a critical history of thought would be an analysis of the conditions under which certain relations of subject to object are formed or modified, insofar as those relations constitute a possible knowledge [*savoir*]" (Foucault, 1984c). Thought is a questioning, a stepping back from what one is doing, a posing of problems. It is what Heidegger called "circumspection." Diverse solutions may be the result, but what is at their root is the problematization (itself the result of social, economic, or political processes) that they attend to.

Finally, the study of problematization is at the same time a critical inquiry. It is a kind of inquiry that looks at specific cases, but in order to learn their general significance: "The study of modes of problematization (that is, of what is neither an anthropological constant nor a chronological variation) is . . . the way to analyze questions of general import in their historically unique form" (Foucault, 1984b). Thought about a problem will have taken a variety of forms. The point of analysis is not to evaluate them but to explore the conditions in which they were possible, the general form of problematization that made them possible. This approach unsettles our tendency to accept a problem as "given" and allows us to see it instead as arising from a specific cultural and historical configuration. In doing this, we recast the problem, examining the situation in which it arose. Critical analysis is itself a practice of problematization and reproblematization. It is "a movement . . . in which one tries to see how the different solutions to a problem have been constructed; but also how these different solutions result from a specific form of problematization" (Foucault, 1984/2006, pp. 118–119).

So the study of problematization is itself a practice of problematization. Foucault considered problematization to be his "way of approaching political questions, as "the development of a domain of acts, practices, and thoughts that seem to me to pose problems for politics." (Foucault, 1984/2006, p. 114). He went on:

> For example, I don't think that in regard to madness and mental illness there is any "politics" that can contain the just and definitive solution. But I think that in madness, in derangement, in behavior problems, there are reasons for questioning politics; and politics must answer these questions, but it never answers them completely. The same is true for crime and punishment: naturally, it would be wrong to imagine that politics [has] nothing to do with the prevention and punishment of crime, and therefore nothing to do with a certain number of elements that modify its form, its meaning, its frequency; but it would be just as wrong to think that there is a political formula likely to resolve the question of crime and put an end to it. The same is true of sexuality: it doesn't exist apart from a relationship to political structures, require-ments, laws, and regulations that have a primary importance for it; and yet one can't expect politics to provide the forms in which sexuality would cease to be a problem. (Foucault, 1984/2006, p. 114)

Thought, then, is a specific practical and concrete way of relating to a situation and considering it as a problem, a way of living a historical and cultural configuration. Thought always arises within a form of life. And the study of thought is itself a kind of thought, the problematization of problematization, with epistemological, political, and ethical aspects. Foucault helped us see the three axes that define any form of life: its members' relations to objects (truth games), their relations to one another (power/knowledge), and their relations to themselves (care of the self). He offered a powerful conception of the project for a critical scientific inquiry and of the appropriate stance for an investigator. This stance is a way of life, a way of living in a form of life – a way of being in a form of life.

15

A Historical Ontology of Ourselves

Foucault's great achievement in my view was to see, and show, how we can have a history of this [self-constituting freedom]: that by describing the different techniques of the self, one can tell the story of different ways in which people have purposefully made themselves into certain kinds of persons, and therefore of the historically specific and definite (and of course always limited) forms which ethical freedom has taken.

Laidlaw, 2002, p. 324

My principal aim in this book has been to consider how social scientists can frame and answer questions that are important and relevant. We have examined typical practices with the tools of qualitative research – interviews, ethnographic fieldwork, and analysis of discourse – with an eye toward their embedded ontological and epistemological assumptions. We have found a recurrent ontological dualism of subject and object, and an epistemology that contrasts subjective experience with objective knowledge. But we have also found important attempts to use these tools in new ways.

DISSOLVING KANT'S PROBLEM

In Part I, I argued that much qualitative research remains stuck in the dualism of subject and object. We began our examination of the practices of qualitative research with the semistructured interview, commonly viewed as central to the objective study of subjectivity, a view that Steiner Kvale summarized well: "If you want to know how people understand their world and their life, why not talk with them?" I questioned the combination of invisibility and confession in the qualitative research interview, and the practices of abstraction and generalization that are central to analysis through coding. The conduit metaphor for language, the view that meaning is contained in concepts and that experience is expressed in talk, contributes to a dualistic opposition between *subjectivity* and *objectivity* that ties this kind of research in knots. This approach maintains the same ontological dualism as empirical-analytic inquiry and ignores intersubjective phenomena: the material practices that form a background to our everyday actions and experiences. We began to consider different ontological assumptions and commitments, looking for the basis of a genuinely new program of inquiry in the social sciences. I suggested that when we interview a person we don't simply test our theories about them or find answers to our questions but encounter them and are challenged by them. We change, we learn, we grow. When we analyze an interview transcript, we should articulate our understanding of it as an invitation to see a different form of life.

In Part II, I traced the dualism of subject/object back to Kant's efforts to solve the problem of how an individual can have genuine knowledge and act ethically. Kant tried to identify those characteristics of human being that make knowledge of truth and goodness possible and concluded that a universal capacity for transcendental reason *constitutes* the world in the form of mental representations. But Kant's model of human being *divided* the individual both from the natural world and from other people, and it *separated* theoretical knowledge from ethical practice. His critical reconstruction of knowledge and ethics effectively doubled both object and subject. The object became *phenomenon* and *noumenon*: thing-for-us and thing-for-itself. The subject became both transcendental subjectivity and empirical subject. The problem can not be solved by "the split introduced by Kantian critique between knowing and willing ... between a language game made of denotations answerable only to the criterion of truth, and a language game governing ethical, social, and political practice that necessarily involves decisions and obligations" (Lyotard, 1979/1997, p. 32).

Kant unwittingly initiated a kind of "social constructionism" that is merely epistemological, that looks only at representation. This approach leads to relativism or nihilism, to the conclusion that we can never truly know reality and that ethical values are merely conventions.

But we saw that there is a different approach to the problem Kant faced, one that avoids the split he introduced. This approach considers constitution to be *ontological*, something that happens in practical activity. The work of those who explored this approach has shown that neither the true nor the good is achievable individually or intellectually. This doesn't mean there is *no* truth or morality, or that thinking is unimportant, but it means these will be multiple, local, and temporal. They have their origins prior to explicit knowledge in the practical techniques in which selves who can know and act are formed. Kant was wrong – consciousness

> does not constitute the world, it divines the world's presence around it as a field not provided by itself; nor does it constitute the word, but speaks as we sing when we are happy; nor again the meaning of the word, which instantaneously emerges for it in its dealing with the world and other men living in it. (Merleau-Ponty, 1945/1962, p. 404)

In the 1970s, calls for a new kind of inquiry in the social sciences proposed that constitution, in this sense of an ontological process within "intersubjective practices," should be the object of inquiry, and that immersion in practices through ethnographic fieldwork would be the central aspect of its method. We saw that unfortunately ethnography has *not* focused on constitution. Instead it has tried to produce veridical descriptions – that is to say, representations – of a cultural order that is generally assumed to be hidden *behind* what people say and do, in their *inner* lives. Ethnography remains stuck with the fiction of the ethnographer as a lone individual who knows the world by representing it. Malinowski invited his reader to: "Imagine yourself suddenly set down surrounded by all your gear, alone on a tropical beach close to a native village, while the launch or dinghy which has brought you sails away out of sight" (Malinowski, 1922/1961, p. 4). *Alone on a tropical beach* is the canonical image of the ethnographer: a heroic, stoic individual. This image persists even in experiments such as autoethnography, insofar as this is "a turning of the ethnographic gaze inward on the self (auto) while maintaining the outward gaze of ethnography, looking at the larger context wherein self experiences occur" (Denzin, 1997, p. 227). It is true that ethnography creates representations that are written and not mental, but its image of knowledge remains Kantian. When it is defined as participant observation, ethnography begins with an unbridgeable division between a cultural order and the ethnographer's description of that order. This description is *removed*, literally and metaphorically, from the form of life it claims to represent.

We need to see fieldwork instead as inquiry where two forms of life make contact. Fieldworkers are dealing with people and objects on, and from, the boundaries. They are never alone, even when they write *of* There, the village, *for* Here, the academy. What they write is not *merely* a representation but a creation and presentation whose meaning and truth are *effects* on readers. Malinowski implicitly acknowledged this: despite all his claims to objectivity, he hoped to foster tolerance and even enlightenment in the people who read his ethnographic accounts. As Foucault said, enlightenment is an attitude of testing the limits imposed on us by culture and history.

In Part III, we explored a key complication to the project of studying a form of life by participating in its practices, namely that participation can lead to *mis*understanding. This brought us back to Kant's notion of critique, to Marx's important transformation of this notion, and then to three different explorations of inquiry with a critical dimension. We saw that the division between person as subject and person as object is not just a bad idea (a contradictory conceptualization) but a cultural and historical reality. In modern forms of life, people have become objects for sciences of knowledge and management. In large part, the traditional human sciences are based on this division and search for techniques to turn people into objects, docile bodies.

The calls for a new kind of research in the human sciences – participatory and interpretive – claimed that it would be able to avoid or escape the dualism that dogged empirical-analytic forms of investigation in psychology, political science, sociology, and anthropology, but we have seen that many qualitative research projects reproduce this dualism rather than resolving it. The notion that qualitative research is "the objective study of subjectivity" *defines* inquiry in dualistic terms. But the program of inquiry Foucault sketched for us, his "historical ontology of ourselves," inspired as it is by the ontological explorations of constitution, does offer a way out of dualism. It does not seek to *solve* Kant's problem so much as *dissolve* it. Like Heidegger, Merleau-Ponty, and Marx, Foucault insisted that the knowing, acting, and judging subject is a product of a more fundamental mode of involvement – practical, social, embodied. Foucault went further than his predecessors in refusing to end his analyses with abstract notions of Space, Time, and Body. Instead he conducted concrete, empirical investigations of specific spaces, times, and bodies. He studied the work that goes into the production of order. As Wolin pointed out, every society "enforces certain types of conduct and discourages others; it, too, defines what sorts of experiments – in the form of individual or group actions – will be encouraged, tolerated, or suppressed" (Wolin, 1968, p. 149, cited in Bernstein, 1976, p. 103). Foucault's "experiments" were designed to test the limits of a form of life.

Habermas, Bourdieu, and Foucault each responded differently to the contradictions of Kant's model of human being and the problems that derailed the Enlightenment project to identify the conditions for valid knowledge, ethical practice, and informed judgment. Habermas has tried to put the project back on track by treating reason not as an innate individual capacity but as a potential lurking within communicative practices. Bourdieu found truth only in the particular practices of science, which he considered able to produce knowledge that transcends its local circumstances and has universal application. Foucault saw all truth claims as local and contingent, and viewed claims to universality as dangerous sources of violence. Rather than foster consensus, he believed we should encourage dissent.

But despite their differences, each of them adopted an ontology of radical realism, in which practices constitute both objects and subjects within particular forms of life. All three learned from Marx's analysis of how the practices of a capitalist economy distort people's lives and consciousness. They saw the Kantian dualisms as consequences of modern forms of life, which divide people from one another and from the fruits of their labor. All three considered it important to investigate these distortions and work to eliminate them, though they had different views of the way a researcher should approach this task.

For Habermas, research requires rational reconstruction from a stance of negotiated impartiality. He distinguishes reflection that reconstructs tacit competence from reflection that undoes the constraints on development, though he doesn't clearly describe the relationship between the two. Research is always interpretation, but this requires theoretical guidance, and Habermas appears to believe that theory somehow stands outside the hermeneutic circle, if not outside every form of life, at least at a "distance." It is not clear how a researcher could achieve the "negotiated impartiality" that Habermas values.

For Bourdieu, research is a reflexive enterprise that turns its techniques of objectification on themselves, exposing symbolic violence in both its own practices and those of the social field that is studied. Reflexivity is a recognition of the fore-structure of research, not an individual activity but part of the collective struggle among scientists in which competition forces selection of the most rational account. Science and the scientist are historically constituted, but Bourdieu insists that scientific knowledge *can* be objective.

For Foucault, research is a matter of problematizing problematization. The researcher works within cultural practices, diagnosing their problems and unmasking political violence. The process is interpretive, but it is neither a depth hermeneutics nor a Romantic hermeneutics. It explores the know-how that underlies know-that. Its program is "a historical ontology of ourselves" that focuses on three aspects of constitution: knowledge, social relations, and selves. A reconstruction of the history of a form of life is necessary, though this will not be a "rational" reconstruction, first because what is recognized as rational changes, and also because the accidental and contingent cannot be sifted out from history to disclose a rational progression.

Habermas maintains that language and communication provide the universal basis for both knowledge and ethics. Bourdieu maintained that science is able to produce knowledge with universal relevance and that the scientist can grasp the whole, albeit not neutrally. Foucault worked from within, insisting that all analysis must be local, yet, as we saw at the end of the last chapter, his local investigations addressed questions of "general import."

These are three different proposals for the stance a critical researcher should adopt, with differing emphases on the dimensions of language, labor, and life. For Habermas, the researcher must be an expert on language, able to decode its distortions. For Bourdieu, the researcher is involved in a special kind of reflexive labor that produces objective knowledge. Foucault first positions the researcher on the axis of language, then that of power/knowledge, and finally proposes that a researcher should be engaged in an *ethical* activity.

COMING TO KNOW WHAT WE HAVE
DONE AND WHO WE ARE

In this final chapter, I will draw together the themes of this book by suggesting how the reforged tools of qualitative research can be used to study the three aspects of constitution in a program like the one Foucault defined. The result – to return to the question of the first chapter: What is science? – would be a social science with political and ethical relevance as well as a new understanding of truth. By posing and answering questions about constitution, we can discover new possibilities for the ways we live and for who we are.

When the way we live in the world becomes a matter of concern, a topic for reflection and debate, and a focus of anxiety, we have what Foucault called a new "problematization." This book grew out of a concern that qualitative research could be more ambitious – that it could and should ask more pressing and grander questions. At the start, I raised the possibility that qualitative research could make a contribution to our planetary problems through an examination of what I called, following Wolin and Bernstein, the "moral paradigms" in which we live. I wanted to explore how qualitative research could contribute to changing an attitude toward the earth and its peoples that is leading precipitously to environmental damage and to war among civilizations.

In a very short period of time, say 10,000 years, humans have created complicated and sophisticated forms of life, with differential and hierarchical divisions of labor, elaborate institutions of governance, and complex systems for the production and exchange of both necessities and luxuries. At the same time we have transformed ourselves using formal institutions of education and training alongside informal practices of socialization and enculturation. We have done all this without explicit knowledge of *how* we do it. For example, we don't have an adequate understanding of how schools function, even though we have been using them for 2,000 years (Packer, 2001a, 2001b; Packer & Greco-Brooks, 1999). The recent global economic crisis makes it clear that we don't fully comprehend how the dominant economic system operates.

Each day on this planet, 6.7 billion people go about their daily labors and half a million more babies are born. Each of us is caught up in forms of life that have evolved – or descended – from origins we didn't witness. None of us, individually, is in a position to grasp completely how they operate, let alone see how to change them. We are playing a game – a set of interlocking games – that we don't understand. We are playing with and against people we don't know, most of whom we will never meet, on a field of play that we can see only dimly and incompletely. We may have a vague sense of the inequities, injustices, contradictions, and conflicts of these games, or perhaps only that play is darned tough, but not enough knowledge to be able to change them. And these games have made us who we are.

We have a nagging sense that things need to change, that these forms of life could be improved. What can we do? We can ask questions. We can try to figure out what we are all doing. We can conduct research. The texts we have looked at in this book provide helpful guidance. Their words offer new ways of seeing, and thinking about, our life on this planet. They open up fresh ways of studying forms of human life. They show that how we live today is not inevitable, not the only way. Alternatives are possible, if we start to look for them.

Qualitative forms of research can examine a level of phenomena that is inaccessible to traditional research. The ontological assumptions of traditional research – preconceptions about the kinds of entities and processes that exist – make this level invisible. Dualism is the key diagnostic feature of these traditional preconceptions. A moral paradigm – let us now call it a "form of life" – is composed neither of subjective beliefs nor of objective objects. It is a shared way of living in the world, material practices in which we live out and transform our biological heritage and in which both objects and subjects are *constituted*. For example, the "individual" of modern democratic cultures is both a product of and a producer of practices that use our planet as a source of raw materials and people as a source of labor and a market of consumers, making and purchasing commodities. We have learned that we must pay attention to three axes of these practical activities: their games of truth and knowledge, their relations of power, and their care and formation of selves.

Concrete and specific studies of constitution are needed because we have gotten things *wrong*, as is clear from the large-scale damage to the planet on which we live and depend, international conflict, and economic instability. The 1970s were a time of existential crisis, a growing sense of lack of meaning. The crises today are more profound, more troubling. We need to change who we are and how we live, and we are the only ones who can make the changes. Research is necessary to help us understand what we have done, how we have done it, and how we might change what we have done. Such research will study embodied, concrete people, in specific material circumstances, not disembodied, abstract subjects and subjectivities.

Stand for a while on a street corner and watch the embodied character of human existence: people gesturing, touching, grasping, inspecting, exploring; people involved in buying, selling, cooking, entertaining, flirting, dancing . . . living. They are clothed, ornamented – equipped with tools, implements, gadgets, and gear – dealing with all kinds of apparatus. They negotiate the spaces of houses, offices, stores, the street, and the pavement. Their bodies have been shaped and formed by a lifetime of such activity. Each one stands, moves, walks, in a distinct way, using specific "techniques of the body" (Mauss, 1935/1973). This bodily comportment, their embodied know-how, is neglected by any study that tries to isolate "experience" with an interview or a questionnaire. To study "concepts" or "subjectivity" is to forget about all of this.

USING THE TOOLS

We can now see how the three central tools of qualitative research can be used to answer questions about constitution in a program of research along the lines of Foucault's "historical ontology of ourselves." Investigation will move from the *truth game* of a form of life, the order of its regional ontology, explored through fieldwork, to the *relations of power* in everyday interaction, fixed and studied in detail, and finally to the *care of the self* in these forms of life, the *ontological complicity* of the participants, evident in interviews, in order to describe how humans become particular kinds of persons by living in a particular form of life. These define the archaeological, genealogical, and ethical phases of investigation (Table 15.1).

Fieldwork

First, fieldwork is an important tool to study the order of a form of life. Fieldwork must be seen as a contact between distinct forms of life, with the fieldworker always a stranger, a newcomer who is complicit (in the sense of collaborating and colluding) (Marcus), negotiating not entry but access (Harrington). They will need to trace the multisited linkages that define the form of life, considered not as bounded, systematic, and integrated but as open, dynamic, and contested (Marcus, Faubian), as a game – contested and dynamic. Researchers write accounts of what they witness to describe and organize, partially and provisionally (Garfinkel). These accounts recount the regional ontology of the form of life: the ways things and people show up. But the knowledge obtained in fieldwork is about entities that appear as a result of the interactions of *two* forms of life, the one studied and that of the researcher, so the researcher must be reflexive, turning around on the form of life they come from to interrogate it, too. Their accounts should be subject to local accountability, perhaps made available to the people they work with, perhaps written with these people, the message a negotiated one. But these accounts should be written for their accountability elsewhere, too, attempts to make visible matters that have been ignored, misrepresented, or covered up. Whether it is Marcus emphasizing that researchers are strangers and representatives of somewhere else, Merleau-Ponty recommending that we slacken the threads that attach us to the world, Habermas arguing for a negotiated impartiality, or Foucault noting that in thinking we establish a distance from the problematized object, it is clear that the researcher is a witness (Bazin) rather than a member. The researcher's analysis will be informed by a grasp of the form of life as a whole, which is not available to the members themselves.

From the start of this book I have emphasized the importance of ontology. We learned from Kuhn that every science has ontological commitments embedded in its practices. These are more than specific *theories* of the world; they *constitute* a specific world. Outside of science, too, people live in what we initially called "moral paradigms," which embody ontological commitments. When we conduct fieldwork we must reflect on the ontology embedded in our practices and the practices of the form of life we study. Such ontologies are not arbitrary; they are tested and proven daily in continuous activity.

TABLE 15.1. *Phases of Investigation in a Historical Ontology*

Phase of Investigation	Research Tool	What Is Studied	What Is Described	The Focus
Archaeological	Fieldwork Contact between distinct forms of life. Issues of access and accountability.	The regional ontology of a form of life, grasped as a whole.	The entities that are recognized; forms of knowledge (*connaissance*) about them.	The order of a game of truth.
Genealogical	Interaction analysis Detailed study of action import. Issues of fixing and interpretation.	Tracing the ontological work accomplished in the everyday practices of a form of life.	How practices constitute the order of a form of life; the know-how (*savoir*) of these practices.	The ordering in relations of power over time.
Ethical	Interviews Dialogue that offers a way of seeing. Issues of *poesis*.	The ontological complicity in which people are bound to their form of life.	How the participants are formed in material practices.	How the orderers are constituted through care of the self.

We have come to see that "on the one hand, man makes his history, on the other he is made by it; he constitutes social reality and is constituted by it. Reflecting on this paradoxical relation is the task of ontology" (Ricoeur, 1979, p. 215).

Foucault's program is built on an ontology of multiple realities and epistemologies, in which subjects and objects are formed in the practical activities of specific forms of life along with ways of knowing. It has been said that the difference "between traditional and critical theory is fundamentally an ontological one: it lies between reflexively disclosed practical performances and empirically known things" (Menke, 1996, p. 61). Traditional inquiry assumes that its task is to describe the relationships among existing independent objects and events that can be known empirically. Critical inquiry, in contrast, presumes an ontology of practical activity that can be known reflexively. This ontological difference has methodological consequences: it is the difference between a method that is directed toward knowing things, toward grasping *being*, what *is*, and one that is directed toward knowing processes, toward fostering *becoming*, what *can be*. "If praxis becomes the object of theory, the theory becomes praxis" (p. 61). This is an ontology of bodies disciplined in social and discursive practices, of power and folded power. It is a radical realism of human action in worlds constituted in embodied and social activity, in which agonism grounds knowledge claims: claims to know objects in the world that are at the same time instructions in how to *see* these objects.

Foucault offered a fresh ontology of human being, built on the work of Heidegger and Merleau-Ponty: "Foucault sought to develop an account of humans as beings-in-the-world situated within an existing web of relations occurring within a context of background practices, all the while possessing an ontological freedom that is not molded by power relations but is instead the condition of possibility of power itself" (Gordeon, 1999, p. 395). We need to attend to power, understood as positive and "constitutive," Foucault insisted. Power is *pouvoir*, meaning "to be able." As Foucault wrote: "In fact, power produces; it produces reality; it produces domains of objects and rituals of truth. The individual and the knowledge that may be gained of him belong to this production."

INTERACTION ANALYSIS

If fieldwork is the first reseach tool, the second is detailed analysis of interaction to articulate the ontological work that produces and reproduces the order of a form of life. An archaeologist reconstructs the order of a form of life on the basis of artifacts, whether pots or statements, but living is not a matter of pots alone. A genealogical step is needed to trace the *practices* in which pots are used. Fieldwork itself moves from the *order* to the *ordering* as the researcher starts to recognize the everyday practices in which this work is done. Now *techniques of objectification* (Bourdieu) are used to fix everyday practices to investigate them closely, suitably transformed (Ricoeur). Detailed descriptions of how interaction is negotiated have explanatory power; they are interpretations that are applications (Gadamer). They are reconstructions (Habermas) of the tacit methods (Garfinkel, Schegloff) used by participants. The concept that unified Foucault's work was surely that of *discourse*, and it might seem that some kind of discourse analysis could help us here. But we saw in Chapter 10 that critical discourse analysis, though its aim is to analyze constitution, typically focuses on representation. We need instead to focus on power, visible as action on others' actions (Foucault), and this is precisely the "action import" of utterances that ethnomethodology and conversation analysis attend to (Garfinkel, Schegloff). The kinds of discourse analysis we looked at in Chapter 10 "are unable to provide adequate means through which to involve the analysis of these material and extra-textual practised forms of powers within their methodology" (Hook, 2001, p. 530). They reduce power to something negative.

The action import of an utterance is the way it influences another person (or oneself). Conversation analysis looks at conversation as "action on another's actions." Each utterance is a response to prior utterances and an action on future ones. This is where the researcher obtains evidence of the way moves in a form of life are understood by the members.

Ethnomethodology and conversation analysis explore how social action accomplishes order. Each of them "calls upon the potential for reflexivity that is intrinsic to such action in order to turn our attention from what we do to how we do it" (Langsdorf, 1995, p. 186). "How we do it" is a matter of power; articulating "how we do it" amounts to describing ontological work.

Interviews

I have been critical of the way interviews are often used as though they were a window on subjectivity, but they are a useful tool to explore how a participant, positioned in a form of life, is ontologically complicit with it – in the sense of being folded together with it, allied and bound. This aspect of constitution will begin to become visible in the detailed analysis of interaction as the techniques for forming selves become apparent, and interviews pick up from there. An interview is a dialogue between members of distinct forms of life (Habermas), objectified (Ricoeur, Bourdieu), and we saw in Chapter 5 that its analysis should be an articulation of the *poesis* (White) of a way of seeing the world that follows from a way of being in the world (Heidegger, Bourdieu, Foucault). The constitution of the participants in a form of life becomes evident as we articulate their perspectives, which correspond to the different positions they occupy, and hence the different interests they have, in its order.

What I have sketched here is a program of research to study the three aspects of constitution: the games of truth in which the *order* of a form of life is known; the relations of power in which subjects and objects are *ordered*; and the care of the self in which the participants, the *orderers*, are also constituted. It is investigation not of a hidden, underlying structure but of the rich, teeming surface of life. The first phase studies how this surface is ordered, the second phase studies the practices of ordering, and the third phase focuses on how humans become orderers of knowledge and action.

In this program of research techniques such as measurement and quantification have their place, though not at the center. Our studies will offer *accounts* of forms of life, accounts that will not be disinterested descriptions but concerned *instructions to see* everyday practices in a new way. These accounts will have *truth* not to the extent that they correspond to how things "really are" but insofar as they point out things we haven't noticed because of the *illusio* (the false consciousness, the alienation) that our practices induce. As such, these accounts will have the potential for *phronesis*: practical relevance that is political in the original sense of guiding our activity.

What is necessary to conduct this kind of inquiry? It is clear that one needs practical know-how, not just formal rules or procedures. Ironically, "how to" books on qualitative research generally offer the latter. It is interesting that in the final volume of *The History of Sexuality* Foucault investigated techniques for the care of the self through the study of ancient "how to manuals," in particular a "handbook-for-daily-living" (Foucault, 1984/1988, p. 6), a "treatise on *how to*" (p. 6, emphasis original). He proceeded by "interpreting the interpretations" (p. 10), looking for "the ethics underlying his analysis" (p. 10). What he found was a *techne* that "constituted the permanent framework of everyday life, making it possible to know at every moment what was to be done and how to do it" (p. 101), a *techne* that implied a particular "perception of the world" (p. 101). *Technique* is a term that Foucault used frequently. He meant not a specialized or esoteric procedure but the practice of a craft, skilled know-how that is directed at producing a final product. Technique has an ethical character, as Aristotle recognized: "Every craft and every inquiry, and similarly every action and project, seems to aim at some good; hence the good has been well defined as that at which everything aims" (Aristotle, 1980, p. 1; translation modified).

BUT IS THIS SCIENCE?

The study of constitution requires inquiry that combines theory, practice, and ethics. But this is a very strange suggestion! Surely this is not what we mean by science? We have come to take for granted Kant's separation of theory, practice, and judgment. Science has come to signify the search for theoretical knowledge without any intrinsic practical relevance, free from aesthetic and ethical judgments. But Habermas, Bourdieu, and Foucault all tried to imagine a kind of inquiry in which these three dimensions were again unified, in which the search for knowledge is not separated from practical relevance or ethical judgment, in which theory has an emancipatory interest. They tried to avoid Kant's separation both in their conception of the object of inquiry and in their conception of inquiry itself.

Is such an inquiry really science? We have certainly arrived at a very different conception of qualitative inquiry than the one we began with, though it is not so different from the account of science that Kuhn offered. Far from being "objective about subjectivity," we have arrived at various reasons for thinking that a researcher should be *involved*; indeed, that there is no way *not* to be involved. Our historicity – our place and participation in human history – makes it possible for us to understand people and events. This understanding – incomplete, partial, practical, and affective though it is – provides the grounding for our more systematic scientific interpretations.

In Chapter 1 we examined Shavelson's assertion that the question, not the method, should drive scientific inquiry. We saw that Shavelson has a very narrow vision of the kinds of questions that social science can ask: (1) What's happening? (2) Is there a causal effect? (3) What is the causal mechanism? This narrow view of social science divides description from explanation, assumes that explanation is solely causal, and excludes all reference to political or ethical values. But I began this book with the proposal that we can, and need to, ask more significant questions. To ask important questions, we need to consider not only our techniques but also our ontologies and, furthermore, our ethics. Over the course of the book I have recommended the general question, "How are we constituted?" Shavelson's committee is surely correct that questions should drive our research designs and choice of techniques. Our studies should indeed be linked to theory, involve direct investigation, have relevance beyond a single case, and we should reason explicitly and coherently, disclosing our data and methods (as Shavelson recommends).

It is easy to assume that scientific research simply accumulates knowledge, and a narrow kind of knowledge at that. But it is possible to imagine that scientific research could lead not only to know-that (*connaitre*) – but also to know-how (*savoir, phronesis*) with political and ethical relevance. Knowledge, ethics, and aesthetics have not always been separated. Life, labor, and language continue to interweave. Kuhn showed us the limitations of the view that science could be a truth game without involving relations of power and requiring the formation of the people who participate in these relations. Now we can say a little more about the kind of social science we can envision.

Science and Truth: The Power of Language

Throughout this book, we have circled around language, never quite able to grasp it. We rejected the conduit metaphor. We examined Garfinkel's development of Wittgenstein's image of multiple language games. We saw how hermeneutics links language and life and how texts do ontological work, inviting the reader to see new worlds. We returned to language again with Habermas's focus on distorted communication, Bourdieu's concept of symbolic violence, and finally with Foucault's attention to discourse. Language has been central, but it has remained mysterious. Merleau-Ponty wrote eloquently of the obscure power of language: "The power possessed by language of bringing the thing expressed into existence, of opening up to thought new ways, new dimensions and new landscapes is, in the last analysis, as obscure for the adult as for the child" (Merleau-Ponty, 1945/1962, p. 401). He pointed out how both Kant and Descartes "begin their meditation in what is already a universe of discourse" (p. 401). At the same time, our certainty that language "is merely the garment and contingent manifestation" for something separate from it "has been implanted in us precisely by language":

> The wonderful thing about language is that it promotes its own oblivion: my eyes follow the lines on the paper, and from the moment I am caught up in their meaning, I lose sight of them. The papers, the letters on it, my eyes and body are there only as the minimum setting of some invisible operation. Expression fades out before what is expressed, and this is why its mediating role may pass unnoticed. (Merleau-Ponty, 1945/1962, p. 401)

This power of language may still be obscure, but it is crucial to the practice of research. We have seen how misleading is the notion that language is transparent. Speech is not a conduit for subjective experiences. The meaning of a word is not a concept that specifies the features belonging to the object the word labels. What is important is the way a word is used in the context of a particular situation. It is the way a thing named shows itself to me, in "a meeting of the human and the non-human" (p. 403). What is general is not a concept but the typicality of the style of the world's behavior.

Garfinkel, too, has been centrally concerned with the link between saying and seeing:

> ... that peculiar way of looking that a member has. The peculiar way of searching, of scanning, of sensing, of seeing finally but not only seeing, but seeing-reporting. It is "observable-reportable." It is available to observation and report. Now I need to run them together. If there was one word in the English language that would run them together, I would use it. There is not, so I have to use the term "accounting" or "accountable" or "account." (Garfinkel, 1974, p. 17)

This power of language is central to qualitative inquiry, though it often goes unrecognized. It is central to interviewing, and our analysis of what an interviewee can show us. It is central to fieldwork, as writing fieldnotes informs and transforms how an ethnographer sees the form of life they are investigating. And it is central to the academy, as the reports we write and the manuscripts we publish inform others about our insights and offer them new ways to see the world we all live in.

Our interest is not simply in discourse but in the dynamic and subtle relationships between speaking and seeing, in the struggle and tension between what we say and what we see. Knowledge *is* this link between the articulable and the visible. We do not simply describe what is visible, putting into words what we see in front of us: "It is not easy to say something new; it is not enough for us to open our eyes, to pay attention, to be aware, for new objects suddenly to light up and emerge out of the ground" (Foucault, 1969/ 1972, pp. 44–45).

Equally, we never simply see what we are told to see. We find ways of viewing the world through ways of writing and talking about it. Discourse analysts study language, but they have neglected the seeing (and thinking and feeling) that language invites. Qualitative researchers employ and deploy language – in their data and in their written and spoken reports – often without acknowledging that to do so is to suggest a way of seeing.

What a scientific investigation seeks is not an objective description of an (underlying) structure but a *new way of seeing* that goes hand in hand with a new way of *saying*. Goodwin's studies of professional vision provide a good illustration of the work that seeing requires. Goodwin (1994) examined the practices of coding, highlighting, and producing material representations in two very different disciplines, archaeological fieldwork and legal argumentation. As he pointed out, "An archaeologist and a farmer see quite different phenomena in the same patch of dirt" (Goodwin, 1994, p. 606). He concluded that "by applying such practices to phenomena in the domain of scrutiny, participants build and contest *professional vision*, which consists of socially organized ways of seeing and understanding events that are answerable to the distinctive interests of a particular social group" (p. 606). Goodwin (1995) cites Ophir and Shapin (1991): science is about forcing the invisible to manifest itelf, and "the invisible appears only to the eyes of those authorized to observe it." He described the "heterotopic" spaces and tools that render a phenomenon of interest visible to those with the authority (of training, funding, etc.) to enter and use them. Goodwin concluded that an object of knowledge emerges when a set of discursive practices is employed in a domain of scrutiny. Practices of recording and transcribing play this role for Goodwin himself as a professional anthropologist. Talk, action, and setting are aspects of a single coherent activity, and learning how to see occurs in structured interactions between expert and learner.

Truth is a matter of pointing out – by offering an account – something unnoticed but relevant to ongoing activity. Truth depends, for the conditions of its possibility, on prior commitment to a paradigm, to a field of thinking, to a language, to a way of saying and seeing, to a way of being in the world, bodily, with others. A "true account" is one that provides a way of seeing what is relevant to our current situation, our problem. From Hegel through Heidegger and Merleau-Ponty to Garfinkel and Foucault, we have traced a new conceptualization of truth that doesn't deny that truths can be found or achieved but does insist that truth can never be absolute and universal. Truths operate within, and depend on, an involvement in a "field of thought," and are evident only to a particular "thinking nature." This means that certainty is never complete, never final. *Doxa* is not provisional and temporary, waiting to be replaced by *episteme*, but original and fundamental:

> Once launched, and committed to a certain set of thoughts, Euclidian space, for example, or the conditions governing the existence of a certain society, I discover evident truths; but these are not unchallengeable, since perhaps this space or this society are not the only ones possible. It is therefore of the essence of certainty to be established only with reservations; there is an *opinion* which is not a provisional form of knowledge destined to give way later to an absolute form, but on the contrary, both the oldest or most rudimentary, and the most conscious or mature form of knowledge – an opinion which is primary in the double sense of "original" and "fundamental." (Merleau-Ponty, 1945/1962, p. 396)

Any truth will be self-evident "because I take for granted a certain acquisition of experience, a certain field of thought, and precisely for this reason it appears to me self-evident for a certain thinking nature, the one which I enjoy and perpetuate, but which remains contingent and given to itself" (p. 396). In particular circumstances, we can say that "we possess a truth, but this experience of truth would be absolute knowledge only if we could thematize every motive, that is, if we ceased to be in a situation" (p. 395).

Isn't this relativism? Is there in fact no solution to the problems Kant struggled to solve? On the contrary, this is a *pluralism* of truth, a *relativity* (Latour, 1988) of truth rather than relativism. A true account is one that has relevance to practical and political problems, one that embodies *phronesis*.

Science and Politics: Informing *Phronesis*

If knowledge is a relationship between saying and seeing, it is also a relationship of forces, the exercise of power in material practices. The knowledge that scientific researchers produce is made possible by the forms of life in which they work, forms of life that integrate particular relations of power.

Social science is not just the production of knowledge but also action on people. Giddens showed us that "every generalization or form of study that is concerned with an existing society constitutes *a potential intervention within that society*" (Giddens, 1979, pp. 244–245, emphasis original). Science is an activity with its own politics, and it is also an activity with political relevance. I am referring here to politics in the original sense of "the process observed in all human (and many non-human) group interactions by which groups make decisions, including activism on behalf of specific issues or causes" (Wikipedia). We saw in the introduction that Aristotle wrote in his *Politics* that man is by nature a political animal, so that ethics and politics are closely linked and a truly ethical life can only be lived by someone who participates in politics.

Qualitative research has tended to reproduce the power strategies of what Foucault called subjectivation, employing strategies to obtain knowledge about people in order to manage them more efficiently. As teachers and researchers, we find ourselves regulated and managed by the institutions that we serve. I am not proposing that we try to work outside these institutions, but we need to think of them also as forms of life in which official knowledge is produced, in which power relations persist, and in which techniques to constitute people are central. As we participate in the truth games, the political power/knowledge, and the ethics of the academy and the research center, we need to be on the lookout for necessary changes.

It is crucial to remember that accounts are *constitutive* of their settings. Scientific accounts act back on the settings, and the people, they describe. When we describe something, our description has relevance to its circumstances – relevance that is practical and political. "Matters of fact and fancy and evidence and good demonstration about the affairs of everyday activities are made a matter for seeing and saying, observing for observation and report. That means then that talk is part of this. Talk is 'a constituent feature of the same setting that it is used to talk about'" (Garfinkel, 1974, p. 17).

As Giddens noted, "Laws in the social sciences are *historical* in character and in principle *mutable* in form" (Giddens, 1979, p. 243) because formulating such laws can change the people they apply to. "Once known – by those to whose conduct they relate – laws may become applied as rules and resources in the duality of structure: the very double meaning (and origin) of 'laws' as both precept of action and generalization about action draws our attention to this" (p. 244). This "reflexive appropriation" of social scientists' accounts by those studied is not a problem but a central feature of inquiry.

If we have an emancipatory interest, this is exactly what we want. At the same time, "we have to avoid the error self-confessedly made by Habermas in *Knowledge and Human Interests*: knowledge acquired in the process of 'self-reflection' is not sufficient condition for social transformation" (Giddens, 1982, p. 15).

Science and Ethics: Asking Ethical Questions

The kind of science we have been building toward in this book has an ethical dimension. It is critical, emancipatory, searching for enlightenment through its focus on constitution. Foucault's approach to critical inquiry was not to take sides in a specific domain – advocating different treatment of criminals, for example. His interest was not in defining radical aims, or proposing political strategies, but in exploring the constitution of various domains. By revealing the contingency and fragility of knowledge, his work made visible the possibility of transformation. The very notion of constitution has a liberatory potential, not only because it has political relevance (though it doesn't advocate a partisan political position) but also because it encourages us to understand how we have become who we are. When we see the fault lines in the way we live, when we map the contingent pathways that we took and those we didn't, and when we recognize how we came to be where we are now, we can see how it could have been otherwise and how it could still be otherwise:

> It is one of my targets to show people that a lot of things that are part of their landscape – that people think are universal – are the result of some very precise historical changes. All my analyses are against the idea of universal necessities in human existence. They show the arbitrariness of institutions and show which space of freedom we can still enjoy and how many changes can still be made. (Foucault in Martin, 1982/1988, p. 11)

A NEW PROBLEMATIC

Anyone who sets out to conduct a research project needs to define the question they are trying to answer. Answering this question ought to make a contribution to the discipline within which the researcher is working. That is to say, it should relate to the problematic of the discipline. We have seen that Foucault viewed the problematics of the contemporary social sciences as having a common basis in a shared episteme, for which Kant was the original and is still the principal spokesperson. If we are now finally escaping from Kant's definition of the problem of knowledge, are we entering a new episteme, with a new problematic? It seems to me that the problematic the Western age currently faces is an *ethical* one. We have largely solved the problem of how to gain technical control of our environment, even our own bodies. We have also learned much about how to influence subjectivity, with advertising and education. Of course, there is much more we can discover, but we have institutions in place that will continue the project of technical mastery. The problem we are faced with now is to what end we should employ this technological wizardry. This amounts to saying that we need to find a new way of relating to ourselves. Along with this will be a new way of relating to others and a new way of relating to the world, with new forms of knowledge.

Qualitative research can contribute to this ethical problematic in precisely the ways Foucault envisioned: investigating alternative modes of relation, for example in indigenous and subjugated practices. As we throw light on the constitution of our taken-for-granted reality, we will see how unnatural and strange it actually is:

We need to anthropologize the West: show how exotic its constitution of reality has been; emphasize those domains most taken for granted as universal (this includes epistemology and economics); make them seem as historically peculiar as possible; show how their claims to truth are linked to social practices and have hence become effective forces in the social world. (Rabinow, 1986, p. 241)

This is research not merely as a search for knowledge but as a way of living and even a way of being. It recognizes that researchers work within specific games of truth and relations of power, that good research inevitably has political and ethical axes. It demands that we not assume that the individual – with personal beliefs and desires – is the only form of subject, or subjectivity, for individuals can exist only within the discursive formations of power/knowledge in which they have been formed, and form themselves, using specific techniques of subjectivation. The objects of our investigation are these multiple formations and techniques, and the subjects and subjectivities, objects and objectivities, formed by them.

There has not been room in this book to provide much detail on *how* to conduct this kind of investigation. As I said at the outset, this is a book on *why* to do qualitative research, and I have concentrated on conceptual, theoretical, and ethical issues that have become either forgotten or ignored. Clearly another volume is needed to provide the *techne*. But I hope that, having examined these aspects, the reader will be clearer about what we are aiming for when we conduct the kind of research we call qualitative. Now that we have decided what we are aiming for, we can figure out in detail what to do and how to do it.

REFERENCES

Adler, P. A., & Adler, P. (1987). *Membership roles in field research*. Newbury Park, CA: Sage.

Adorno, T. W. (1950). *The authoritarian personality*. New York: Harper.

Adorno, T. W. (1964/1973). *The jargon of authenticity*. London: Routledge and Kegan Paul.

Allen, A. (2003). Foucault and Enlightenment: A critical reappraisal. *Constellations, 10*(2), 180–198.

Allison, H. E. (1983). *Kant's transcendental idealism: An interpretation and defense*. New Haven, CT: Yale University Press.

Althusser, L. (1970/1971). Ideology and ideological state apparatuses. In *Lenin and philosophy and other essays* (pp. 127–186). New York: Monthly Review Press.

Anderson, G. J., & Herr, K. (1999). The new paradigm wars: Is there room for rigorous practitioner knowledge in schools and universities? *Educational Researcher, 28*(5), 12–21.

Aristotle. (1962). *On interpretation*. Milwaukee, WI: Marquette University Press.

Aristotle. (1967). *Poetics* (G. F. Else, Trans.). Ann Arbor: University of Michigan Press.

Aristotle. (1980). *The Nicomachean ethics* (D. Ross, Trans.). Oxford: Oxford University Press.

Aristotle. (1988). *Metaphysics* (J. Annas, Trans.). Oxford: Oxford University Press.

Aristotle. (1995). *Politics* (T. J. Saunder, Trans.). Oxford: Clarendon Press.

Arnheim, R. (1967). *Toward a psychology of art*. Berkeley: University of California Press.

Atkinson, P. (1990). *The ethnographic imagination: Textual construction of reality*. London: Routledge.

Atkinson, P. (2005). Qualitative research: Unity and identity. *Forum: Qualitative Social Research, 6*(3).

Atkinson, P. (2006). Rescuing autoethnography. *Journal of Contemporary Ethnography, 35*(4), 400–404.

Atkinson, P., & Hammersley, M. (1994). Ethnography and participant observation. In N. K. Denzin & Y. S. Lincoln (Eds.), *Handbook of qualitative research* (pp. 248–261). Thousand Oaks, CA: Sage.

Atkinson, P., & Silverman, D. (1997). Kundera's *Immortality*: The interview society and the invention of self. *Qualitative Inquiry, 3*, 304–325.

Atkinson, P., Coffey, A., Delamont, S., Lofland, J., & Lofland, L. (Eds.). (2001). *Handbook of ethnography*. London: Sage.

Auerbach, C. F., & Silverstein, L. B. (2003). *Qualitative data: An introduction to coding and analysis*. New York: New York University Press.

Augustine. (397/2002). *The confessions of St. Augustine* (A. C. Outler, Trans.). Mineola, NY: Dover.

Austin, J. L. (1975). *How to do things with words* (2nd ed.). Cambridge, MA: Harvard University Press.

Ayer, A. J. (1980). *Hume*. New York: Hill and Wang.

Badillo, R. P. (1991). *The emancipatory theory of Jurgen Habermas and metaphysics*. Washington, DC: The Council for Research in Values and Philosophy.

Bakhtin, M. (1994). *Speech genres and other essays*, quoted in P. Willeman, *Looks and frictions: Essays in cultural studies and film theory* (p. 199). Bloomington: Indiana University Press.

Bakhurst, D. (1991). *Consciousness and revolution in Soviet philosophy: From the Bolsheviks to Evald Ilyenkov*. Cambridge: Cambridge University Press.

Barthes, R. (1973/1975). *The pleasure of the text* (R. Miller, Trans.). New York: Harper Collins.

Bateson, G., Jackson, D. D., Haley, J., & Weakland, J. (1956). Toward a theory of schizophrenia. *Behavioral Science, 1*, 251.

Bauman, Z. (1981). *Hermeneutics and social science*. New York: Columbia University Press.

Bazin, J. (2003). Questions of meaning. *Anthropological Theory, 3*(4), 418–434.

Becker, C. L. (1932/1961). *The heavenly city of the eighteenth-century philosophers*. New Haven, CT: Yale University Press.

Behar, R. (1995). Introduction: Out of exile. In R. Behar & D. A. Gordon (Eds.), *Women writing culture* (pp. 1–29). Berkeley: University of California Press.

Behar, R., & Gordon, D. (Eds.). (1995). *Women writing culture*. Berkeley: University of California Press.

Bennett, J. W. (1998). Classic anthropology. *American Anthropologist, 100*(4), 951–956.

Benzecri, J. P. (1992). *Correspondence analysis handbook*. New York: Marcel Dekker.

Berger, P. L., & Luckmann, T. (1966). *The social construction of reality: A treatise in the sociology of knowledge*. Garden City, NY: Anchor Press.

Bernstein, R. J. (1971). *Praxis and action: Contemporary philosophies of human action*. Philadelphia: University of Pennsylvania Press.

Bernstein, R. J. (1976). *The restructuring of social and political theory*. Philadelphia: University of Pennsylvania Press.

Bernstein, R. J. (1983). *Beyond objectivism and relativism: Science, hermeneutics, and praxis*. Philadelphia: University of Pennsylvania Press.

Billig, M. (1999a). Whose terms? Whose ordinariness? Rhetoric and ideology in conversation analysis. *Discourse and Society, 10*(4), 543–558.

Billig, M. (1999b). Conversation analysis and the claims of naivety. *Discourse and Society, 10*(4), 572–576.

Bittner, E. (1973). Objectivity and realism in sociology. In G. Psathas (Ed.), *Phenomenological sociology* (pp. 109–125). New York: Wiley.

Blauner, R. (1964). *Alienation and freedom: The factory worker and his industry*. Chicago: University of Chicago Press.

Blaxter, L., Hughes, C., & Tight, M. (2001). *How to research* (2nd ed.). Buckingham: Open University Press.

Bogdan, R. (1983). Teaching fieldwork to educational researchers. *Anthropology and Education Quarterly, 14*(3), 171–178.

Bogdan, R. C., & Biklen, S. K. (1992). *Qualitative research for education: An introduction to theory and methods* (2nd ed.). Boston: Allyn and Bacon.

Bologh, R. W. (1979). *Dialectical phenomenology: Marx's method.* Boston: Routledge and Kegan Paul.

Borgatti, S. (n.d.). Introduction to grounded theory (http://www.analytictech.com/mb870/introtoGT.htm). Last accessed October 2009.

Bourdieu, P. (1972/1977). *Outline of a theory of practice* (R. Nice, Trans.). Cambridge: Cambridge University Press.

Bourdieu, P. (1979/1984). *Distinction: A social critique of the judgement of taste* (R. Nice, Trans.). Boston: Harvard University Press.

Bourdieu, P. (1980/1991). *The logic of practice* (R. Nice, Trans.). Cambridge: Polity Press.

Bourdieu, P. (1981/1991). *Language and symbolic power.* Cambridge: Polity Press.

Bourdieu, P. (1984/1990). *Homo academicus* (P. Collier, Trans.). Stanford, CA: Stanford University Press.

Bourdieu, P. (1988/1991). *The political ontology of Martin Heidegger.* Stanford, CA: Stanford University Press.

Bourdieu, P. (1991). The peculiar history of scientific reason. *Sociological Forum, 6*(1), 3–26.

Bourdieu, P. (1992/1996). *The rules of art: Genesis and structure of the literary field.* Stanford, CA: Stanford University Press.

Bourdieu, P. (1997/2000). *Pascalian meditations* (R. Nice, Trans.). Stanford, CA: Stanford University Press.

Bourdieu, P. (2003). Participant objectivation: The Huxley Medal Lecture. *Journal of the Royal Anthropological Institute, 9*(2), 281–294.

Bourdieu, P., & Passeron, J.-C. (1970/1990). *Reproduction in education, society and culture* (R. Nice, Trans.). London: Sage.

Bourdieu, P., & Wacquant, L. J. D. (1992). *An invitation to reflexive sociology.* Chicago: University of Chicago Press.

Bowen, E. S. [Bohannan, L.] (1964). *Return to laughter: An anthropological novel.* New York: Doubleday.

Boyatzis, R. E. (1998). *Transforming qualitative material: Thematic analysis and code development.* Thousand Oaks, CA: Sage.

Brenner, N. (1994). Foucault's new functionalism. *Theory and Society, 23*(5), 679–709.

Bridgman, P. W. (1945). Some general principles of operational analysis. *Psychological Review, 52*, 246–249, 281–284.

Bridgman, P. W. (1959). *The way things are.* Cambridge, MA: Harvard University Press.

Briggs, J. (1970). *Never in anger: Portrait of an Eskimo family.* Cambridge, MA: Harvard University Press.

Broady, D. (2002). French prosopography: Definition and suggested readings. *Poetics, 30* (5–6), 381–385.

Bruner, J. (1986). *Actual minds, possible worlds.* Cambridge, MA: Harvard University Press.

Bruner, J. (1987). Life as narrative. *Social Research, 54*, 11–32.

Bührmann, A. D., Diaz-Bone, R., Gutiérrez-Rodríguez, E., Schneider, W., Kendall, G., & Tirado, F. (2007). Editorial: From Michel Foucault's theory of discourse to empirical discourse research. *Forum: Qualitative Social Research, 8*(2).

Burgess, E. W. (1927). Statistics and case studies as methods of sociological research. *Sociology and Social Research, 12*, 103–120.

Burwood, S., Gilbert, P., & Lennon, K. (1999). *Philoso*
Queen's University Press.

Calhoun, C. (1993). Habitus, field of power, and cap*ita*
specificity. In C. Calhoun, E. LiPuma, & M. Posto*ne*
perspectives (pp. 61–88). Chicago: University of Chicag*o*

Calhoun, C. (2003). Pierre Bourdieu. In G. Ritzer (Ed.), *The*
major social theorists (pp. 696–730). Cambridge, MA: Blac*k*

Caputo, J. (1993). On not knowing who we are: Madness, her*me*
of truth in Foucault. In J. Caputo & M. Yount (Eds.), *Fouca*
institutions (pp. 233–262). University Park: Pennsylvania State

Caputo, J. D. (1987). *Radical hermeneutics: Repetition, deconstruc*
neutic project. Bloomington: Indiana University Press.

Carbone, M. (2004). *The thinking of the sensible: Merleau-Pont*
Evanston, IL: Northwestern University Press.

Carnap, R. (1937/2002). *The logical syntax of language*. Chicago: Open

Carr, W., & Kemmis, S. (1986). *Becoming critical: Education, knowledg*
research. London: Falmer Press.

Charmaz, K. (2000). Grounded theory: Objectivist and constructivist *me*
N. K. Denzin & Y. S. Lincoln (Eds.), *Handbook of qualitative research*
(pp. 509–535). Thousand Oaks, CA: Sage.

Chi, M. T. H. (1997). Quantifying qualitative analyses of verbal data: A practic*al*
The Journal of the Learning Sciences, 6(3), 271–315.

Chomsky, N. (1957). *Syntactic structures*. The Hague: Mouton.

Clifford, J. (1983). On ethnographic authority. *Representations*, 2, 118–146.

Clifford, J. (1986). Introduction: Partial truths. In J. Clifford & G. Marcus (*Eds.*
Writing culture: The poetry and politics of ethnography (pp. 1–26). Berkel*ey*
University of California Press.

Clifford, J. (1988). *The cultural predicament*. Cambridge, MA: Harvard University Press.

Clifford, J. (1990). Notes on (field) notes. In R. Sanjek (Ed.), *Fieldnotes: The making of
anthropology* (pp. 47–70). Ithaca, NY: Cornell University Press.

Clifford, J., & Marcus, G. E. (Eds.). (1986). *Writing culture: The poetics and politics of
ethnography*. Berkeley: University of California Press.

Collins, A. (1999). *Possible experience: Understanding Kant's 'Critique of Pure Reason.'*
Berkeley: University of California Press.

Collins, J. (1998). Language, subjectivity, and social dynamics in the writings of Pierre
Bourdieu. *American Literary History*, 10(4), 725–732.

Connelly, F. M., & Clandinin, D. J. (1990). Stories of experience and narrative inquiry.
Educational Researcher, 19(5), 2–14.

Corbin, J. & Strauss, A. (2008). *Basics of qualitative research: Techniques and procedures
for developing grounded theory* (3rd ed.). Thousand Oaks, CA: Sage.

Culler, J. (1976). *Ferdinand de Saussure*. Harmondsworth: Penguin.

Dallmayr, F. R., & McCarthy, T. A. (1977). *Understanding and social inquiry*. Notre
Dame, IN: University of Notre Dame Press.

Darwin, C. (1871). *The descent of man and selection in relation to sex*. London: John
Murray.

Davidson, A. L. (1986). Archaeology, genealogy, ethics. In D. C. Hoy (Ed.), *Foucault:
A critical reader* (pp. 221–233). Oxford: Blackwell.

Deegan, M. J. (2001). The Chicago School of ethnography. In P. Atkinson, A. Coffey,
S. Delamont, J. Lofland, & L. Lofland (Eds.), *Handbook of ethnography* (pp. 11–25).
London: Sage.

Deleuze, G. (1986/1988). *Foucault* (S. Hand, Trans.). Minneapolis: University of Minnesota Press.

Denzin, N. K. (1988). Review of Strauss, A. L., 'Qualitative Analysis for Social Scientists.' *Contemporary Sociology, 17*(3), 430–432.

Denzin, N. K. (1997). *Interpretive ethnography: Ethnographic practices for the 21st century*. Thousand Oaks, CA: Sage.

Descartes, R. (1637, 1641/2003). *Discourse on method* and *Meditations* (E. Haldane & G. R. T. Ross, Trans.). Mineola, NY: Dover.

Descombes, V. (2002). A confusion of tongues. *Anthropological Theory, 2*(4), 433–446.

Dillard, J. L. (1973). *Black English: Its history and usage in the United States*. New York: Vintage Books.

Dilthey, W. (1964/1990). The rise of hermeneutics. In G. L. Ormiston & A. D. Schrift (Eds.), *The hermeneutic tradition: From Ast to Ricoeur* (pp. 101–114). Albany: State University of New York Press.

Dilthey, W. (1914/1960). *Weltanschauung und Analyse des Menschen seit Renaissance und Reformation. Gesammelte Schriften Vol. II*. Stuttgart: B. G. Teubner.

Dingwall, R. (1988). Oration in awarding Harold Garfinkel an honorary doctorate from the University of Nottingham. (http://therapy.massey.ac.nz/175771/garfinkel/garfinkel.html).

Donmoyer, R. (1996). Educational research in an era of paradigm proliferation: What's a journal editor to do? *Educational Researcher, 25*(2), 19–25.

Dreyfus, H. L. (1984). Beyond hermeneutics: Interpretation in late Heidegger and recent Foucault. In G. Shapiro & A. Sica (Eds.), *Hermeneutics: Questions and prospects* (pp. 66–83). Amherst: University of Massachusetts Press.

Dreyfus, H. L. (1991). *Being-in-the-world: A commentary on Heidegger's Being and time, Division 1*. Cambridge, MA: MIT Press.

Dreyfus, H. L., & Hall, H. (Eds.). (1982). *Husserl, intentionality and cognitive science*. Cambridge, MA: MIT Press.

Dreyfus, H. L., & Rabinow, P. (1982). *Michel Foucault: Beyond structuralism and hermeneutics*. Chicago: University of Chicago Press.

Dryden, J. (1672/1968). *The conquest of Granada by the Spaniards*. In M. Summers, *Dryden: The dramatic works*. New York: Gordian Press.

Duranti, A., & Goodwin, C. (1992). *Rethinking context: Language as an interactive phenomenon*. Cambridge: Cambridge University Press.

Durkheim, E. (1895/1982). *The rules of sociological method* (W. D. Halls, Trans.). New York: Free Press.

Durkheim, E. (1912/1995). *The elementary forms of the religious life*. New York: Free Press.

Eagleton, T. (1983). *Literary theory: An introduction*. Minneapolis: University of Minnesota Press.

Eagleton, T. (1991). *Ideology: An introduction*. London: Verso.

Edgar, A. (2006). *Habermas: The key concepts*. London: Routledge.

Emerson, R. M., Fretz, R. I., & Shaw, L. L. (1995). *Writing ethnographic fieldnotes*. Chicago: University of Chicago Press.

Engels, F. (1845/2008). *The condition of the working class in England in 1844*. New York: Cosimo Classics.

Engels, F. (1926/1960). *Dialectics of nature*. New York: International Publishers.

Erickson, F. (2002). Culture and human development. *Human Development, 45*, 299–306.

Ericsson, K. A., & Simon, H. A. (1984). *Protocol analysis: Verbal reports as data*. Cambridge, MA: MIT Press.

Fabian, J. (1995). Ethnographic misunderstanding and the perils of context. *American Anthropologist, New Series*, 97(1), 41–50.

Fairclough, N. (1989). *Language and power*. London: Addison-Wesley Longman.

Fairclough, N. (1992). *Discourse and social change*. Cambridge: Polity Press.

Fairclough, N. L., & Wodak, R. (1997). Critical discourse analysis. In T. A. van Dijk (Ed.), *Discourse studies: A multidisciplinary introduction. Vol. 2. Discourse as social interaction* (pp. 258–284). London: Sage.

Faubion, J. D. (2001). Currents of cultural fieldwork. In P. Atkinson, A. Coffey, S. Delamont, J. Lofland & L. Lofland (Eds.), *Handbook of ethnography* (pp. 39–59). London: Sage.

Ferraris, M. (1988/1996). *History of hermeneutics* (L. Somigli, Trans.). Atlantic Highlands, NJ: Humanities Press.

Feuer, M. J., Towne, L., & Shavelson, R. J. (2002). Scientific culture and educational research. *Educational Researcher*, 31(8), 4–14.

Fontana, A. (2002). Postmodern trends in interviewing. In J. F. Gubrium & J. A. Holstein (Eds.), *Handbook of interview research: Context and method* (pp. 161–180). Thousand Oaks, CA: Sage.

Fontana, A., & Frey, J. H. (2000). The interview: From structured questions to negotiated text. In N. K. Denzin & Y. S. Lincoln (Eds.), *Handbook of qualitative research* (2nd ed.) (pp. 645–672). Thousand Oaks, CA: Sage.

Forbath, W. E. (1998). Habermas's constitution: A history, guide, and critique. *Law and Social Inquiry*, 23(4), 969–1016.

Foucault, M. (1961/1988). *Madness and civilization: A history of insanity in the Age of Reason* (R. Howard, Trans.). London: Vintage Books.

Foucault, M. (1963/1975). *The birth of the clinic: An archaeology of medical perception* (A. M. Sheridan Smith, Trans.). New York: Vintage/Random House.

Foucault, M. (1966/1973). *The order of things: An archaeology of the human sciences*. New York: Vintage Books.

Foucault, M. (1967/1999). Nietzsche, Freud, Marx. In J. Faubion (Ed.), *Aesthetics, method, and epistemology: Essential works of Foucault, 1954–1984, Volume 2* (pp. 269–278). New York: New Press.

Foucault, M. (1969/1972). *The archaeology of knowledge* (A. M. Sheridan Smith, Trans.). London: Tavistock.

Foucault, M. (1969/2006). Candidacy presentation: Collége de France, 1969. In P. Rabinow (Ed.), *Essential Works of Michel Foucault, 1954–1984, Volume 1. Ethics: Subjectivity and truth* (pp. 5–11). New York: Pantheon.

Foucault, M. (1971/1984). Nietzsche/genealogy/history. In P. Rabinow (Ed.), *The Foucault reader* (pp. 76–100). New York: Random House.

Foucault, M. (Ed.). (1973/1975). *I, Pierre Riviere, having slaughtered my mother, my sister, and my brother. . .: A case of parricide in the 19th century*. New York: Random House.

Foucault, M. (1974). Human nature: Justice versus power. In E. Fons (Ed.), *Reflexive water: The basic concerns of mankind* (pp. 135–197). London: Souvenir Press.

Foucault, M. (1975/1977). *Discipline and punish* (A. M. Sheridan Smith, Trans.). Cambridge: Cambridge University Press.

Foucault, M. (1976/1980). *The history of sexuality, Volume 1: An introduction*. New York: Vintage Books.

Foucault, M. (1980). *Power/Knowledge: Selected interviews & other writings 1972–1977* (C. Gordon, Ed.) New York: Pantheon.

Foucault, M. (1982). Afterword: The subject and power. In H. L. Dreyfus & P. Rabinow (Eds.), *Michel Foucault: Beyond structuralism and hermeneutics* (pp. 208–226). Chicago: University of Chicago Press.

Foucault, M. (1983/2006). On the genealogy of ethics: An overview of work in progress. In P. Rabinow (Ed.), *Essential works of Michel Foucault 1954–1984, Volume 1. Ethics: Subjectivity and truth* (pp. 253–280). New York: New Press.

Foucault, M. (1984a). The order of discourse. In M. Shapiro, (Ed.). *Language and politics*. Oxford: Oxford University Press.

Foucault, M. (1984b). What is Enlightenment? In P. Rabinow (Ed.), *The Foucault reader* (pp. 32–50). New York: Pantheon Books.

Foucault, M. [using the pseudonym Maurice Florence] (1984c). Foucault. In D. Huisman & J. F. Braunstein (Eds.), *Dictionnaire des philosophes* (pp. 942–944). Paris: Presses Universitaires de France.

Foucault, M. (1984d). On the genealogy of ethics: An overview of work in progress. In P. Rabinow (Ed.), *The Foucault reader* (pp. 340–372). New York: Pantheon Books.

Foucault, M. (1984/1986). *The history of sexuality, Volume 2: The use of pleasure* (R. Hurley, Trans.). New York: Vintage Books.

Foucault, M. (1984/1988). *The history of sexuality, Volume 3: The care of the self* (R. Hurley, Trans.). New York: Vintage Books.

Foucault, M. (1984/2006). Polemics, politics and problematizations [Interview with Paul Rabinow, 1984]. In Paul Rabinow (Ed.), *Essential works of Michel Foucault 1954–1984, Volume 1. Ethics: Subjectivity and truth* (pp. 111–120). New York: New Press.

Foucault, M. (1985/2006). Preface to *The history of sexuality, Volume 2*. In P. Rabinow (Ed.), *Essential works of Michel Foucault 1954–1984, Volume 1. Ethics: Subjectivity and truth* (pp. 199–206). New York: New Press.

Foucault, M. (1991). Governmentality. In G. Burchell, C. Gordon, & P. Miller (Eds.), *The Foucault effect: Studies in governmentality* (pp. 87–104). Chicago: University of Chicago Press.

Foucault, M. (1994/2006). Penal theories and institutions. In P. Rabinow (Ed.), *Essential works of Michel Foucault 1954–1984, Volume 1. Ethics: Subjectivity and truth* (pp. 17–22). New York: New Press.

Foucault, M. (1997/2006). The will to knowledge. In Paul Rabinow (Ed.), *Essential works of Michel Foucault 1954–1984, Volume 1. Ethics: Subjectivity and truth* (pp. 11–16). New York: New Press.

Fraser, M., & Greco, M. (2005). Introduction. In M. Fraser & M. Greco (Eds.), *The body: A reader* (pp. 1–42). London: Routledge.

Freud, S. (1963). *General psychological theory: Papers on metapsychology*. New York: Macmillan.

Friedman, J. (2003). Prefatory note on Jean Bazin, the author of 'Questions of meaning.' *Anthropological Theory, 3*(4), 416–434.

Gadamer, H.-G. (1960/1976). *Philosophical hermeneutics*. Berkeley: University of California Press.

Gadamer, H.-G. (1960/1986). *Truth and method*. New York: Crossroad Publishing Company.

Gadamer, H.-G. (1979). The problem of historical consciousness. In P. Rabinow & W. M. Sullivan (Eds.), *Interpretive social science: A reader*. Berkeley: University of California Press.

Gage, N. L. (1989). The paradigm wars and their aftermath: A 'historical' sketch of research on teaching. *Educational Researcher, 18*(7), 4–10.

Garfinkel, H. (1964). Studies of the routine grounds of everyday activities. *Social Problems, 11*(3), 225–250.

Garfinkel, H. (1967). *Studies in ethnomethodology*. Englewood Cliffs, NJ: Prentice-Hall.

Garfinkel, H. (1974). The origins of the term 'ethnomethodology.' In R. Turner (Ed.), *Ethnomethodology* (pp. 15–18). Harmondsworth: Penguin Education.

Garfinkel, H. (1988). Evidence for locally produced, naturally accountable phenomena of order, logic, reason, meaning, method, etc. in and as of the essential quiddity of immortal ordinary society (I of IV): An announcement of studies. *Sociological Theory, 6*(1), 103–109.

Garfinkel, H. (1996). Ethnomethodology's program. *Social Psychology Quarterly, 59*(1), 5–21.

Garfinkel, H. (2002). *Ethnomethodology's program: Working out Durkheim's aphorism*. London: Rowman and Littlefield.

Garfinkel, H., & Wieder, L. (1992). Two incommensurable, asymmetrically alternate technologies of social analysis. In J. C. McKinney & E. A. Tiryakian (Eds.), *Theoretical sociology: Perspectives and developments* (pp. 337–366). New York: Appleton-Century-Crofts.

Gay, P. (1969). *The Enlightenment: An interpretation, Vol. 1. The rise of modern paganism*. New York: Norton.

Gay, P. (1977). *The Enlightenment: An interpretation, Vol. 2. The science of freedom*. New York: Norton.

Geertz, C. (1972). Deep play: Notes on the Balinese cockfight. *Daedalus, 101*(1), 1–37.

Geertz, C. (1973). *The interpretation of cultures*. New York: Harper and Row.

Geertz, C. (1976/1979). From the native's point of view: On the nature of anthropological understanding. In P. Rabinow & W. M. Sullivan (Eds.), *Interpretive social science: A reader* (pp. 225–241). Berkeley: University of California Press.

Geertz, C. (1988). *Works and lives: The anthropologist as author*. Stanford, CA: Stanford University Press.

Gerard, P. (1996). *Creative nonfiction: Researching and crafting stories of real life*. Cincinnati: Story Press.

Gergen, K. J. (2001). Construction in contention: Toward consequential resolutions. *Theory and Psychology, 11*(3), 419–432.

Giddens, A. (1976). *New rules of sociological method: A positive critique of interpretative sociologies*. New York: Basic Books.

Giddens, A. (1977). *Studies in social and political theory*. New York: Basic Books.

Giddens, A. (1979). *Central problems in social theory: Action, structure and contradiction in social analysis*. Berkeley: University of California Press.

Giddens, A. (1982). *Profiles and critiques in social theory*. Berkeley: University of California Press.

Giddens, A. (1984). *Positivism and sociology*. London: Heinemann.

Giddens, A. (1987). *Sociology: A brief but critical introduction* (2nd ed.). San Diego: Harcourt Brace Jovanovich.

Giorgi, A. (1985). Sketch of a psychological phenomenological method. In A. Giorgi (Ed.), *Phenomenology and psychological research* (pp. 8–21). Pittsburgh: Duquesne University Press.

Glaser, B., & Strauss, A. (1964). The social loss of dying patients. *American Journal of Nursing, 64*, 119–121.

Glaser, B. G., & Strauss, A. L. (1967). *The discovery of grounded theory: Strategies for qualitative research*. Chicago: Aldine.

Goodwin, C. (1994). Professional vision. *American Anthropologist, 96*(3), 606–633.

Goodwin, C. (1995). Seeing in depth. *Social Studies of Science*, 25(2), 237–274.

Goodwin, C., & Duranti, A. (1992). Rethinking context: Language as an interactive phenomenon. In A. Duranti & C. Goodwin (Eds.), *Rethinking context: An introduction* (pp. 1–42). Cambridge: Cambridge University Press.

Goodwin, C., & Heritage, J. (1990). Conversation analysis. *Annual Review of Anthropology*, 19, 283–307.

Gorden, R. (1980). *Interviewing strategy, techniques and tactics*. New York: Dorsey.

Gordeon, N. (1999). Foucault's subject: An ontological reading. *Polity*, 31(3), 395–414.

Gordon, D. (1988). Writing culture, writing feminism: The poetics and politics of experimental ethnography. *Inscriptions, 3/4*.

Grice, H. P. (1975). Logic and conversation. In D. Davison & G. Harman (Eds.), *The logic of grammar* (pp. 64–75). Encino, CA: Dickenson.

Grunbaum, A. (1984). *The foundations of psychoanalysis: A philosophical critique*. Berkeley: University of California Press.

Guba, E. G. (Ed.). (1990). *The paradigm dialog*. Newbury Park, CA: Sage.

Guba, E., & Lincoln, Y. (1994). Competing paradigms in qualitative research. In N. Denzin & Y. Lincoln (Eds.), *Handbook of qualitative research* (pp. 105–177). Thousand Oaks, CA: Sage.

Gubrium, J. F., & Holstein, J. A. (2002). From the individual interview to the interview society. In J. F. Gubrium & J. A. Holstein (Eds.), *Handbook of interview research: Context and method* (pp. 3–32). Thousand Oaks, CA: Sage.

Guignon, C. B. (1983). *Heidegger and the problem of knowledge*. Indianapolis: Hackett.

Gumperz, J. (1982). *Discourse strategies*. Cambridge: Cambridge University Press.

Gutting, G. (2008). Foucault. In E. N. Zalta (Ed.), *The Stanford Encyclopedia of Philosophy* (http://plato.stanford.edu/archives/win2003/entries/davidson/).

Habermas, J. (1967/1988). *On the logic of the social sciences* (S. W. Nicholson & J. A. Stark, Trans.). Cambridge, MA: MIT Press.

Habermas, J. (1968/1971). *Knowledge and human interests* (J. Shapiro, Trans.). Boston: Beacon Press.

Habermas, J. (1973). *Theory and practice*. Boston: Beacon Press.

Habermas, J. (1977). A review of Gadamer's "Truth and method." In F. R. Dallmayr & T. McCarthy (Eds.), *Understanding and social inquiry* (pp. 335–363). Notre Dame, IN: University of Notre Dame Press.

Habermas, J. (1978). *Legitimation crisis* (T. McCarthy, Trans.). Boston: Beacon Press.

Habermas, J. (1979). *Communication and the evolution of society* (T. McCarthy, Trans.). Boston: Beacon Press.

Habermas, J. (1981/1984). *The theory of communicative action, Vol. 1. Reason and the rationalization of society* (T. McCarthy, Trans.). Boston: Beacon Press.

Habermas, J. (1981/1989). *The theory of communicative action, Vol. 2. Lifeworld and system: A critique of functionalist reason* (T. McCarthy, Trans.). Boston: Beacon Press.

Habermas, J. (1983/1990). *Moral consciousness and communicative action* (C. Lenhardt & S. W. Nicholson, Trans.). Cambridge, MA: MIT Press.

Habermas, J. (1990). Discourse ethics: Notes on a program of philosophical justification. In *Moral Consciousness and Communicative Action* (pp. 43–115) (C. Lenhart & S. W. Nicholson, Trans.). Cambridge, MA: MIT Press.

Hacking, I. (1983). *Representing and intervening: Introductory topics in the philosophy of natural science*. Cambridge: Cambridge University Press.

Hahn, H. (1933/1959). Logic, mathematics and knowledge of nature. In A. J. Ayer (Ed.), *Logical positivism* (pp. 147–161). New York: The Free Press.

Hanfling, O. (Ed.). (1981). *Essential readings in logical positivism*. Oxford: Blackwell.

Harding, S. (1991). *Whose science? Whose knowledge?* Ithaca, NY: Cornell University Press.

Harrington, A. (1999). Objectivism in hermeneutics? Gadamer, Habermas, Dilthey. *Philosophy of the Social Sciences, 30*(4), 491–507.

Harrington, B. (2003). The social psychology of access in ethnographic research. *Journal of Contemporary Ethnography, 32*(5), 592–626.

Harris, R. (1987). *Reading Saussure: A critical commentary on the "Cours de linguistique générale."* La Salle, IL: Open Court.

Harry, B., Sturges, K., & Klingner, J. K. (2005). Mapping the process: An exemplar of process and challenge in grounded theory analysis. *Educational Researcher, 34*(2), 3–13.

Hartsock, N. (1987). The feminist standpoint: Developing the ground for a specifically feminist historical materialism. In S. Harding (Ed.), *Feminism and methodology: Social science issues*. Bloomington: Indiana University Press.

Hayano, D. M. (1979). Auto-ethnography: Paradigms, problems, and prospects. *Human Organization, 38*, 99–104.

Hegel, G. W. F. (1807/1977). *Phenomenology of spirit* (A. V. Miller, Trans.). Oxford: Oxford University Press.

Heidegger, M. (1927/1962). *Being and time* (J. M. E. Robinson, Trans.). New York: Harper and Row.

Heidegger, M. (1975/1982). *The basic problems of phenomenology* (A. Hofstadter, Trans.). Bloomington: Indiana University Press.

Heidegger, M. (1980/1988). *Hegel's phenomenology of spirit* (P. Emad & K. Maly, Trans.). Bloomington: Indiana University Press.

Held, D. (1980). *Introduction to critical theory: Horkheimer to Habermas*. Berkeley: University of California Press.

Heller, A. (1982). Habermas and Marxism. In John B. Thompson & David Held (Eds.), *Habermas: Critical debates* (pp. 21–41). London: Macmillan.

Hempel, C. G. (1935). On the logical positivists' theory of truth. *Analysis, 2*, 49–59.

Hempel, C. G., & Oppenheim, P. (1948). The logic of explanation. *Science, 15*, 135–175.

Heritage, J. (1984). *Garfinkel and ethnomethodology*. Cambridge: Polity Press.

Heritage, J. (1998). Conversation analysis and institutional talk: Analyzing distinctive turn-taking systems. In S. Cmejrkovà, J. Hoffmannovà, O. Müllerovà, & J. Svetlà (Eds.), *Proceedings of the 6th international congress of IADA (International Association for Dialog Analysis)* (pp. 3–17). Tubingen: Niemeyer.

Holman Jones, S. (2005). Autoethnography: Making the personal political. In N. Denzin & Y. Lincoln (Eds.), *The Sage handbook of qualitative research* (3rd ed.) (pp. 763–791). Thousand Oaks, CA: Sage.

Holmes, D. R., & Marcus, G. E. (2005). Refunctioning ethnography: The challenge of an anthropology of the contemporary. In N. K. Denzin & Y. S. Lincoln (Eds.), *The Sage handbook of qualitative research* (3rd ed.) (pp. 1099–1113). Thousand Oaks, CA: Sage.

Holstein, A., & Gubrium, J. F. (1995). *The active interview*. Thousand Oaks, CA: Sage.

Honey, M. A. (1987). The interview as text: Hermeneutics considered as a model for analyzing the clinically informed research interview. *Human Development, 30*, 69–82.

Hook, D. (2001). Discourse, knowledge, materiality, history: Foucault and discourse analysis. *Theory and Psychology, 11*(4), 521–547.

Horkheimer, M., Fromm, E., & Marcuse, H. (1936). *Studien über Autorität und Familie*. Paris: Librairie Felix Alcan.

Horkheimer, M. (1972). *Critical theory*. New York: Herder and Herder.

Horkheimer, M. (1982). *Critical theory: Selected essays*. New York: Continuum.

Horkheimer, M., & Adorno, T. W. (1944/1989). *Dialectic of enlightenment* (J. Cumming, Trans.). New York: Continuum.

Howarth, D., & Stavrakakis, Y. (2000). Introducing discourse theory and political analysis. In D. Howarth, A. J. Norval, & Y. Stavrakakis (Eds.), *Discourse theory and political analysis: Identities, hegemonies and social change* (pp. 1–23). Manchester: Manchester University Press.

Howe, K. R. (2005). The question of education science: *Experimentism versus experimentalism. Educational Theory*, 55(3), 307–321.

Hume, D. (1739–1740/1978). *A treatise on human nature* (L. A. Selby-Bigge, Ed.). Oxford: Oxford University Press.

Hume, D. (1748/2000). *An enquiry concerning human understanding* (T. Beauchamp, Ed.). Oxford: Oxford University Press.

Husserl, E. (1900/1913). *Logical investigations* (2 volumes) (J. N. Findlay, Trans.). London: Routledge.

Husserl, E. (1913/1983). *Ideas pertaining to a pure phenomenology and to a phenomenological philosophy. Book One: General introduction to a pure phenomenology* (F. Kersten, Trans.). The Hague: Nijhoff.

Husserl, E. (1917/1981). Pure phenomenology: Its method and its field of investigation (Inaugural lecture at Freiburg im Breisgau) (R. Welsh Jordan, Trans.). In P. McCormick & F. A. Elliston (Eds.), *Husserl: Shorter Works* (pp. 10–17). Notre Dame, IN: University of Notre Dame Press.

Husserl, E. (1936/1970). *The crisis of European sciences and transcendental phenomenology: An introduction to phenomenology* (D. Carr, Trans.). Evanston, IL: Northwestern University Press.

Husserl, E. (1999). *The essential Husserl: Basic writings in transcendental phenomenology*. (D. Welton, Ed.). Bloomington: Indiana University Press.

Hymes, D. (Ed.) (1964). *Language in culture and society*. New York: Harper and Row.

Hymes, D. (1972). The use of anthropology: Critical, political, personal. In D. Hymes (Ed.), *Reinventing anthropology* (pp. 3–79). New York: Pantheon Books.

Hyppolite, J. (1946/1974). *Genesis and structure of Hegel's "Phenomenology of spirit"* (S. C. J. Heckman, Trans.). Evanston, IL: Northwestern University Press.

Iser, W. (1980). *The act of reading: A theory of aesthetic response* (Johns Hopkins Paperbacks ed.). Baltimore: Johns Hopkins University Press.

Jackson, P. W. (1968). *Life in classrooms*. New York: Holt, Rinehart and Winston.

Jameson, F. (1979/1997). Foreword. In J. F. Lyotard, *The postmodern condition: A report on knowledge* (pp. vii–xxi). Minneapolis: University of Minnesota Press.

Janik, A., & Toulmin, S. (1973). *Wittgenstein's Vienna*. New York: Simon and Schuster.

Jay, M. (1973). *The dialectical imagination: A history of the Frankfurt School and the Institute for Social Research 1923–1950*. Boston: Little, Brown and Co.

Johnson, R. B., & Onwuegbuzie, A. J. (2004). Mixed methods research: A research paradigm whose time has come. *Educational Researcher*, 33(7), 14–26.

Jorgensen, D. L. (1989). *Participant observation: A methodology for human studies* (Vol. 15). Newbury Park, CA: Sage.

Jules-Rosette, B. (1985). Interview with Harold Garfinkel. *Sociétés*, 5(1), 35–39.

Kant, I. (1781/1965). *The critique of pure reason* (N. K. Smith, Trans.). New York: St. Martin's Press.

Kant, I. (1783/1977). *Prolegomena to any future metaphysics* (P. Carus & J. Ellington, Trans.). Indianapolis: Hackett.

Kant, I. (1784/2000). What is enlightenment? In B. Gupta & J. Mohanty (Eds.), *Philosophical questions: East and West* (pp. 401–407). Oxford: Rowman and Littlefield.

Kant, I. (1788/1956). *The critique of practical reason.* (T. K. Abbot, Trans.). Bobbs-Merrill.

Kant, I. (1790/1952). *The critique of judgement* (J. C. Meredith, Trans.). Oxford: Clarendon Press.

Kellner, D. (1989). *Critical theory, Marxism and modernity.* Baltimore: Johns Hopkins University Press.

Kellner, H. (2005). White, Hayden. In M. Groden & M. Kreiswirth (Eds.), The Johns Hopkins Guide to Literary Theory and Criticism (http://litguide.press.jhu.edu/index.html). Last accessed 2010.

Kohlberg, L. (1981). *Essays on moral development. The philosophy of moral development: Moral stages and the idea of justice* (Vol. 1). New York: Harper and Row.

Koschmann, T., Stahl, G., & Zemel, A. (2004). The video analyst's manifesto (or the implications of Garfinkel's policies for studying practice within design-based research). In Y. B. Kafai, W. A. Sandoval, N. Enyedy, A. S. Nixon, & F. Herrera (Eds.), *Proceedings of the 6th international conference on Learning sciences* (pp. 278–285). Mahwah, NJ: Lawrence Erlbaum Associates.

Kroeber, A. L., & Kluckhohn, C. (1952). *Culture: A critical review of concepts and definitions.* New York: Vintage Books.

Kuhn, T. (1962). *The structure of scientific revolutions.* Chicago: University of Chicago Press.

Kuhn, T. S. (1970). *The structure of scientific revolutions* (2nd ed.). Chicago: University of Chicago Press.

Kuhn, T. S. (1974/1977). Second thoughts on paradigms. In T. S. Kuhn (Ed.), *The essential tension: Selected studies in scientific tradition and change* (pp. 293–319). Chicago: University of Chicago Press.

Kvale, S. (1996). *InterViews: An introduction to qualitative research interviewing.* Thousand Oaks, CA: Sage.

Laboratory of Comparative Human Cognition (LCHC). (1983). Culture and cognitive development. In P. H. Mussen & W. Kessen (Eds.), *Handbook of child psychology, Vol. 1: History, theory, and methods* (pp. 295–356). New York: Wiley.

Laclau, E. (2000). Foreword. In D. Howarth, A. J. Norval, & Y. Stavrakakis (Eds.), *Discourse theory and political analysis: Identities, hegemonies and social change* (pp. x–xi). Manchester: Manchester University Press.

Laclau, E., & Mouffe, C. (1985). *Hegemony and socialist strategy: Towards a radical democratic politics.* London: Verso.

Laidlaw, J. (2000). A free gift makes no friends. *Journal of the Royal Anthropological Institute, 6,* 617–634.

Laidlaw, J. (2002). For an anthropology of ethics and freedom. *Journal of the Royal Anthropological Institute, 8*(2), 311–332.

Lakoff, G., & Johnson, M. (1980). *Metaphors we live by.* Chicago: University of Chicago Press.

Langsdorf, L. (1995). Treating method and form as phenomena: An appreciation of Garfinkel's phenomenology of social action. *Human Studies, 18,* 177–188.

Lather, P. (1986). Research as praxis. *Harvard Educational Review, 56,* 257–277.

Lather, P. (2004). This *is* your father's paradigm: Government intrusion and the case of qualitative research in education. *Qualitative Inquiry, 10*(1), 15–34.

Latour, B. (1988). A relativistic account of Einstein's relativity. *Social Studies of Science, 18*(1), 3–44.

Latour, B. (1996). On interobjectivity. *Mind, Culture, and Activity, 3*(4), 228–245.

Latour, B. (2005). *Reassembling the social: An introduction to Actor-Network-Theory.* Oxford: Oxford University Press.

Latour, B., & Woolgar, S. (1979/1986). *Laboratory life: The construction of scientific facts.* Princeton, NJ: Princeton University Press.

Levi-Strauss, C. (1958/1963). *Structural anthropology.* New York: Basic Books.

Levinson, S. C. (1983). *Pragmatics.* Cambridge: Cambridge University Press.

Lewis, O. (1951). *Life in a Mexican village: Tepoztlàn re-studied.* Urbana: University of Illinois Press.

Lieblich, A., Tuval-Mashiach, R., & Zilber, T. (1998). *Narrative research: Reading, analysis, and interpretation.* Thousand Oaks, CA: Sage.

Lincoln, Y. S., & Guba, E. G. (2000). Paradigmatic controversies, contradictions, and emerging confluences. In N. K. Denzin & Y. S. Lincoln (Eds.), *Handbook of qualitative research* (2nd ed.) (pp. 163–188). Thousand Oaks, CA: Sage.

Locke, J. (1690/1975). *An essay concerning human understanding.* Oxford: Clarendon Press.

Lukàcs, G. (1923/1988). *History and class consciousness: Studies in Marxist dialectics.* Cambridge, MA: MIT Press.

Luke, A. (1995). Text and discourse in education: An introduction to critical discourse analysis. In M. W. Apple (Ed.), *Review of research in education* (Vol. 21) (pp. 3–48). Washington, DC: AERA.

Lyotard, J.-F. (1979/1997). *The postmodern condition: A report on knowledge* (G. Bennington & B. Massumi, Trans.). Minneapolis: University of Minnesota Press.

MacIntyre, A. (1966). *A short history of moral philosophy: A history of moral philosophy from the Homeric age to the twentieth century.* New York: Macmillan.

MacIntyre, A. (1984). *After virtue: A study in moral theory* (2nd ed.). South Bend, IN: University of Notre Dame Press.

MacKinnon, C. (1999). *Toward a feminist theory of the state.* Cambridge, MA: Harvard University Press.

Malinowski, B. (1922/1961). *Argonauts of the western Pacific.* New York: Dutton.

Malinowski, B. (1967). *A diary in the strict sense of the term.* London: Routledge and Kegan Paul.

Mallin, S. B. (1979). *Merleau-Ponty's philosophy.* New Haven, CT: Yale University Press.

Manning, P. (2004). Ethno's threads [Review of Harold Garfinkel, 'Ethnomethodology's Program: Working out Durkheim's aphorism']. *Contemporary Sociology, 33*(3), 278–281.

Marcus, G. E. (1994). After the critique of ethnography: Faith, hope, and charity, but the greatest of these is charity. In R. Borofsky (Ed.), *Assessing cultural anthropology* (pp. 40–53). New York: McGraw-Hill Humanities, Social Sciences and World Languages.

Marcus, G. E. (1995). Ethnography in/of the world system: The emergence of multi-sited ethnography. *Annual Review of Anthropology, 24,* 95–117.

Marcus, G. E. (1997). The uses of complicity in the changing mise-en-scene of anthropological fieldwork. *Representations, 59,* 85–108.

Marcus, G. E. (1998). *Ethnography through thick and thin.* Princeton, NJ: Princeton University Press.

Marcus, G., & Fischer, M. (Eds.). (1986a). *Anthropology as cultural critique: An experimental moment in the human sciences.* Chicago: University of Chicago Press.

Marcus, G., & Fischer, M. (1986b). A crisis of representation in the human sciences. In G. Marcus & M. Fischer (Eds.), *Anthropology as cultural critique: An experimental moment in the human sciences* (pp. 7–16). Chicago: University of Chicago Press.

Margolis, J. (1993). Redeeming Foucault. In J. Caputo & M. Yount (Eds.), *Foucault and the critique of institutions* (pp. 41–59). University Park: Pennsylvania State University Press.

Martin, J. R. (1993). Becoming educated: A journey of alienation or integration? In H. S. Shapiro & D. E. Purpel (Eds.), *Critical social issues in American education: Toward the 21st century* (pp. 137–148). New York: Longman.

Martin, R. (1982/1988). Truth, power, self: An interview with Foucault. In L. H. Martin, H. Gutman, & P. Hutton (Eds.), *Technologies of the self: A seminar with Michel Foucault* (pp. 9–15). Amherst: University of Massachusetts Press.

Marx, K. (1844/1983). Economico-philosophical manuscripts of 1844. In E. Kamenka (Ed.), *The portable Karl Marx* (pp. 131–152). New York: Penguin.

Marx, K. (1857–1858/1973). *Grundrisse: Foundations of the critique of political economy* (M. Nicolaus, Trans.). New York: Penguin.

Marx, K. (1867/1977). *Capital: A critique of political economy, Vol. 1* (B. Fowkes, Trans.). New York: Vintage Books.

Marx, K. (1888/1983). Theses on Feuerbach. In E. Kamenka (Ed.), *The portable Karl Marx* (pp. 155–158). New York: Penguin.

Marx, K., & Engels, F. (1845/1988). *The German ideology*. New York: International Publishers.

Mauss, M. (1935/1973). Techniques of the body. *Economy and Society, 2*, 71–88.

Mauss, M. (1960). *The gift: Forms and functions of exchange in archaic societies*. New York: Norton.

Maxwell, J. A. (2005). *Qualitative research design: An interpretive approach* (2nd ed.) Thousand Oaks, CA: Sage Publications.

Maynard, D. W. (1986). New treatment for an old itch [review of John Heritage, 'Garfinkel and ethnomethodology,' 1984]. *Contemporary Sociology, 15*(3), 346–349.

Maynard, D. W., & Clayman, S. E. (1991). The diversity of ethnomethodology. *Annual Review of Sociology, 17*, 385–418.

Maynard, D. W., & Clayman, S. E. (2003). Ethnomethodology and conversation analysis. In L. T. Reynolds and N. J. Herman-Kinney (Eds.), *The handbook of symbolic interactionism* (pp. 173–202). Walnut Creek, CA: Altamira Press.

Mayrl, W. M. (1973). Ethnomethodology: Sociology without society? *Catalyst, 7*, 15–28.

McCarthy, T. (1978). *The critical theory of Jurgen Habermas*. Cambridge, MA: MIT Press.

McCracken, G. (1988). *The long interview*. Newbury Park, CA: Sage.

McPhee, J. (1969). *Levels of the game*. New York: The Noonday Press, Farrar, Straus and Giroux.

McPhee, J. (2000). *Annals of the former world*. New York: Farrar, Straus and Giroux.

Mehan, H., & Wood, H. (1975). *The reality of ethnomethodology*. New York: Wiley.

Mendelson, J. (1979). The Habermas–Gadamer debate. *New German Critique, 18*, 44–73.

Menke, C. (1996). Critical theory and tragic knowledge. In D. M. Rasmussen (Ed.), *The handbook of critical theory* (pp. 57–73). Oxford: Blackwell.

Merleau-Ponty, M. (1942/1963). *The structure of behaviour*. London: Methuen.

Merleau-Ponty, M. (1945/1962). *Phenomenology of perception*. London: Routledge and Kegan Paul.

Merleau-Ponty, M. (1964). *The primacy of perception*. Evanston, IL: Northwestern University Press.

Merleau-Ponty, M. (1964/1968). *The visible and the invisible.* Evanston, IL: Northwestern University Press.

Michell, J. (2003). The quantitative imperative: Positivism, naive realism and the place of qualitative methods in psychology. *Theory and Psychology,* 13(1), 5–31.

Miles, M. B., & Huberman, A. M. (1994). *Qualitative data analysis: An expanded sourcebook* (2nd ed.). Thousand Oaks, CA: Sage.

Mill, J. S. (1843/1987). *The logic of the moral sciences.* London: Duckworth.

Miller, A. I. (1962/1983). Introduction: P. W. Bridgman and the special theory of relativity. In P. W. Bridgman, *A sophisticate's primer of relativity* (pp. xii–xlv). Middletown, CT: Wesleyan University Press.

Misgeld, D. (1976). Critical theory and hermeneutics: The debate between Habermas and Gadamer. In J. O'Neill (Ed.), *On critical theory.* New York: Seabury Press.

Mishler, E. G. (1986). *Research interviewing: Context and narrative.* Cambridge, MA: Harvard University Press.

Mohanty, J. (1993). Foucault as philosopher. In J. Caputo & M. Yount (Eds.), *Foucault and the critique of institutions* (pp. 27–40). University Park: Pennsylvania State University Press.

Morgan, D. L. (2007). Paradigms lost and pragmatism regained: Methodological implications of combining qualitative and quantitative methods. *Journal of Mixed Methods Research,* 1(1), 48–76.

Morris, C. (1938). Foundations of the theory of signs. In O. Neurath, R. Carnap, & C. Morris (Eds.), *International encyclopedia of unified science* (pp. 77–138). Chicago: University of Chicago Press.

Moser, K. S. (2000). Metaphor analysis in psychology: Method, theory, and fields of application. *Forum: Qualitative Social Research,* 1(2).

Mruck, K. (2000). Qualitative research in Germany. *Forum: Qualitative Social Research,* 1(1).

Mruck, K., & Breuer, F. (2003). Subjectivity and reflexivity in qualitative research: The FQS issues. *Forum: Qualitative Social Research,* 4(2).

Neurath, O. (1938). Unified science as encyclopedic integration. In O. Neurath, R. Carnap, & C. Morris (Eds.), *Foundations of the unity of science: Toward an international encyclopedia of unified science* (pp. 1–27). Chicago: University of Chicago Press.

Neurath, O., Carnap, R., & Morris, C. (1938). *Foundations of the unity of science: Toward an international encyclopedia of unified science.* Chicago: University of Chicago Press.

Nietzsche, F. (1877/1998). *On the genealogy of morality* (M. Clark & A. Swensen, Trans.). Indianapolis: Hackett.

Nofsinger, R. E. (1991). *Everyday conversation.* Newbury Park, CA: Sage.

Oakley, A. (1981). Interviewing women: A contradiction in terms. In H. Roberts (Ed.), *Doing feminist research* (pp. 30–61). Boston: Routledge and Kegan Paul.

Ochs, E. (1979). Transcription as theory. In E. Ochs & B. B. Schieffelin (Eds.), *Developmental pragmatics* (pp. 43–72). New York: Academic Press.

Ollman, B. (1976). *Alienation* (2nd ed.). Cambridge: Cambridge University Press.

Ollman, B. (1990). Putting dialectics to work: The process of abstraction in Marx's method. *Rethinking Marxism,* 3, 26–74.

Ollman, B. (2003). *Dance of the dialectic: Steps in Marx's method.* Urbana: University of Illinois Press.

Ophir, A., & Shapin, S. (1991). The place of knowledge: A methodological survey. *Science in Context,* 4(1), 3–21.

Ormiston, G. L., & Schrift, A. D. (1990a). *The hermeneutic tradition: From Ast to Ricoeur*. Albany: State University of New York Press.

Ormiston, G. L., & Schrift, A. D. (1990b). *Transforming the hermeneutic context: From Nietzsche to Nancy*. Albany: State University of New York Press.

O'Rourke, B. K., & Pitt, M. (2007). Using the technology of the confessional as an analytical resource: Four analytical stances towards research interviews in discourse analysis. *Forum: Qualitative Social Research*, 8(2).

Packer, M. J. (1989). Tracing the hermeneutic circle: Articulating an ontical study of moral conflicts. In M. J. Packer & R. B. Addison (Eds.), *Entering the circle: Hermeneutic investigation in psychology* (pp. 95–117). Albany: State University of New York Press.

Packer, M. J. (2001a). *Changing classes: School reform and the new economy*. Cambridge: Cambridge University Press.

Packer, M. (2001b). Changing classes: Shifting the trajectory of development in school. In M. Packer & M. B. Tappan (Eds.), *Cultural and critical perspectives on human development*. Albany: State University of New York Press.

Packer, M., & Greco-Brooks, D. (1999). School as a site for the production of persons. *Journal of Constructivist Psychology*, 12, 133–149.

Packer, M. J. (2008). Is Vygotsky relevant? Vygotsky's Marxist psychology. *Mind, Culture, and Activity*, 15(1), 8–31.

Paget, M. A. (1983). Experience and knowledge. *Human Studies*, 6, 67–90.

Palmer, R. E. (1969). *Hermeneutics: Interpretation theory in Schleiermacher, Dilthey, Heidegger and Gadamer*. Evanston, IL: Northwestern University Press.

Park, R. (1915/1997). The city: Suggestions for the investigation of human behavior in the urban environment, Reprinted in K. Gelder & S. Thornton (Eds.), *The subcultures reader*. London: Routledge.

Pepper, S. C. (1942). *World hypotheses: A study of evidence*. Berkeley: University of California Press.

Pettit, P. (1975). *The concept of structuralism: A critical analysis*. Berkeley: University of California Press.

Phillips, D. C. (1995). The good, the bad, and the ugly: The many faces of constructivism. *Educational Researcher*, 24(7), 5–12.

Phillips, L., & Jörgensen, M. W. (2002). *Discourse analysis as theory and method*. London: Sage.

Piaget, J. (1937/1955). *The construction of reality in the child*. London: Routledge and Kegan Paul.

Piaget, J. (1945/1962). *Play, dreams and imitation in childhood* (C. Gattegno & F. M. Hodgson, Trans.). New York: Norton.

Piaget, J. (1970/1972). *The principles of genetic epistemology* (W. Mays, Trans.). New York: Basic Books.

Piaget, J. (1970/1988). *Structuralism* (C. Maschler, Trans.). New York: Harper and Row.

Pike, K. (1954). *Language in relation to a unified theory of the structure of human behavior*. Glendale, CA: Summer Institute of Linguistics.

Polanyi, M. (1967). *The tacit dimension*. New York: Doubleday.

Pollner, M., & Emerson, R. M. (2001). Ethnomethodology and ethnography. In P. Atkinson, A. Coffey, S. Delamont, J. Lofland, & L. Lofland (Eds.), *Handbook of ethnography* (pp. 118–135). London: Sage.

Popper, K. R. (1934/1959). *The logic of scientific discovery*. New York: Basic Books.

Popper, K. R. (1963). *Conjectures and refutations: The growth of scientific knowledge* (4th ed.). New York: Harper and Row.

Poster, M. (1989). *Critical theory and poststructuralism*. Ithaca, NY: Cornell University Press.

Poster, M. (1993). Foucault and the problem of self-constitution. In J. Caputo & M. Yount (Eds.), *Foucault and the critique of institutions* (pp. 63–80). University Park: Pennsylvania State University Press.

Potter, J., & Hepburn, A. (2005). Qualitative interviews in psychology: Problems and possibilities. *Qualitative Research in Psychology, 2*(4), 281–306.

Powdermaker, H. (1966). *Stranger and friend: The way of an anthropologist*. New York: Norton.

Powdermaker, H. (1967). An agreeable man. *New York Review of Books, 9*(8).

Prince, G. (1987). *A dictionary of narratology*. Lincoln: University of Nebraska Press.

Prior, L. (1997). Following in Foucault's footsteps: Text and context in qualitative research. In D. Silverman (Ed.), *Qualitative research: Theory, method and practice* (pp. 63–79). London: Sage.

Propp, V. (1928/1977). *Morphology of the folk tale*. Austin: University of Texas Press.

Rabinow, P. (1986). Representations are social facts: Modernity and post-modernity in anthropology. In J. Clifford & G. E. Marcus (Eds.), *Writing culture: The poetics and politics of ethnography* (pp. 234–261). Berkeley: University of California Press.

Rabinow, P., & Rose, N. (2003). Foucault today. In P. Rabinow & N. Rose (Eds.), *The essential Foucault: Selections from the essential works of Foucault, 1954–1984* (pp. vii–xxxv). New York: New Press.

Rabinow, P., & Sullivan, W. M. (Eds.). (1979). *Interpretive social science: A reader*. Berkeley: University of California Press.

Rapport, N. (1990). 'Surely everything has already been said about Malinowski's diary!' *Anthropology Today, 6*(1), 5–9.

Rasmussen, D. M. (1996). *The handbook of critical theory*. Oxford: Blackwell.

Rawls, A. W. (1996). Durkheim's epistemology: The neglected argument. *The American Journal of Sociology, 102*(2), 430–482.

Rawls, A. W. (1997). Durkheim and pragmatism: An old twist on a contemporary debate. *Sociological Theory, 15*(1), 5–29.

Rawls, A. W. (1998). Durkheim's challenge to philosophy: Human reason explained as a product of enacted social practice. *American Journal of Sociology, 104*(3), 887–901.

Rawls, A. W. (2006). Respecifying the study of social order: Garfinkel's transition from theoretical conceptualization to practices in detail. In H. Garfinkel & A. W. Rawls (Eds.), *Seeing sociologically: The routine grounds of social action* (pp. 1–98). Boulder, CO: Paradigm Publishers.

Reddy, M. J. (1979). The conduit metaphor: A case of frame conflict in metaphor and thought. In A. Ortony (Ed.), *Language and metaphor* (pp. 284–324). Cambridge: Cambridge University Press.

Redfield, R. (1930). *Tepoztlan, a Mexican village: A study of folk life*. Chicago: University of Chicago Press.

Rennie, D. L., Watson, K. D., & Monteiro, A. (2000). Qualitative research in Canadian psychology. *Forum: Qualitative Social Research, 1*(2).

Ricoeur, P. (1971/1979). The model of the text: Meaningful action considered as a text [Reprinted from *Social Research, 38*]. In P. Rabinow & W. M. Sullivan (Eds.), *Interpretive social science: A reader* (pp. 73–101). Berkeley: University of California Press.

Ricoeur, P. (1973). Ethics and culture: Habermas and Gadamer in dialogue. *Philosophy Today, 17*, 153–165.

Ricoeur, P. (1973/1990). Hermeneutics and the critique of ideology. In G. L. Ormiston & A. D. Schrift (Eds.), *The hermeneutic tradition: From Ast to Ricoeur* (pp. 298–334.). Albany: State University of New York Press.

Ricoeur, P. (1976). *Interpretation theory: Discourse and the surplus of meaning.* Fort Worth: The Texas Christian University Press.

Ricoeur, P. (1979). *Main trends in philosophy.* New York: Holmes and Meier.

Ricoeur, P. (1981). The narrative function. In J. B. Thompson (Ed.), *Hermeneutics and the human sciences* (pp. 274–296). Cambridge: Cambridge University Press.

Rizo, F. M. (1991). The controversy about quantification in social research: An extension of Gage's "'Historical' sketch." *Educational Researcher, 20*(9), 9–12.

Robben, A. C. G. M., & Sluka, J. A. (Eds.). (2007). *Ethnographic fieldwork: An anthropological reader.* Malden, MA: Blackwell.

Rock, P. (2001). Symbolic interactionism and ethnography. In P. Atkinson, A. Coffey, S. Delamont, J. Lofland, & L. Lofland (Eds.), *Handbook of ethnography* (pp. 26–38). London: Sage.

Rockmore, T. (1997). *Cognition: An introduction to Hegel's Phenomenology of Spirit.* Berkeley: University of California Press.

Rogers, C. (1945). The nondirective method as a technique for social research. *American Journal of Sociology, 50* (4), 279–283.

Rosaldo, R. (1989). Imperialist nostalgia. *Representations, 26,* 107–122.

Rosaldo, R. (1993). After objectivism. In S. During (Ed.), *The cultural studies reader* (pp. 104–117). London: Routledge.

Ryan, G. W., & Bernard, H. R. (2000). Data management and analysis methods. In N. K. Denzin & Y. S. Lincoln (Eds.), *Handbook of qualitative research* (2nd ed.) (pp. 769–802). Thousand Oaks, CA: Sage.

Ryan, G. W., & Bernard, H. R. (2003). Techniques to identify themes. *Field Methods, 15* (1), 85–109.

Ryle, G. (1968). The thinking of thoughts: What is 'Le penseur' doing? In *University Lectures* #18. Saskatoon: University of Saskatchewan.

Sacks, H., Schegloff, E. A., & Jefferson, G. (1974). A simplest systematics for the organization of turn-taking in conversation. *Language, 50,* 696–735.

Sanjek, R. (1990). On ethnographic validity. In R. Sanjek (Ed.), *Fieldnotes: The making of anthropology* (pp. 385–418). Ithaca, NY: Cornell University Press.

Saussure, F. de (1915/1959). *Course in general linguistics* (W. Baskin, Trans.). New York: Philosophical Library.

Sayer, D. (1987). *The violence of abstraction: The analytic foundations of historical materialism.* Oxford: Blackwell.

Schacht, R. (1970). *Alienation.* New York: Anchor Books.

Schegloff, E. A. (1995). Discourse as an interactional achievement III: The omnirelevance of action. *Research on Language and Social Interaction, 28*(3), 185–211.

Schegloff, E. A. (1997). Whose text? Whose context? *Discourse and Society, 8*(2), 165–187.

Schegloff, E. A. (1998). Reply to Wetherell. *Discourse and Society, 9*(3), 413–416.

Schegloff, E. A. (1999). 'Schegloff's texts' as 'Billig's data': A critical reply. *Discourse and Society, 10*(4), 558–572.

Schiffrin, D. (1994). *Approaches to discourse.* Oxford: Blackwell.

Schiffrin, D., Tannen, D., & Hamilton, H. E. (Eds.). (2001). *The handbook of discourse analysis.* Oxford: Blackwell.

Schleiermacher, F. D. E. (1810/1990). The aphorisms on hermeneutics from 1805 and 1809/10. In G. L. Ormiston & A. D. Schrift (Eds.), *The hermeneutic tradition: From Ast to Ricoeur* (pp. 57–84). Albany: State University of New York Press.

Schleiermacher, F. D. E. (1819/1990). The hermeneutics: Outline of the 1819 lectures. In G. L. Ormiston & A. D. Schrift (Eds.), *The hermeneutic tradition: From Ast to Ricoeur* (pp. 85–100). Albany: State University of New York Press.

Schlick, M. (1935). Facts and propositions. *Analysis*, 2, 65–70.

Schlick, M. (1936). Meaning and verification. *The Philosophical Review*, 4, 339–369.

Schutz, A. (1954). *Collected papers II: Studies in social theory*, A. Brodersen (Ed.). The Hague: Martinus Nijhoff.

Schutz, A. (1963a). Concept and theory formation in the social sciences. In M. Natanson (Ed.), *Philosophy of the social sciences* (pp. 231–249). New York: Random House.

Schutz, A. (1963b). Common-sense and scientific interpretation of human action. In M. Natanson (Ed.), *Philosophy of the social sciences* (pp. 302–346). New York: Random House.

Schutz, A. (1970). *On phenomenology and social relations*: Chicago: University of Chicago Press.

Searle, J. (1969). *Speech acts: An essay in the philosophy of language*. Cambridge: Cambridge University Press.

Searle, J. (1997). *The construction of social reality*. New York: Free Press.

Seidman, I. (1998). *Interviewing as qualitative research: A guide for researchers in education and the social sciences* (2nd ed.). New York: Teachers College Press.

Selman, R. (1980). *The growth of interpersonal understanding*. New York: Academic Press.

Sharrock, W. (2004). What Garfinkel makes of Schutz: The past, present and future of an alternate, asymmetric and incommensurable approach to sociology. *Theory and Science*, 5(1) (http://theoryandscience.icaap.org/).

Shavelson, R. J., & Towne, L. (Eds.) (2002). *Scientific research in education*. Washington, DC: National Academy Press.

Shavelson, R. J., & Towne, L. (2004). What drives scientific research in education? *American Psychological Society Observer*, 17(4) (http://www.psychologicalscience. org/observer/).

Shweder, R. A. (1991). *Thinking through cultures: Expeditions in cultural psychology*. Cambridge, MA: Harvard University Press.

Shweder, R. A. (1996). True ethnography: The lore, the law, and the lure. In R. Jessor, A. Colby, & R. A. Shweder (Eds.), *Ethnography and human development: Context and meaning in social inquiry* (pp. 15–52). Chicago: University of Chicago Press.

Shweder, R. A. (2007). Something else: The resolute irresolution of Clifford Geertz. *Common Knowledge*, 13(2/3), 191–205.

Simon, J. (1971). A conversation with Michel Foucault. *Partisan Review*, 38(2), 192–201.

Sluka, J. A., & Robben, A. C. G. M. (2007). Fieldwork in cultural anthropology: An introduction. In A. C. G. M. Robben & J. A. Sluka (Eds.), *Ethnographic fieldwork: An anthropological reader* (pp. 1–28). Malden, MA: Blackwell.

Smart, B. (1983). *Foucault, Marxism and critique*. London: Routledge and Kegan Paul.

Smith, J. A. (2004). Reflecting on the development of interpretative phenomenological analysis and its contribution to qualitative research in psychology. *Qualitative Research in Psychology*, 1, 39–54.

Smith, J. K., & Heshusius, L. (1986). Closing down the conversation: The end of the quantitative–qualitative debate among educational inquirers. *Educational Researcher*, 15, 4–12.

Solomon, R. C. (1983). *In the spirit of Hegel: A study of G. W. F. Hegel's 'Phenomenology of Spirit.'* New York: Oxford University Press.

Spencer, J. (2001). Ethnography after postmodernism. In P. Atkinson, A. Coffey, S. Delamont, J. Lofland, & L. Lofland (Eds.), *Handbook of ethnography* (pp. 443–452). London: Sage.

Spradley, J. (1980). *Participant observation.* New York: Holt.

Strauss, A., & Corbin, J. (1990). *Basics of qualitative research: Grounded theory procedures and techniques.* Newbury Park, CA: Sage.

Strauss, A., & Corbin, J. (1998). *Basics of qualitative research: Techniques and procedures for developing grounded theory* (2nd ed.). Thousand Oaks, CA: Sage.

Stroud, B. (1977). *Hume.* London: Routledge and Kegan Paul.

Suchman, L., & Jordan, B. (1990). Interactional troubles in face-to-face survey interviews. *Journal of the American Statistical Association, 85*, 232–253.

Sudnow, D. (1974). *Ways of the hand: The organization of improvised conduct.* Cambridge, MA: Harvard University Press.

Sudnow, D. (1979). *Talk's body: A meditation between two keyboards.* New York: Knopf.

Sullivan, R. J. (1989). *Immanuel Kant's moral theory.* Cambridge: Cambridge University Press.

Swanson, G. E. (1968). Review of Harold Garfinkel, 'Studies in Ethnomethodology.' *American Sociological Review, 33*(1), 122–124.

Taylor, C. (1971). Interpretation and the sciences of man. *The Review of Metaphysics, 25*, 3–34, 45–51.

Taylor, C. (1975). *Hegel.* Cambridge: Cambridge University Press.

Taylor, C. (1980). Understanding in human science. *The Review of Metaphysics, 34*, 3–23.

Taylor, C. (1983/1985). Understanding and ethnocentricity. In *Philosophical papers 2: Philosophy and the human sciences.* Cambridge: Cambridge University Press.

Taylor, C. (1989). *Sources of the self: The making of the modern identity.* Cambridge, MA: Harvard University Press.

Taylor, C. (1993). Engaged agency and background in Heidegger. In C. B. Guignon (Ed.), *The Cambridge companion to Heidegger* (pp. 317–336). Cambridge: Cambridge University Press.

ten Have, P. (2002). The notion of member is the heart of the matter: On the role of membership knowledge in ethnomethodological inquiry. *Forum: Qualitative Social Research, 3*(3).

ten Have, P. (2004). *Understanding qualitative research and ethnomethodology.* London: Sage.

Thompson, J. (1991). Introduction. In P. Bourdieu, *Language and symbolic power* (pp. 1–31). Cambridge: Polity Press.

Turner, V., & Bruner, E. (Eds.). (1986). *The anthropology of experience.* Urbana: University of Illinois Press.

Urry, J. (1972). "Notes and Queries on Anthropology" and the development of field methods in British anthropology, 1870–1920. *Proceedings of the Royal Anthropological Institute of Great Britain and Northern Ireland, 1972*, 45–57.

U. S. Department of Education. (n.d.). Evidence-based education. Power Point presentation consisting of 24 slides prepared by Grover J. Whitehurst, Assistant Secretary, Office of Educational Research and Improvement (www.ed.gov/offices/OERI/presentations/evidencebase.html).

van Dijk, T. A. (Ed.). (1985). *Handbook of discourse analysis* (Vols. 1–4). London: Academic Press.

van Dijk, T. A. (Ed.). (1997a). *Discourse studies: A multidisciplinary introduction, Vol. 1: Discourse as structure and process.* London: Sage.

van Dijk, T. A. (Ed.). (1997b). *Discourse studies: A multidisciplinary introduction, Vol. 2: Discourse as social interaction.* London: Sage.

van Dijk, T. A. (2001). Critical discourse analysis. In D. Schiffrin, D. Tannen, & H. E. Hamilton (Eds.), *The handbook of discourse analysis* (pp. 352–371). Oxford: Blackwell.

Van Maanen, J. (1988). *Tales from the field.* Chicago: University of Chicago Press.

Vandenberghe, F. (1999). "The real is relational": An epistemological analysis of Pierre Bourdieu's generative structuralism. *Sociological Theory, 17*(1), 32–67.

Vidich, A. J., & Lyman, S. M. (2000). Qualitative methods: Their history in sociology and anthropology. In N. K. Denzin & Y. S. Lincoln (Eds.), *Handbook of qualitative research* (2nd ed.) (pp. 37–84). Thousand Oaks, CA: Sage.

Von Mises, R. (1956). *Positivism: A study in human understanding.* New York: George Braziller, Inc.

Vygotsky, L. S. (1987). *The collected works of L. S. Vygotsky, Vol. 1: Problems of general psychology.* New York: Plenum Press.

Wacquant, L. (1989). Towards a reflexive sociology: A workshop with Pierre Bourdieu. *Sociological Theory, 7*(1), 26–63.

Wacquant, L. (1998). A fleshpeddler at work: Power, pain, and profit in the prize-fighting economy. *Theory and Society, 27*(1), 1–42.

Wacquant, L. (2004a). Taking Bourdieu into the field. *Berkeley Journal of Sociology, 46,* 180–186.

Wacquant, L. (2004b). *Body & soul: Notebooks of an apprentice boxer.* Oxford: Oxford University Press.

Wacquant, L. (2005). Shadowboxing with ethnographic ghosts: A rejoinder. *Symbolic Interaction, 28*(3), 441–447.

Warren, C. A. B. (2002). Qualitative interviewing. In J. F. Gubrium & J. A. Holstein (Eds.), *Handbook of interview research: Context and method* (pp. 83–101). Thousand Oaks, CA: Sage.

Watzlawick, P., Beavin, J. H., & Jackson, D. D. (1967). *Pragmatics of human communication.* New York: Norton.

Wetherell, M. (1998). Positioning and interpretative repertoires: Conversation analysis and post-structuralism in dialogue. *Discourse and Society, 9*(3), 387–412.

White, H. V. (1969/1990). *The content of the form: Narrative discourse and historical representation.* Baltimore: Johns Hopkins University Press.

White, H. V. (1973a). *Metahistory: The historical imagination in nineteenth-century Europe.* Baltimore: Johns Hopkins University Press.

White, H. V. (1973b). Foucault decoded: Notes from underground. *History and Theory, 12*(1), 23–54.

Wieder, D. L. (1974). Telling the code. In R. Turner (Ed.), *Ethnomethodology* (pp. 144–172). Harmondsworth: Penguin Education.

Wiesenfeld, E. (2000, June). Between prescription and action: The gap between the theory and practice of qualitative inquiries. *Forum: Qualitative Social Research* [On-line Journal], *1*(2). Available at http://qualitative-research.net\fqs\fqs-e/2–00 inhalt-e.htm.

Willis, P. (1981). Cultural production is different from cultural reproduction is different from social reproduction is different from reproduction. *Interchange, 12*(2–3), 48–67.

Winch, P. (1956). Social science. *The British Journal of Sociology, 7*(1), 18–33.

References

Winch, P. (1958). *The idea of a social science and its relation to philosophy.* London: Routledge and Kegan Paul.

Wittgenstein, L. (1922). *Tractatus-Logico-Philosophicus* (C. K. Ogden & F. P. Ramsey, Trans.). London: Kegan Paul, Trench and Truber.

Wittgenstein, L. (1953). *Philosophical investigations* (G. E. M. Anscombe, Trans.). Oxford: Blackwell.

Wittgenstein, L. (1953/2001). *Philosophical investigations: The German text, with a revised English translation* (E. Anscombe, Trans.). Oxford: Blackwell.

Wittgenstein, L. (1958/1969). *The blue and brown books.* New York: Barnes and Noble.

Wolin, S. S. (1968). Paradigms and political theories. In P. King & B. C. Parekh (Eds.), *Politics and experience.* Cambridge: Cambridge University Press.

Wuthnow, R. (1987). *Meaning and moral order: Explorations in cultural analysis.* Berkeley: University of California Press.

Zizek, S. (1993). *Tarrying with the negative: Kant, Hegel, and the critique of ideology.* Durham, NC: Duke University Press.

NAME INDEX

Adorno, Theodore, 284, 285–287, 295
Aristotle, 10, 23, 70, 83, 288, 340, 389, 394
Austin, John, 252, 326

Bakhtin, Mikhail, 234
Barthes, Roland, 325
Bazin, Jaques, 220–221, 222–223,
    266, 385
Behar, Ruth, 228
Benjamin, Walter, 284
Berger, Peter, 159–163, 164–166, 202
Bernstein, Richard, 37–38, 39, 383
Biklen, Sari, 47, 58, 114
Boas, Franz, 210
Bogdan, Robert, 42, 47, 49, 58, 114
Bohannan, Laura, 228
Bourdieu, Pierre, 12, 272, 288, 309–312,
    316–340, 342, 374, 381, 382, 388,
    389, 391
Bridgman, Percy, 22
Briggs, Joan, 228
Bruner, Jerome, 52, 103
Burgess, Ernest, 210

Carnap, Rudolf, 29
Chomsky, Noam, 115, 307, 308
Clifford, James, 225–226, 227–228, 229, 236, 237,
    240, 340
Comte, Auguste, 21
Corbin, Janet, 66–67, 68, 72, 73

Darwin, Charles, 176, 356
Descartes, Rene, 21, 141–143, 391
Dewey, John, 210
Dilthey, Wilhelm, 9, 82, 83, 88–92, 96, 112–113, 217,
    220, 296
Dreyfus, Hubert, 355
Durkheim, Emile, 132, 169, 170

Eagleton, Terence, 83, 231
Einstein, Albert, 21–22

Fairclough, Norman, 246–248
Foucault, Michel, 6, 12–13, 288, 305, 309–312, 336,
    340, 354, 377, 380, 381, 382, 385, 387, 388, 389, 391,
    393, 394–395
Freud, Sigmund, 285, 299–302, 303, 311
Fromm, Erich, 284

Gadamer, Hans-Georg, 9, 83, 97, 217, 220, 223,
    224, 231, 234, 295–299, 305, 310, 388
Garfinkel, Harold, 9, 74–78, 80, 112, 169, 170,
    190–207, 223, 224, 233, 241, 261, 319, 344, 348,
    385, 388, 390, 391, 393
Geertz, Clifford, 10, 213–218, 219–220, 222–224,
    227, 228–229, 232, 234, 237, 241, 242,
    267, 340
Giddens, Anthony, 10, 128–134, 304, 393, 394
Glaser, Bernard, 64, 72
Goffman, Erving, 200–201
Goodwin, Charles, 392
Gordon, Deborah, 228, 239
Grice, Herbert, 252
Guba, Egon, 40

Habermas, Jurgen, 5, 12, 98, 272, 288, 289–315, 316,
    335, 338, 342, 367, 374, 381, 382, 385, 388, 389, 391,
    394
Hegel, Georg, 150, 168–176, 177, 201, 273–275, 277,
    279, 305, 343, 360, 392
Heidegger, Martin, 70, 87, 112, 150, 201–207, 223,
    224, 261, 295, 296, 298, 303, 327, 343, 353, 376,
    381, 387, 388, 392
Honey, Margaret, 102
Horkheimer, Max, 284, 285–287, 295
Hume, David, 21, 141–143, 144, 170
Husserl, Edmund, 149–153, 164–166, 202, 242, 343,
    354, 355

Iser, Wolfgang, 9, 101, 104–107, 110, 113, 220

Jefferson, Gail, 255
Jordan, Brigitte, 44–46, 50, 51

419

SUBJECT INDEX

DATE DUE
